Springer Tracts in Modern Physics
Volume 192

Springer
New York
Berlin
Heidelberg
Hong Kong
London
Milan
Paris
Tokyo

Physics and Astronomy ONLINE LIBRARY

http://www.springer.de

Springer Tracts in Modern Physics

Springer Tracts in Modern Physics provides comprehensive and critical reviews of topics of current interest in physics. The following fields are emphasized: elementary particle physics, solid-state physics, complex systems, and fundamental astrophysics.
Suitable reviews of other fields can also be accepted. The editors encourage prospective authors to correspond with them in advance of submitting an article. For reviews of topics belonging to the above mentioned fields, they should address the responsible editor, otherwise the managing editor.
See also springeronline.com

Thomas Ihn

Electronic Quantum Transport in Mesoscopic Semiconductor Structures

With 90 Illustrations, 5 in Full Color

Springer

Thomas Ihn
Solid-State Laboratory
ETH Hoenggerber
CH-8093 Zurich, Switzerland
ihn@phys.ethz.ch

Physics and Astronomy Classification Scheme (PACS): 71.30.+h, 73.23.-b, 73.23.Hk, 68.37.Ps, 73.43.Fj, 73.63.Nm

Library of Congress Cataloging-in-Publication Data
Ihn, Thomas.
Electronic quantum transport in mesoscopic semiconductor structures / Thomas Ihn.
p. cm. – (Springer tracts in modern physics)
Includes bibliographical references and index.
ISBN 0-387-40096-6 (alk. paper)
1. Electron transport. 2. Semiconductors. 3. Mesoscopic phenomena (Physics)
4. Quantum theory. I. Title. II. Series.
QC176.8.E4I36 2003
537.6′22-dc21 2003050501

ISSN print edition: 0081-3869
ISSN electronic edition: 1615-0430
ISBN 0-387-40096-6 Printed on acid-free paper.

Printed in the United States of America. (SBA)

9 8 7 6 5 4 3 2 1 SPIN 10928523

Springer-Verlag is a part of *Springer Science+Business Media*

springeronline.com

Preface

The physics of semiconductors has seen an enormous evolution within the last fifty years. Countless achievements have been made in scientific research and device applications have revolutionized everyday life. We have learned how to customize materials in order to tailor their optical as well as electronic properties. The ongoing trend toward device miniaturization has been the driving force on the application side and it has fertilized fundamental research. Nowadays, advanced processing techniques allow the fabrication of sub-micron semiconductor structures in many university research laboratories. At the same time, experiments down to millikelvin temperatures allow researchers to anticipate the observation of quantum phenomena, so far hidden at room temperature by the large thermal energy and strong dephasing.

The field of mesoscopic physics deals with systems under experimental conditions where several quantum length scales for electrons such as system size and phase coherence length, or phase coherence length and elastic mean free path, are comparable. Intense research over the last twenty years has revealed an enormous richness of quantum effects in mesoscopic semiconductor physics, which is typically characterized by an interplay of quantum interference and many-body interactions. The most famous phenomena are probably the integer and fractional quantum Hall effects, the quantization of conductance through a quantum point contact, the Aharonov–Bohm effect, and single-electron charging of quantum dots.

The sheer unlimited versatility of materials and processing techniques does not only sustain a flourishing field of physics but still inspires a large community of researchers to develop more and more challenging visions that may be realized in semiconductor systems. Some top issues are strongly correlated mesoscopic electronic systems, spins in mesoscopic systems and spintronics, phase-coherence, controlled entanglement, controlled dephasing, and quantum information processing.

This book tries to give some insight into recent research work in the field of electron transport in mesoscopic semiconductors by covering three selected topics. Since no selection of topics in this broad field could ever be representative, it was mainly dictated by my own research pursued over the last years. However, an introductory historical part I was added in order to give the non-specialist a chance to see which developments form the basis for current research.

Part II of this book is devoted to the possible metal-insulator transition in two-dimensional systems at zero magnetic field and low temperatures. Although the existence of novel interaction-dominated metallic ground states is still under debate, the topic has initiated substantial research on the experimental side as well as in theory over the last years.

In part III electron transport through semiconductor quantum dots is introduced and certain aspects of recent research are discussed. In particular, it treats electron transport through a Coulomb blockaded quantum ring system in which Aharonov–Bohm-type effects allow a surprisingly detailed understanding of the energy spectrum. In addition, the question of the ground-state spins in conventional singly connected quantum dots is addressed.

Part IV gives an overview over novel setups and experiments that aim at the local investigation of mesoscopic system's interiors by utilizing scanning probe techniques at cryogenic temperatures. The "marriage" of scanning probes, which are subject of a research field on their own, and mesoscopic semiconductor physics has turned out to be an extremely demanding enterprise for two reasons: first, it is experimentally very challenging to operate scanning force microscopes at temperatures where mesoscopic physics is observed in semiconductors and second, there are almost no detailed theories that allow the interpretation of scanned images or that teach the experimentalists how exactly to perform experiments in order to get meaningful results. However, exactly these two aspects have made research in this direction so exciting.

Sincere thanks are given to all those who have contributed to the presented work, in particular to those talented young scientists who have devoted some years of their career to this fascinating field of mesoscopic electronics. Although many more people are acknowledged at the end of this book, already at this point I wish to thank the key-players in our team at ETH Zurich, Volkmar Senz, Andreas Fuhrer, Stefan Lindemann, Jörg Rychen, Tobias Vancura, and Klaus Ensslin.

Zurich, March 2003 *Thomas Ihn*

Contents

Part I

Introduction to Electron Transport

1

Electrical conductance: Historical account from Ohm to the semiclassical Drude–Boltzmann theory

At the heart of the phenomenon of electrical conductance is *Ohm's law*

$$U = RI \tag{1.1}$$

empirically found by Georg Simon Ohm[1] and published in Refs. [1, 2] dating back to the years 1826 and 1827. It states that the application of a voltage U to a conductor gives rise to a current I proportional to U or, equivalently, if a current is driven through a conductor, a voltage will build up proportional to the current (see Fig. 1.1). The proportionality constant is called electrical resistance R, its inverse $G = R^{-1}$ is called electrical conductance. Over the last 175 years tremendous insight has been gained into the microscopic mechanisms leading to the appearance of a resistance in conducting materials.

The Hall effect, discovered by Edwin Herbert Hall[2] in 1879 [3, 4] opened the field of the so-called magnetotransport (or galvanomagnetic) phenomena meaning all the effects an external homogeneous magnetic field causes in a sample through which a current flows. Hall found that in a magnetic field normal to the current flow a voltage can be measured between two points whose connecting line is perpendicular to current and magnetic field (Hall effect, see Fig. 1.1). The Hall voltage U_{H} is proportional to the applied magnetic field B and the current I through the sample, i.e.

$$U_{\mathrm{H}} = R_{\mathrm{H}} B I \tag{1.2}$$

The constant of proportionality, R_{H}, is called Hall coefficient. In contrast, the longitudinal resistance of Hall's samples measured between points along the direction of the current flow was independent of the applied magnetic field.

[1] Georg Simon Ohm, ⋆ 16 March 1789, Erlangen, Germany, † 6 July 1854, Munich, Germany.

[2] Edwin Herbert Hall, ⋆ 7 November 1855, Gorham (Maine), USA, † 20 November 1938, Cambridge (Massachusetts), USA.

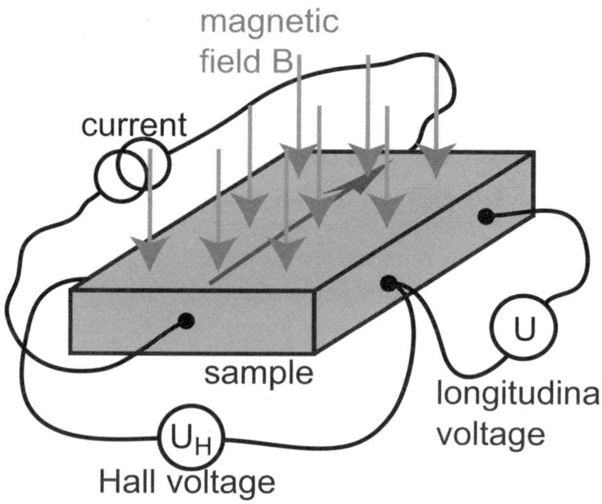

Fig. 1.1. Schematic drawing of a resistance measurement. A current source supplies the current I to the sample. The voltage drop U measured along the current flow is a measure of the ohmic resistance of the sample. If a magnetic field is applied normal to the current flow, a Hall voltage U_H builds up in the direction perpendicular to both the current and the magnetic field in the sample.

Historically, the discovery of the electron in 1897 by Sir Joseph John Thomson[3] offered an obvious mechanism for conduction in metals. The corresponding model, the *Drude theory of electrical and thermal conduction*, was presented by Paul Drude[4] at the end of the 19th century [5, 6] (see Ref. [7] for an excellent introduction). This model was capable of explaining Ohm's law and tracing the conductance back to a few material specific quantities (such as the electron density n and the scattering rate $1/\tau$ of the electrons) and to the geometry of a specific sample, e.g., its width W, its length L, and its thickness d in a rectangular solid. For this geometry, a geometry-independent specific resistivity (or simply, the resistivity) ϱ can be defined by

$$R = \varrho \times \frac{L}{Wd}.$$

The inverse of the resistivity $\sigma = \varrho^{-1}$ is called the conductivity. The Drude theory expression for the conductivity is

$$\sigma = \frac{ne^2\tau}{m}, \tag{1.3}$$

where m is the mass of the charge carriers. The Hall effect could also be described within this theory and became a method to determine the density and the sign of the

[3] Sir Joseph John Thomson, ⋆ 18 December 1856, Cheetham Hill (Manchester), England, † 30 August 1940, Cambridge, England; Nobel prize in physics 1906.

[4] Paul Drude, ⋆ 12 July 1863, Braunschweig, Germany, † 5 July 1906, Berlin, Germany.

charge carriers in conductors. The expression for the Hall coefficient [c.f. eq. (1.2)] is

$$R_{\mathrm{H}} = \frac{1}{end}. \tag{1.4}$$

The central idea of the Drude model is the notion that electrical resistance arises due to the presence of electron scattering in conducting materials characterized by the scattering time τ.

At that time, the kinetic theory of gases and Boltzmann's statistical theory including the kinetic equation for the distribution function (Boltzmann equation) were already available and concepts like the relaxation time approximation of the Boltzmann equation could be applied. However, it was still the classical Maxwell–Boltzmann distribution function that had to be used. It was, for example, predicted by Hendrik Antoon Lorentz[5] [8] that an energy-dependent relaxation time would have serious consequences for the resistance. It would lead to a temperature dependence and a magnetic field dependence of the resistance and to a magnetic field and temperature-dependent Hall coefficient. These predictions turned out later to be of considerable importance for electronic conduction in semiconductors.

With the advent of quantum mechanics, the whole theory of solids was revolutionized and all the concepts within solid-state physics familiar to physicists today, such as the band structure and the theory of phonons, were developed. The development of a quantum theory of electrical conduction started with the introduction of the Fermi–Dirac distribution function

$$f(E, \mu_{\mathrm{ch}}, T) = \frac{1}{\exp\left[(E-\mu_{\mathrm{ch}})/kT\right]+1} \tag{1.5}$$

in Boltzmann's kinetic equation by Arnold Sommerfeld[6] [9–12]. Here E is the energy of a quantum state, μ_{ch} is the chemical potential of the system under consideration, k is Boltzmann's constant, and T is the temperature. With this improvement, the theory of conduction in metals had the form still taught at universities today. The presence of the chemical potential μ_{ch} in the distribution function made the theory applicable to low temperatures and allowed the investigation of the so-called residual resistivity of metals. Introductions to the use of Boltzmann's theory for the description of electrical conduction can be found in Refs. [7, 13, 14].

Bloch's[7] theory of electrons moving in a periodic crystal lattice [15] made it necessary to revise the basics of the Drude–Boltzmann theory of conduction again. The concept of electronic conduction by electrons and holes was introduced within the semiclassical model of electron dynamics. In this model the classical momentum (velocity) is replaced by the crystal momentum, i.e., by the wave vector of a Bloch electron (or hole) times Planck's constant $\hbar$ in Newton's equations of motion. In a

[5] Hendrik Antoon Lorentz, ⋆ 18 July 1853, Arnheim, Netherlands, † 4 February 1928, Haarlem, Netherlands; Nobel prize in physics 1902 (with P. Zeeman).

[6] Arnold Sommerfeld, ⋆ 5 December 1868, Königsberg, Prussia, † 26 April 1951, Munich, Germany.

[7] Felix Bloch, ⋆ 23 October 1905, Zurich, Switzerland, † 10 September 1983, Zurich, Switzerland, Nobel prize in physics 1952 (with E.M. Purcell).

similar fashion, the wave vector also enters the distribution function for the charge carriers. In cases where electrons or holes have energies close to the minimum or maximum of a Bloch-band, a situation frequently occurring in semiconductor structures, an effective electron or hole mass $m^\star$ can be introduced and the charge carriers move effectively according to Newton's laws of motion. In summary, the Drude model of electrical conduction is still valid within the semiclassical description of electron dynamics if the electron mass m in eq. (1.3) is replaced by an effective electron or hole mass $m^\star$.

Within the semiclassical theory the question of the sources of electron scattering in solids remains unsettled. According to Bloch's theory an electron in the Bloch-state of a perfect crystal lattice does not lead to an electrical resistance. As a consequence, electron scattering important for the occurrence of a finite resistance can only occur by (1) impurities or lattice defects in the crystal, (2) by intrinsic deviations from perfect periodicity due to thermal motion of the ions in the lattice (phonon scattering), or (3) by scattering at the boundaries of a sample. It becomes obvious that the theory of electrical resistance must be intimately related to the scattering theory of electrons (see e.g. Ref. [14] for an overview over scattering mechanisms in semiconductors).

2

Toward the microscopic understanding of conductance on a quantum mechanical basis

2.1 Quantum transport in metals

When Lew Dawydowitsch Landau[1] published his paper on the diamagnetism of metals [16], he was probably not aware of the importance of his calculation for quantum transport. In this publication he determined the quantized energy levels of free electrons in a homogeneous magnetic field

$$E_n = \hbar\omega_c \left(n + \frac{1}{2}\right),$$

where $n \geq 0$ is an integer and $\omega_c = eB/m$ is the cyclotron frequency. Similar calculations emphasizing the role of the sample edges were performed by V. Fock[2] [17], C.G. Darwin [18], and Edward Teller[3] [19]. The energy levels that we now call Landau levels played their first important role in experiments on the magnetoresistance of bismuth performed by L. Shubnikov and Wander Johannes de Haas[4] in 1930 [20]. They found that the measured magnetoresistance oscillated as a function of magnetic field, an effect now called *Shubnikov–de Haas effect*. A theoretical description of this effect was, for example, given in Ref. [21].

A full quantum mechanical description of electron transport was developed between 1950 and 1960 as a special case of quantum mechanical linear response theory. Kubo [22, 23] and Greenwood [24] were the first to propose a quantum theory of the conductance culminating in what is today known as the *Kubo–Greenwood formula* for the electrical conductance. It is based on linear response theory and takes at zero frequency and zero magnetic field the form [25]

---

[1] Lew Dawydowitsch Landau, ⋆ 22 January 1908, Baku, USSR, † 1 April 1968, Moscow, USSR; Nobel prize in physics 1962.

[2] Vladimir Alexandrovich Fock, ⋆ 22 December 1898, St Petersburg, Russia, † 27 December 1974, Leningrad, Russia.

[3] Edward Teller, ⋆ 15 January 1908, Budapest, Hungary

[4] Wander Johannes de Haas, ⋆ 2 March 1878, Lisse, Holland, † 26 April 1960, Bilthoven, Netherlands.

$$\boldsymbol{j}(\boldsymbol{x}) = \int \mathrm{d}^3 x' \; \sigma(\boldsymbol{x}, \boldsymbol{x}') \boldsymbol{\mathcal{E}}(\boldsymbol{x}'),$$

where

$$\sigma(\boldsymbol{x}, \boldsymbol{x}') = h \int \mathrm{d}\alpha \int \mathrm{d}\beta \left[-\frac{\mathrm{d}f(E_\alpha)}{\mathrm{d}E} \right] \delta(E_{\beta\alpha}) \boldsymbol{j}_{\beta\alpha}(\boldsymbol{x}) \otimes \boldsymbol{j}_{\alpha\beta}(\boldsymbol{x}'). \qquad (2.1)$$

Here the notion is that on the application of an electric field $\boldsymbol{\mathcal{E}}(\boldsymbol{x})$ the current created is described by the local current density $\boldsymbol{j}(\boldsymbol{x})$, which is a non-local microscopic reformulation of Ohm's law in eq. (1.1). The zero-frequency limit excludes time-dependent effects such as phonon scattering. We have introduced the non-local conductivity tensor $\sigma(\boldsymbol{x}, \boldsymbol{x}')$, which can be calculated from the knowledge of the matrix elements of the current density $\boldsymbol{j}_{\alpha\beta}(\boldsymbol{x})$. The $\otimes$ symbol represents the tensor product of the two vectors. The basis states $|\alpha\rangle$ used for this representation are the exact eigenstates of the system's Hamiltonian including the disorder potential created by residual impurities and crystal imperfections at zero applied electric field. The quantities $E_{\alpha\beta} = E_\alpha - E_\beta$ are differences of eigenenergies. No impurity averaging has been performed in eq. (2.1). The temperature dependence of the conductivity enters through the Fermi–Dirac distribution function $f(E)$ [see eq. (1.5)]. Since in typical conduction problems the local electric $\boldsymbol{\mathcal{E}}(\boldsymbol{x})$ field is not known, it is frequently assumed to be homogeneous and the spatially averaged current density is calculated, resulting again in a local conductivity tensor.

It was shown, for example, by Kohn[5] and Luttinger [26] that such a quantum mechanical linear response theory for the conductance leads to the Drude–Boltzmann–Sommerfeld result in lowest order, if the electric field is homogeneous and *the average is taken over impurity configurations* in the system. The distribution function of the Boltzmann equation can be identified with the diagonal matrix elements of the density matrix in momentum representation. For macroscopic systems the familiar Boltzmann result for the energy-dependent elastic scattering rate [7, 27]

$$\frac{1}{\tau(E)} = N_\mathrm{i} \frac{m^\star}{(2\pi)^{d-1}\hbar^3} \left(\frac{2m^\star E}{\hbar^2} \right)^{d/2-1} \int \mathrm{d}\Omega_d \left| V(q) \right|^2 (1 - \cos\theta) \qquad (2.2)$$

is recovered, in which d is the dimension of the system, Ω_d is the solid angle in d dimensions, N_i denotes the volume density of the scatterers, $V(q)$ is the Fourier transform of the potential of an individual scatterer and $q = 2k\sin(\theta/2)$ and $k = \sqrt{2m^\star E/\hbar^2}$. The scattering time τ enters the expression for the conductivity in eq. (1.3). Equation (2.2) will be of particular importance in the analysis of our experimental results in part II of this book. Kohn and Luttinger recognized that beyond the semiclassical theory interference effects between scattered electron waves would lead to higher-order corrections. An important aspect of this approach to the description of transport is the fact that only properties of states at the Fermi-energy

[5] Walter Kohn, ⋆ 9 March 1923, Vienna, Austria, Nobel prize in chemistry 1998 (with J.A. Pople).

of the system enter the calculation of the conductance at zero temperature. The low-temperature conductance can therefore give some insight in the nature of the ground state of electronic systems.

At about the same time Rolf Landauer proposed a different theoretical view on the appearance of the resistance [28]. He considered the flow of electrons past an individual scattering center and predicted the presence of a resistivity dipole, today known as *Landauer's resistivity dipole*, around the scatterer. Such a resistivity dipole is essentially made up of electrons accumulating upstream and lacking downstream of the impurity, i.e., the electron gas is polarized around the impurity similar to a polarized atom in a dielectric material. For a one-dimensional conductor supporting a single mode, he found a formula for the conductance in the linear transport regime today known as the *Landauer formula* [29–32]

$$G = \frac{2e^2}{h}\frac{T}{1-T}, \tag{2.3}$$

in which T is the transmission probability for electrons through the conductor. This approach was later extended to the case of multimode and multiprobe conductors by M. Büttiker [33], leading to the Landauer–Büttiker formalism of the conductance. Its relation to the Kubo–Greenwood approach was clarified by Baranger and Stone in Ref. [34].

The quantum mechanical transmission is also of central importance for tunneling phenomena and Landauer's view on the resistance can also be applied to tunneling transport. Historically, already in 1928, i.e., soon after the formulation of quantum mechanics, Fowler and Nordheim investigated the field emission of electrons from metals [35]. The concept of inter-band tunneling in solids was proposed by Zener in 1943 [36] and in 1957, the Esaki diode[6] [37] made use of this concept. Tunneling through thin insulating layers from normal metals into superconductors [38] or between superconductors [39] studied around 1960 by I. Giaever[7] was the experimental technique used for verifying the existence of a superconducting energy gap and for measuring the density of states in superconductors. On the theoretical side, the transfer Hamiltonian approach was developed by John Bardeen[8] in Ref. [40]. The early work on tunneling in solids has been reviewed in Ref. [41].

[6] After Leo Esaki, $\star$ 12 March 1925, Osaka, Japan; Nobel prize in physics 1973 (with I. Giaever and B.D. Josephson)

[7] Ivar Giaever, $\star$ 5 April 1929, Bergen, Norway; Nobel price in physics 1973 (with L. Esaki and B.D. Josephson).

[8] John Bardeen, $\star$ 23 May 1908, Madison (Wisconsin), USA, † 30 January 1991, Boston, USA; Nobel prizes in physics 1956 (with W.H. Brattain and W. Shockley) and in 1972 (with L.N. Cooper and J.R. Schrieffer).

2.2 Transistors and two-dimensional electron gases in semiconductors

Parallel to the developments in fundamental research on conductance in the late 1950s, the remarkable technological success of microelectronics based on semiconducting materials began. In 1947/48 the transistor[9] was developed by John Bardeen, Walter Houser Brattain,[10] and William Shockley[11] at the Bell Laboratories on the basis of germanium. In 1954 the silicon technology started, the first transistor-based computer worked in 1955 in the United States and the concept of the integrated circuits was invented in 1958 by Jack St. Claire Kilby, engineer of Texas Instruments.

2.2.1 Two-dimensional electron gases in field-effect transistors

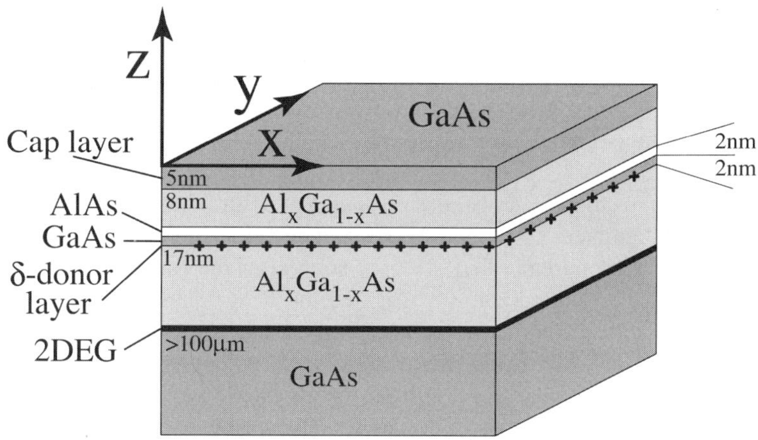

Fig. 2.1. Layer sequence in a GaAs/AlGaAs heterostructure. The two-dimensional electron gas (2DEG) forms at the interface between the two materials, here 34 nm below the sample surface.

From these early days on, fruitful mutual interactions began between the semiconductor industry and fundamental physics research. With the field-effect transistors,[12] two-dimensional electron gases became available for fundamental research

[9] From: **trans**fer res**istor**; the planar bipolar transistor was patented in 1948, mass production started in 1951; already in 1963 the worldwide production of transistors was 1 billion pieces, higher than the established electron tube.

[10] Walter Houser Brattain, ⋆ 10 February 1902, Amoy (today Xiamen, China), † 13 October 1987, Seattle, USA; Nobel prize in physics 1956 (with J. Bardeen and W. Shockley).

[11] William Shockley, ⋆ 13 February 1910, London, England, † 12 August 1989, Stanford (California), USA; Nobel prize in physics 1956 (with J. Bardeen and W.H. Brattain).

[12] the most famous variant is the silicon based MOSFET – **M**etal-**O**xide **F**ield **E**ffect **T**ransistor

in the 1960s and 1970s, initially based on the elemental semiconductors, silicon and germanium, later also on III-V compounds like gallium arsenide, aluminium arsenide, and others. These systems opened a new research area, namely, the physics of the electronic properties of two-dimensional systems (see Refs. [42, 43]). Figure 2.1 shows an example of a GaAs/AlGaAs heterostructure similar to the ones used in parts III and IV of this book for experiments. The two-dimensional electron gas (2DEG) accumulates at the interface between the two materials. Structures with very high crystal purity (typically less than 10^{14} cm^{-3} residual impurity concentration) can, for example, be grown by molecular beam epitaxy (MBE). Growth is based on GaAs (100) wafers on top of which a thick layer of GaAs is grown. Subsequently, the growth continues with an AlGaAs alloy that is almost perfectly lattice-matched to the GaAs cystal. Atomically sharp interfaces between these materials can be realized without introducing lattice distortions or interrupting the lattice periodicity. A lattice plane remote from the GaAs/AlGaAs interface is doped with silicon incorporated mainly on gallium lattice sites (n-type δ-doping). The layer sequence is completed by burying the doping plane in AlGaAs and capping the structure with a thin GaAs layer in order to prevent strong surface oxidation of the AlGaAs.

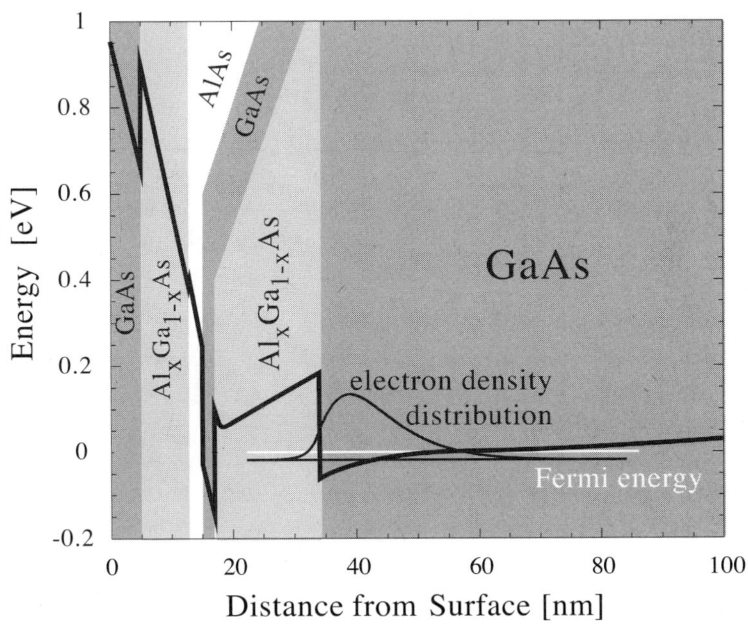

Fig. 2.2. Self-consistent conduction band profile of the GaAs/AlGaAs heterostructure

The resulting self-consistent conduction band structure of such a sample is shown in Fig. 2.2. The electrochemical potential of the system is pinned deep inside the GaAs substrate layer, typically close to the middle of the bandgap. At the interface between GaAs and AlGaAs, a triangular potential well gives rise to a single occupied quantum state confining electrons to a thin sheet (about 10 nm thick). The thickness of the sheet in growth direction is given by the extent of the quantum state and can in principle not be made thinner due to arguments based on the uncertainty principle. The conduction band minimum rises strongly toward the surface, since at the surface the electrochemical potential is pinned by surface states in the bandgap of GaAs. The presence of this built-in barrier together with the insulating character of the AlGaAs barrier material allows the application of electric fields between the two-dimensional electron gas and a metallic gate electrode evaporated onto the surface of such a sample. The electron density can then be controlled via the field effect.

Electron transport in two-dimensional electron gases is routinely characterized and described in terms of the Drude–Boltzmann theory of transport in metals described above. Magnetotransport experiments in the spirit of Fig. 1.1 apply a current I to the two-dimensional system patterned into a Hall bar structure and the longitudinal and transverse voltages, U and U_{H}, respectively, are measured. If W is the width of the Hall bar and L is the separation between longitudinal voltage probes, the components of the two-dimensional resistivity tensor

$$\varrho = \begin{pmatrix} \varrho_{xx} & \varrho_{xy} \\ -\varrho_{xy} & \varrho_{xx} \end{pmatrix}$$

can be determined to be

$$\varrho_{xx} = \frac{W}{L}\frac{U}{I} \quad \text{and} \quad \varrho_{xy} = \frac{U_{\mathrm{H}}}{I}.$$

The corresponding conductivity tensor σ is obtained from ϱ by tensor inversion. The electron density (Hall density) n_{H} can then be determined from the low magnetic field Hall resistivity, which is linear in magnetic field B

$$n_{\mathrm{H}} = \frac{1}{e\,\mathrm{d}\varrho_{xy}(B)/\mathrm{d}B|_{B=0}}$$

and the Hall mobility $\mu_{\mathrm{H}} = e\tau/m^{\star}$ of the electrons follows from

$$\mu_{\mathrm{H}} = \frac{\mathrm{d}\varrho_{xy}(B)/\mathrm{d}B|_{B=0}}{\varrho_{xx}(0)}.$$

Typical values for electron densities in such two-dimensional systems at liquid helium temperatures range from $10^{14}\,\mathrm{m}^{-2}$ to $10^{16}\,\mathrm{m}^{-2}$, mobilities are typically between $0.1\,\mathrm{m}^2$/(Vs) to $1000\,\mathrm{m}^2$/(Vs). From the mobility μ_{H} one finds the scattering time τ, if the effective mass of the charge carriers is known. For electrons in GaAs, for example, the effective mass $m^{\star} = 0.067$ and a mobility of $100\,\mathrm{m}^2$/(Vs) corresponds to a scattering time $\tau = 38\,\mathrm{ps}$. The Hall density allows the determination of the Fermi wave vector of the electrons via

$$k_{\mathrm{F}} = \sqrt{2\pi n_{\mathrm{H}}}$$

and the related Fermi wavelength λ_F, Fermi velocity v_F, and Fermi energy E_F. For an electron density of $5 \times 10^{15}\,\mathrm{m}^{-2}$ in GaAs, one obtains $\lambda_\mathrm{F} = 2\pi/k_\mathrm{F} = 35\,\mathrm{nm}$, $v_\mathrm{F} = \hbar k_\mathrm{F}/m^\star = 3 \times 10^5$ m/s, and $E_\mathrm{F} = \hbar^2 k_\mathrm{F}^2/(2m^\star) = 17.9\,\mathrm{meV}$. Combining information about the electron density and the mobility, the elastic mean free path can be determined from

$$l_\mathrm{e} = v_\mathrm{F}\tau$$

and the diffusion constant is given by

$$D = \frac{1}{2}v_\mathrm{F}^2\tau = \frac{1}{2}v_\mathrm{F}l_\mathrm{e}.$$

In a high-quality two-dimensional electron gas in GaAs with $n_\mathrm{H} = 5 \times 10^{15}\,\mathrm{m}^{-2}$ and a mobility $\mu_\mathrm{H} = 100\,\mathrm{m}^2/(\mathrm{Vs})$, one has $D = 1.8\,\mathrm{m}^2/\mathrm{s}$ and $l_\mathrm{e} = 11.7\,\mu\mathrm{m}$, which is already a macroscopic length scale. Similar electron gases form the basis for the measurements discussed in parts III and IV of this book.

The Shubnikov–de Haas effect in two-dimensional systems is another common measurement allowing the determination of the carrier density. The observed quantum oscillations in the resistance are periodic in $1/B$ [42]. The period $\Delta(1/B)$ is related to the Shubnikov–de Haas density via

$$n_\mathrm{SdH} = \frac{2e}{h\Delta(1/B)},$$

if spin-degenerate Landau levels are assumed. If the two-dimensional electron or hole gas is in the quantum limit (only one subband is occupied), then the Shubnikov–de Haas density and the Hall density are typically measured to be the same within a few percent. A more detailed analysis of the Shubnikov–de Haas effect allows the extraction of a quantum life-time of an electron in a Landau level and of the effective mass $m^\star$ [44, 45].

2.2.2 Resonant tunneling in semiconductors

In addition to the field-effect transistors, the advances in semiconductor material quality made the development of double-barrier resonant tunneling structures possible. In these structures, a quantum well extended in two dimensions is coupled via tunneling barriers to emitter and collector electrodes. The weak coupling leads to well separated quantized energy levels in the quantum well. If on the application of an external bias voltage occupied emitter states become resonant with quantum well states, resonant tunneling transport can occur through the structure. A further increase in the bias voltage detunes the resonant levels and as a consequence a negative differential resistance arises. This was first demonstrated by Leo Esaki and coworkers in the early 1970s [46, 47]. The resonant tunneling effect has to be mentioned here, because it is at the heart of tunneling transport through semiconductor quantum dot structures, which will be discussed in part III of this book.

2.2.3 Integer and fractional quantum Hall effect

In the early 1980s research on magnetotransport in two-dimensional electron gases was greatly stimulated by the experimental observation of important quantum phenomena: the integer quantum Hall effect found in silicon MOSFETs by Klaus von Klitzing[13] in 1980 [48] impressively showed the precise quantization of the Hall resistance U_{H}/I in integer fractions of the resistance quantum h/e^2 [c.f. eqs. (1.2) and (1.4)]:

$$\frac{U_{\mathrm{H}}}{I} = \frac{h}{e^2}\frac{1}{\nu},$$

where $\nu > 0$ is an integer called the "filling factor" of Landau levels. Figure 2.3

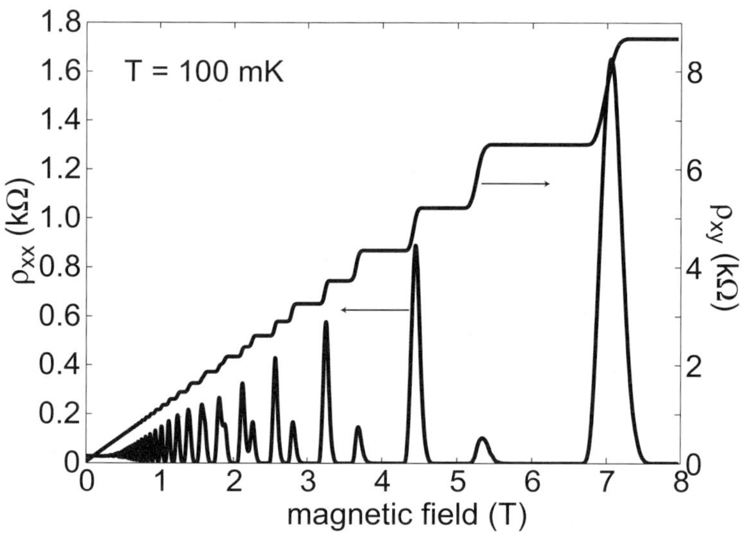

Fig. 2.3. Quantum Hall effect measured in a GaAs/AlGaAs heterostructure at a temperature of 100 mK. At magnetic fields where the Hall resistivity $\varrho_{xy} = U_{\mathrm{H}}/I$ shows plateaus, the longitudinal resistivity ϱ_{xx} is exceedingly small.

shows the quantum Hall effect measured on a high-quality GaAs/AlGaAs heterostructure at a temperature of 100 mK. The quantization of the Hall resistance at integer fractions of h/e^2 can be seen in the Hall resistivity ϱ_{xy}. At magnetic fields where the Hall resistivity shows plateaus, the longitudinal resistivity of the sample, ϱ_{xx}, is close to zero. Büttiker suggested a phenomenological and transparent description of the integer quantum Hall effect in terms of the transmission of edge channels [49]. The self-consistent nature of edge channels predicted by Chklovskii and coworkers in Ref. [50] will be further discussed in the scanning gate experiments described in chapter 16.

[13] Klaus von Klitzing, ⋆ 28 June 1943, Schroda, Poland; Nobel prize in physics 1985.

The fractional quantum Hall effect, where plateaus in ϱ_{xy} develop at fractional values of the filling factor ν, was discovered in gallium arsenide heterostructures by Tsui[14] and Störmer[15] [51] is based on the Coulomb interaction between electrons in two-dimensional systems leading to novel ground states. For example, an excellent approximate wave function for the $\nu = 1/3$ correlated many-body ground state was suggested by R. Laughlin[16] [52]. The fractional quantum Hall effect can be described as the integer quantum Hall effect of novel quasiparticles called composite fermions (see e.g. Refs. [53–56] for an introduction to the fractional quantum Hall effect).

2.2.4 Weak localization

In the late 1970s, two-dimensional electron gases were predicted to become insulating at zero magnetic field as the temperature approaches absolute zero. The predictions were based on the quantum interference of electrons leading to localization no matter how small the disorder potential due to residual crystal imperfections, an effect known as "weak localization" [57]. It turns out that the effect is closely related to the interference of elastically back-scattered electron waves. As a result of this finding, experimentalists working on two-dimensional systems were able to study phase-coherence phenomena — a still insufficiently understood topic reaching into fundamental quantum mechanics. The weak localization effect has a logarithmic temperature dependence, and it can be suppressed by applying a magnetic field normal to the plane of the two-dimensional system. This is shown in Fig. 2.4 for a p-SiGe sample measured at a temperature of 100 mK. The weak localization effect

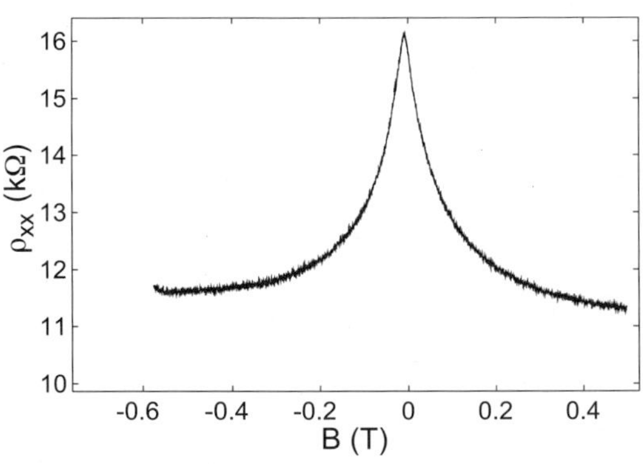

Fig. 2.4. Measurement of the suppression of the weak localization effect in a magnetic field. The sample is a p-SiGe quantum well with a two-dimensional hole gas.

will be discussed in much more detail in part II of this book where it plays a very

[14] Daniel C. Tsui, ⋆ 1939, Henan, China; Nobel prize in physics 1998 (with H.L. Störmer and R.B. Laughlin).

[15] Horst L. Störmer, ⋆ 6 April 1949, Frankfurt am Main, Germany; Nobel prize in physics 1998 (with R.B. Laughlin and D.C. Tsui).

[16] Robert Laughlin, ⋆ 1 November 1950, Visalia, California, USA; Nobel prize in physics 1998 (with H. Störmer and D.C. Tsui).

important role in connection with a possible metal–insulator transition in strongly interacting two-dimensional systems at zero magnetic field.

2.3 Basic phenomena in semiconductor structures of reduced size and dimensionality

2.3.1 The Aharonov–Bohm effect and conductance fluctuations

Already in 1939 it was pointed out by Franz [58] that two electron beams enclosing a magnetic flux Φ between them would acquire a relative phase given by

$$\Delta\varphi = 2\pi\frac{\Phi}{\Phi_0},$$

where $\Phi_0 = h/e$ is the flux quantum. Later, Aharonov and Bohm[17] realized, that in contrast to the classical motion of a charged particle in an electromagnetic field where only the fields play a role, in the quantum description the electromagnetic potentials cannot be eliminated from the equations of motion [59]. They suggested interference experiments between electron waves that could prove the physical significance of the electromagnetic potentials. One of these experiments is schematically depicted in Fig. 2.5. This fundamental effect, nowadays called Aharonov–Bohm ef-

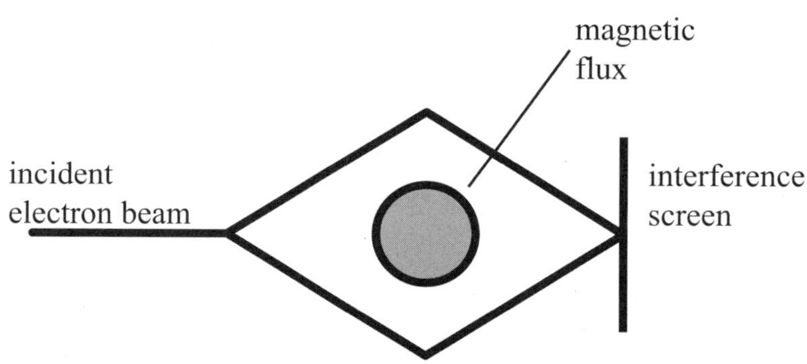

Fig. 2.5. Scheme of the interference experiment suggested by Aharonov and Bohm

fect, is not only the basis for the microscopic understanding of the weak localization phenomenon, but it became of great importance in the field of semiconductor structures with reduced dimensionality, which developed rapidly in the 1980s, certainly also inspired by the industrial requirements of smaller and faster electronic devices. In 1985 first indications of the Aharonov–Bohm effect were observed in

[17] David Joseph Bohm, ⋆ 26 December 1917, Wilkes-Barre, Pennsylvania, USA, † 27 October 1992.

double-quantum well structures [60] and later also in laterally defined semiconducting ring-shaped systems [61–64], at a time when this phenomenon was already well studied in metallic systems [65, 66]. A good introduction to the phenomenon is given in Ref. [67]. Figure 2.6a shows the Aharonov–Bohm effect measured in a ring structure of $d = 1\,\mu\text{m}$ diameter, realized on a GaAs/AlGaAs heterostructure. Periodic oscillations of the resistance as a function of magnetic field are clearly visible. The period of about 4 mT corresponds to the magnetic flux $\Phi = B\pi d^2/4$ through the ring being an integer number times the flux quantum Φ_0. This Aharonov–Bohm period will play an important role in chapter 12, where we show measurements of the Coulomb-blockade effect (see below) in a ring geometry.

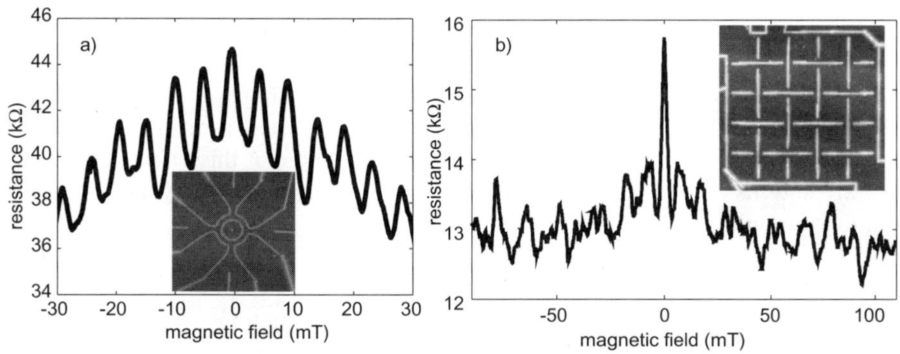

Fig. 2.6. a) Aharonov–Bohm effect measured at a temperature of 30 mK in the quantum ring structure shown in the inset. The current driven through the ring was 1.7 nA. The period of the oscillations of about 4 mT corresponds to the area of the ring with radius $r = 0.5\,\mu\text{m}$. b) Conductance fluctuations measured in a sample in which 1 µm square cavities are arranged in a 5×5 array (see inset). Neighboring cavities are coherently coupled via small openings. The characteristic field scale of the fluctuations is again 4 mT corresponding to 1 µm^2, the area of an individual cavity.

The search of the Aharonov–Bohm effect in metals and also in semiconductor structures was hindered by the presence of another related interference effect, namely, the conductance fluctuations, which manifest themselves as aperiodic but reproducible fluctuations in the resistance of a sample as a function of an external parameter such as the magnetic field or the electron density. Conductance fluctuations occur in samples whose characteristic size L is of the order of the phase coherence length l_φ of the electrons. They occur in systems where the elastic mean free path of the electrons l_e is much smaller than L (then called universal conductance fluctuations, or UCF), or in classically chaotic systems if $l_\text{e} \gg l_\varphi$ (then called ballistic conductance fluctuations). A substantial discussion of conductance fluctuations can be found in Refs. [30, 32]. An example of ballistic conductance fluctuations is depicted in Fig. 2.6b [68]. They were measured at 100 mK on a sample in which an array of 5×5 ballistic cavities of size $1 \times 1\,\mu\text{m}^2$ was realized. Neighboring cavities were interconnected by narrow constrictions. The sample was based on a GaAs/AlGaAs heterostructure.

In general, structures in which the electronic phase coherence length l_φ is comparable to the characteristic system size L but much larger than the Fermi wavelength λ_F are called *mesoscopic systems*. Depending on the relation between the mean free path of the electrons l_e and the system size L, one talks about ballistic ($l_\mathrm{e} \gg l_\varphi$) or diffusive ($l_\mathrm{e} \ll l_\varphi$) systems.

2.3.2 Conductance quantization in semiconductor quantum point contacts

In 1988 the quantization of the conductance in very narrow constrictions in a two-dimensional electron gas was discovered by van Wees and coworkers [69] and Wharam and coworkers [70]. They were able to vary the width of the constriction from less than one to many Fermi wavelengths by applying voltages to a split-gate. They found that with increasing gate voltage, i.e., increasing width of the constriction, the conductance did not increase linearly, as expected from purely classical arguments, but it increased in steps of $2e^2/h$. An example of this effect is shown in Fig. 2.7. It turns out that the effect can most naturally be described by the Landauer–Büttiker formula

$$G = \frac{2e^2}{h} NT, \tag{2.4}$$

where N is the number of transverse modes in the constriction with average transmission $T = 1$. This formula differs from eq. (2.3) in the appearance of the factor N and in the fact that only the transmission T appears instead of $T/(1-T)$. In fact, the latter goes to infinity as T approaches 1. The reason for a finite conduc-

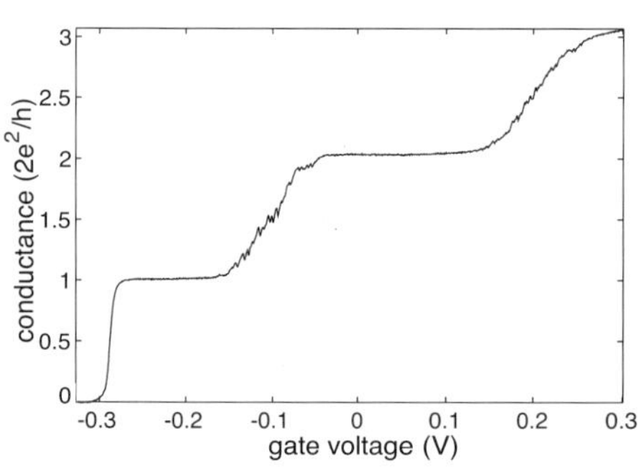

Fig. 2.7. Conductance quantization in a quantum point contact realized on a GaAs/AlGaAs-heterostructure. The plateaus in the conductance are reminiscent of the plateaus in the quantum Hall effect in Fig. 2.3.

tance despite the absence of scattering is the fact that the electron gas has to couple adiabatically from a very large (infinitely large) reservoir into a narrow constriction, which gives rise to the "contact" resistance $R_\mathrm{c} = h/(2e^2N)$. If this resistance is coupled in series with the pure resistance of the constriction in analogy to eq. (2.3), $R_\mathrm{s} = [h/(2e^2N)](1-T)/T$, then we indeed recover eq. (2.4) [31]. Further details about the conductance quantization can be found in Refs. [30–32]. A very clear discussion of the effect in terms of the adiabatic approximation was given in Refs. [32, 71]. Although quantum point contacts are interesting research objects themselves (see e.g. Ref. [72] for an interesting recent publication), they also form the basic element for the fabrication of more complex nanostructures such as quantum dots (see below in this chapter and part III).

If constrictions are made long and thin, the resulting structures are called quantum wires. These systems are interesting because they are tunable one-dimensional systems. In the terminology of mesoscopic physics, a system becomes one dimensional as soon as its width W becomes smaller than the phase-coherence length l_φ. Such wires will be investigated with scanning gate techniques in chapter 17. A strictly one-dimensional system in the quantum mechanical sense, however, is reached only if W is comparable to $\lambda_{\mathrm{F}}/2$, which means that only one mode is occupied. Experimentally, such systems are difficult to realize. Even in wider mesoscopically one-dimensional systems, the quantization of the conductance is lost because imperfections in the wires do not allow T to approach unity. The resulting backscattering in such wires leads to the observation of conductance fluctuations. Exceptions are very pure quantum wire fabricated by cleaved-edge overgrowth, which indeed show conductance quantization in spite of their length [73].

2.3.3 Semiconductor quantum dots and artificial atoms

At about the same time as the discovery of conductance quantization, another important effect was found in mesoscopic semiconductor structures and in mesoscopic metallic devices. If a small conducting island of charge is coupled to two large electron reservoirs (source and drain contacts) via quantum point contacts driven into pinch-off, the conductance of the device becomes much smaller than the conductance quantum $2e^2/h$. The current–voltage characteristics of such devices are highly nonlinear at very low temperatures (about 100 mK) showing an exponential suppression of the current at low bias. When a gate electrode is placed next to the island, the conductance as a function of gate voltage exhibits a series of thermally broadened peaks with extremely small conductance in between. Such a measurement made on a quantum dot realized on a GaAs heterostructure is shown in Fig. 2.8. The suppression of the conductance in the valleys between peaks is called the Coulomb-blockade effect. Its origin is the classical Coulomb repulsion of electrons. It leads to a charging energy of the almost isolated island characterized by the energy scale

$$\Delta E_{\mathrm{c}} = \frac{e^2}{2C}.$$

Here, C is the capacitance of the island. The separation of neighboring conductance peaks is closely related to the charging energy. While in the Coulomb-blockade regions the number of electrons on the island is constant, on a conductance peak it can change by one thus allowing a current to flow. On a conductance peak, a single electron enters and leaves the island before the next electron can enter. In this sense, such islands are called *single-electron transistors* (SETs). Although experiments on metallic grains embedded in thin insulating films were already investigated

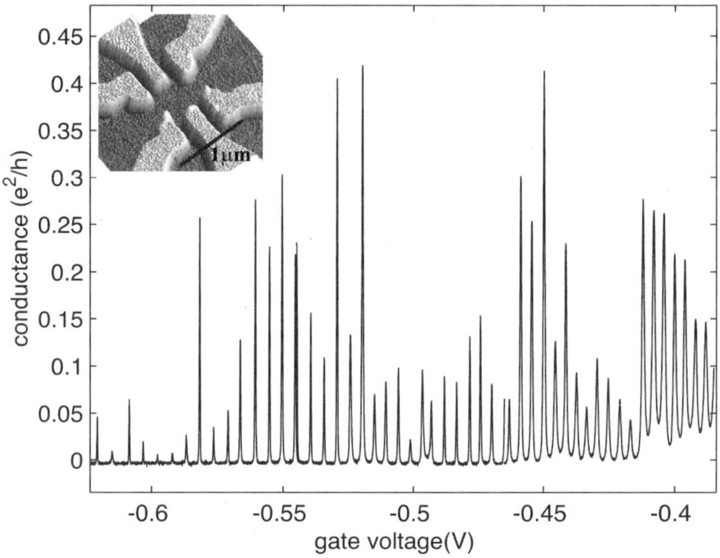

Fig. 2.8. Coulomb blockade in the conductance of a semiconductor quantum dot. The inset shows a scanning electron microscope image of the gate-defined quantum dot structure.

in the 1950s and 1960s [74–76] and a transport theory was developed by Kulik and Shekhter [77], the first indication for the Coulomb-blockade effect in an individual artificially fabricated metallic SET was reported by Fulton and Dolan in Ref. [78]. First results in semiconducting samples were obtained by Scott-Thomas and coworkers [79] looking at inhomogeneous wires in Si-MOSFET inversion channels. In the 1990s, semiconducting islands containing very few (down to one) electrons could be fabricated by Tarucha and coworkers [80, 81]. These quantum dots showed shell filling effects and are therefore referred to as *artificial atoms*. In 1997 the Kondo effect in an artificial quantum dot was found by Goldhaber-Gordon and coworkers [82]. This effect, which is well known from dilute magnetic alloys (see Ref. [83]) had been predicted to occur in quantum dots with an unpaired spin in the regime of relatively strong coupling to source and drain [84–86].

Electron transport through quantum dots is discussed in a number of books [32, 87–90] and review articles [81, 91–95]. An introduction to the subject is given in chapter 11 as a preparation for the subsequent description of measurements on quantum dots and quantum rings.

Part II

Conductance in Strongly Interacting and Disordered Two-Dimensional Systems

The physics of electron transport in two-dimensional systems has been intensely investigated in the 1970s and early 1980s. With the availability of Si-MOSFETs (Silicon Metal-Oxide-Field-Effect-Transistors) and GaAs/AlGaAs heterostructures, two-dimensional electron gases could be probed at low temperatures and in magnetic fields. The effects of disorder and interactions could be demonstrated convincingly. The theoretical description has been built on Fermi-liquid concepts and found to agree well with experiments.

With increasing sample quality (in both material systems, Si and GaAs) and the advent of other high-quality two-dimensional systems like p-GaAs heterostructures or n- and p-SiGe structures, a new range of parameters became accessible. Carrier densities could be driven to very low values, where interactions among electrons or holes become more and more important, without entering the strongly localized regime. In the mid-1990s experiments were performed that had an enormous impact on the community and lead to speculations about a metal–insulator transition in two-dimensional systems at zero magnetic field.

In the following chapters a perspective of this very active field of research is presented and results obtained on the p-SiGe system are discussed in detail. In chapter 3 to 7 we introduce the well-established theoretical concepts based on Fermi-liquid theory and beyond. In chapters 8 and 9 an overview is given of experiments and theories related to the possible metal–insulator transition in two-dimensions. Chapter 10 will be devoted to experiments performed on p-SiGe quantum wells which exhibit features of the metal–insulator transition, their analysis, and interpretation.

3

The concept of metals and insulators

According to the quantum mechanical concept of metals and insulators introduced by Wilson [96, 97], metallic conductivity is expected for systems in which the density of states at the Fermi level remains finite at zero temperature. In contrast, an insulator exhibits a gap in the density of states at the Fermi level. This situation is schematically depicted in Fig. 3.1. It was realized later that this condition is not suf-

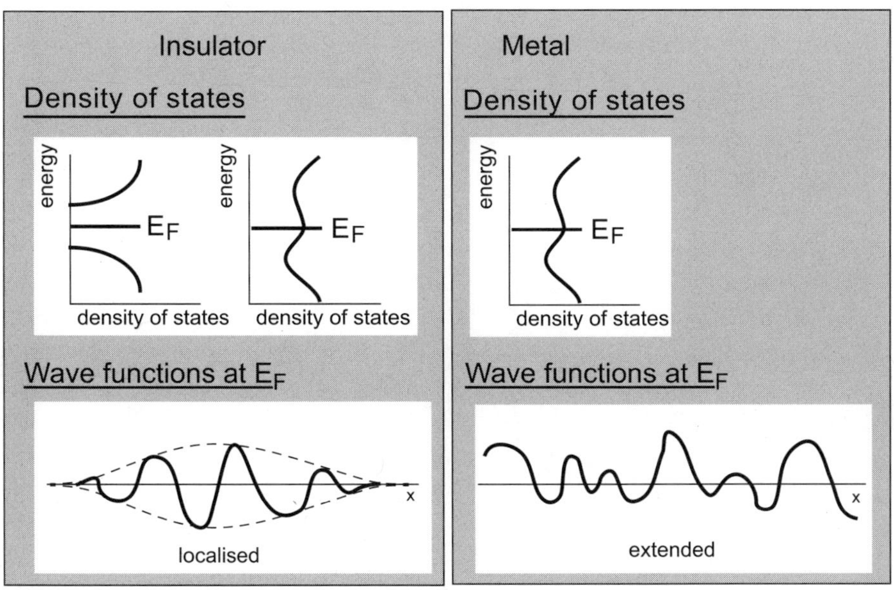

Fig. 3.1. Concept of metals and insulators. If at zero temperature the density of states at the Fermi level is finite, the character of the wave function at the Fermi energy is decisive for metallic or insulating behavior. Localized (extended) wave functions lead to insulating (metallic) behavior.

ficient in disordered conductors, where the nature of the wave function at the Fermi level is decisive [98]. Extended wave functions are typical for a metal, while localized wave functions lead to insulating behavior [99] (see Fig. 3.1). These concepts, however, rely on a single-particle description and it will be discussed in later sections how interactions change this notion of metals and insulators.

Measurements of the resistivity have proven to give experimental access to the nature of the ground state at low temperatures. In the framework of the Kubo–Greenwood formula for the conductivity based on linear response [c.f. eq. (2.1)] it can be shown that localized wave functions lead to zero conductivity, whereas extended wave functions lead to zero or finite resistivity at zero temperature. However, because the absolute zero is not accessible for experiment, the criterion $\mathrm{d}\varrho/\mathrm{d}T > 0$ measured at the lowest accessible temperatures is taken as an indication for a metallic ground state, whereas $\mathrm{d}\varrho/\mathrm{d}T < 0$ hints toward an insulating ground state.

4

Scaling theory of localization

In two-dimensional (2D) systems it was suggested theoretically by Abrahams and coworkers [57] that the ground state at zero temperature is of an insulating nature. This prediction was based on a picture of non-interacting electrons. In their paper these authors present scaling arguments indicating that the conductance G of a disordered electronic system depends on its size L in a universal way. They introduce the scaling function

$$\beta(g) = \frac{\mathrm{d}\ln g}{\mathrm{d}\ln L},$$

where $g = G/(e^2/\hbar)$ is the dimensionless conductance and derive asymptotic forms for this function. The scaling function β describes the evolution of the conductance, when the size of the system is increased. For any dimension d, at large g the ohmic limit must be reached leading to [57, 100]

$$\beta(g) \rightarrow d - 2 \text{ for } g \rightarrow \infty.$$

For small g, however, exponential localization is expected and [57, 100]

$$\beta(g) \rightarrow \ln(g/g_0) \text{ for } g \rightarrow 0,$$

where g_0 is some constant of order unity. Abrahams and coworkers argued that within an Anderson model in two dimensions, a perturbation analysis in the disorder will give corrections to the ohmic limit that guarantee

$$\beta(g) \rightarrow -g_{\mathrm{a}}/g < 0 \tag{4.1}$$

for two dimensions. This implies that no matter what the conductance of a small square of a two-dimensional system is, increasing the size of the conductor will always decrease the conductance g. Therefore, all (infinitely extended) two-dimensional systems of non-interacting electrons are insulators at zero temperature. The corresponding $\beta(g)$ function is depicted in Fig. 4.1 together with the results for three and one dimension as solid lines.

In the limit of $g \rightarrow \infty$, the limiting form of the scaling function, eq. (4.1) can be integrated, leading to the conductance [101, 102]

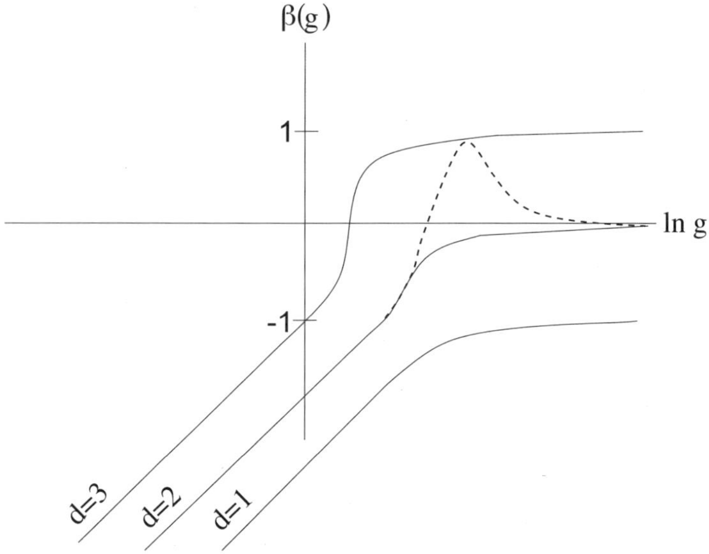

Fig. 4.1. Scaling function $\beta(g)$ for three- ($d = 3$), two- ($d = 2$), and one-dimensional ($d = 1$) systems. Solid lines represent the behavior obtained by Abrahams and coworkers in Ref. [57]. The dashed line shows how the scaling function would have to be modified, e.g., by interaction effects, in order to produce a metal–insulator transition in two dimensions.

$$G(L) = G_0 - \frac{e^2}{\hbar\pi^2} \ln\left(\frac{L}{l}\right), \tag{4.2}$$

where G_0 is the conductance at the lower cutoff l_D, i.e., the Drude conductivity $G_0 = e^2/h \cdot (k_\mathrm{F} l_\mathrm{D})$.

5

Electron–electron interactions within the Fermi-liquid concept

Interactions between electrons can be incorporated in this line of thought. The theory of Fermi liquids initiated by Landau [103] was well developed at that time, describing the elementary excitations of an interacting Fermi liquid by quasiparticles, the so-called quasi-electrons, which behave like non-interacting particles [104]. In simple terms, the quasi-electron is the bare electron dressed by a screening cloud formed by all the other electrons in the Fermi sea. This leads, for example, to a renormalized electron mass and to an effective screening of the electron–electron interaction and makes the success of the picture of non-interacting electrons at least plausible. While the renormalization of physical quantities such as the specific heat or the magnetic susceptibility in the Fermi-liquid theory can be large, they remain finite. In this framework, a transport equation for quasi-electrons can be derived, called the Landau–Silin equation [104], which is the analog of the Boltzmann equation for classical gases. The arguments given above for the insulating nature of a non-interacting two-dimensional electron gas at zero temperature can then be directly transferred to the case of interacting electrons if quasiparticle interactions are neglected.

Interacting electron gases are commonly characterized by the so-called interaction parameter r_s defined as

$$r_\mathrm{s} = \frac{1}{\sqrt{\pi n_\mathrm{s} {a_\mathrm{B}^{\star}}^2}} \tag{5.1}$$

in two-dimensional systems. Given the sheet density of electrons (or holes) n_s, r_s is the radius (normalized to $a_\mathrm{B}^{\star}$) of a circle within which a single electron is found on average.

Beyond the non-interacting quasiparticle picture it was shown by Altshuler and coworkers [105, 106] in a perturbative approach treating r_s as a small parameter that interactions between quasi-electrons lead to strong singularities near the Fermi level in disordered Fermi liquids, for example, in the tunneling density of states.

On the same perturbative level the effect of interactions on the conductivity in disordered systems has been shown to be twofold. On the one hand, they lead to direct corrections to the Drude conductivity with a logarithmic temperature dependence [101]. On the other hand, they cause dephasing of the electronic wave functions [67, 101, 107, 108].

Yet another effect of interactions of electrons on transport properties has been known for a long time. Local scattering sites like ionized impurities or remote donors, which cause resistance at low temperatures, are known to be screened by the Fermi sea of the electrons. The temperature dependence of the screening leads to a temperature-dependent conductivity [42, 109, 110].

5.1 Dephasing in diffusive two-dimensional systems

In connection with the dephasing effect, we return briefly to the prediction of localization in two dimensions. On a microscopic level, the localization of a two-dimensional electron gas at zero temperature is believed to be due to phase coherent backscattering of electrons [30, 108]. The corresponding theory of weak localization valid for systems with weak disorder was first presented in Refs. [111, 112] using diagrammatic techniques and later interpreted as coherent backscattering by Bergmann [113, 114] and others [108, 115, 116]. Connection to experiments, where the system size is typically not an easily variable parameter, is made by noting that phase coherent backscattering is limited at finite temperatures by the finite dephasing rate $\hbar/\tau_\varphi(T)$ [or dephasing length $l_\varphi(T) = \sqrt{D\tau_\varphi(T)}$, with the diffusion constant $D = v_{\mathrm{F}} l/2$] of the wave functions. This leads to a replacement of the system size L by the phase coherence length $l_\varphi(T)$ in the scaling result eq. (4.2)

$$G(T) = G_0 - \frac{e^2}{\hbar\pi^2} \ln\left(\frac{l_\varphi(T)}{l}\right),$$

i.e., the effective system size can now be tuned by changing the temperature T of the sample provided that $L \gg l_\varphi$. The theory of dephasing relies on the effect that an individual electron feels the fluctuating electromagnetic field produced by all the other electrons in the Fermi sea, which leads to a loss of the phase information over time. The phase-breaking rate depends linearly on temperature following [67, 107, 108]

$$\frac{\hbar}{\tau_\varphi} = kT\frac{e^2/h}{G} \ln\frac{kT}{\hbar/\tau_\varphi} \overset{G \gg e^2/h}{\approx} kT\frac{e^2/h}{G} \ln\frac{G}{2e^2/h}. \tag{5.2}$$

5.2 Interaction corrections to the conductivity

In addition to the dephasing mechanism, interactions between quasiparticles lead to direct corrections of the zero-temperature Drude conductivity. The theoretical aspects of these corrections have been reviewed for example by Lee and Ramakrishnan in Ref. [101] and by Altshuler and Aronov in Ref. [107]. Recently, the theoretical background was very intuitively interpreted and extended to higher temperatures by Zala and coworkers [117]. They emphasize that the interaction corrections to the conductivity known to date, i.e., the effects of temperature-dependent screening [109, 110, 118] and the interaction corrections calculated by Altshuler and coworkers [107], originate from the same microscopic physics. They are the two limits of the same processes, namely, elastic scattering of carriers by the self-consistent potential created by all the other electrons. In the following we briefly discuss the physics behind the two corrections trying to make their common origin transparent.

5.2.1 Temperature-dependent screening

Following the qualitative discussion in Ref. [117] we recall that potential inhomogeneities and impurities in a two-dimensional electron gas are screened. The induced electron density has an oscillatory contribution — the well-known Friedel oscillations — with a wavelength of half the Fermi wavelength λ_{F} and decaying with $1/r$ in two dimensions (see e.g. [25]). The total induced electron density is the origin of a self-consistently induced *Hartree* potential [118] with an oscillatory contribution of wavelength $\lambda_{\mathrm{F}}/2$. Scattering of electrons at these Friedel oscillations of the screening cloud of a single impurity (see Fig. 5.1) result in an energy-dependent scattering rate that increases linearly with $E - E_{\mathrm{F}}$ for $E > E_{\mathrm{F}}$ while it is constant for $E < E_{\mathrm{F}}$ [117]. Increasing temperature makes these higher energies more and more accessible, leading to a linear metallic temperature dependence as long as $\hbar/\tau < kT \ll E_{\mathrm{F}}$ (in this context called the ballistic limit) , i.e.,

$$G(T) = G_0 \left[1 - c\left(\frac{kT}{E_{\mathrm{F}}}\right)\right] \quad \text{for } \hbar/\tau < kT \ll E_{\mathrm{F}}, \tag{5.3}$$

where c is a positive constant that depends on the scattering mechanisms and the carrier density. Such a linear temperature dependence has already been anticipated by Stern in 1980 [110]. Scattering at Friedel oscillations is the microscopic description of the physics behind transport theories that include temperature-dependent screening in the random-phase approximation (Lindhard screening) that have been widely used for the description of transport in two-dimensional systems [42, 110, 119–124]. For the special case of p-Si/SiGe quantum wells this theory was applied by Laikhtman and Kiehl [125]. The microscopic picture used here is intimately related to the Kohn anomaly in the polarization function of Stern [118] (see Fig. 5.2) which leads to the same energy dependence of the scattering rate.

An extension of the theory of temperature-dependent screening including the effect of collision broadening, i.e., an extension into the region where $kT < \hbar/\tau$ (called the diffusive limit in this context) has been presented by Das Sarma [121]. The disorder potential removes the zero-temperature Kohn anomaly from the polarization function and results in temperature-independent screening in this limit:

$$G(T) = \text{const.} \quad \text{for } kT < \hbar/\tau.$$

It was pointed out by Zala and coworkers that exchange effects are neglected in the standard random-phase description of temperature-dependent screening and that

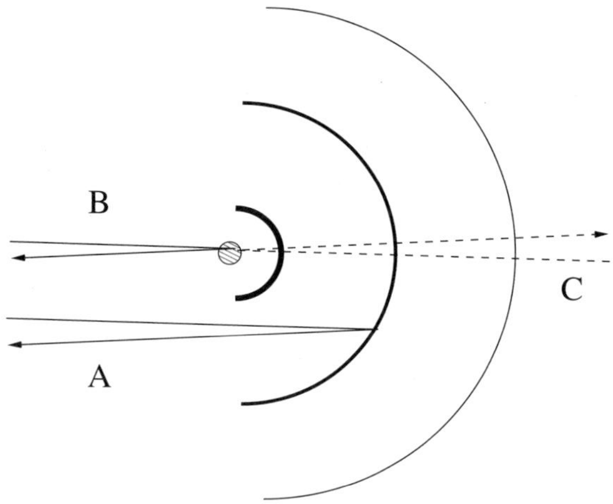

Fig. 5.1. Scattering off Friedel oscillations originating from a single impurity lead to a temperature-dependent screening correction to the $T = 0$ Drude conductivity. Interference between the two paths A and B contributes mainly to an enhanced backscattering (after Ref. [117]).

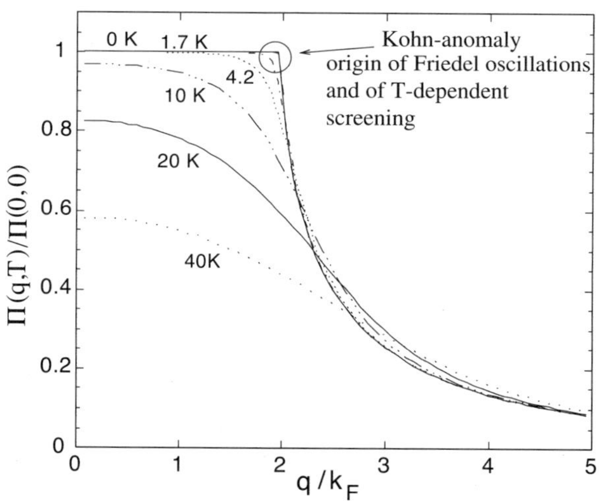

Fig. 5.2. Polarisation function after Stern in Ref. [118]. The temperature dependence of this function leads to an implicit temperature dependence of the scattering rate of an electron and thereby to an interaction correction of the conductivity linear in kT/E_{F}.

their inclusion decreases c compared to the Hartree result. Using a diagrammatic approach these authors find that with a very special choice of parameters the exchange contribution can even overpower the Hartree result and $c < 0$ in this case, leading to a linear insulating temperature dependence of the conductance [117].

In a different approach exchange and correlation effects can also be incorporated into the standard random-phase approximation via the so-called local field correction [126]. As mentioned in [127], this has been done in the analytic T-dependent screening theory by Gold and Dolgopolov [122] in the Hubbard approximation including exchange corrections only. This theory is therefore applicable to cases where the interaction parameter $r_s < 1$. In Ref. [127] the range of validity is extended to larger values of r_s by considering correlation effects. It is found that with increasing r_s values the constant c in eq. (5.3) increases monotonously.

5.2.2 Interaction corrections due to interference of multiply scattered paths

Similar to the weak localization theory, successive scattering events involving Friedel oscillations and more than a single impurity can interfere and lead to additional logarithmic corrections to the conductivity [107, 117]. These corrections have been

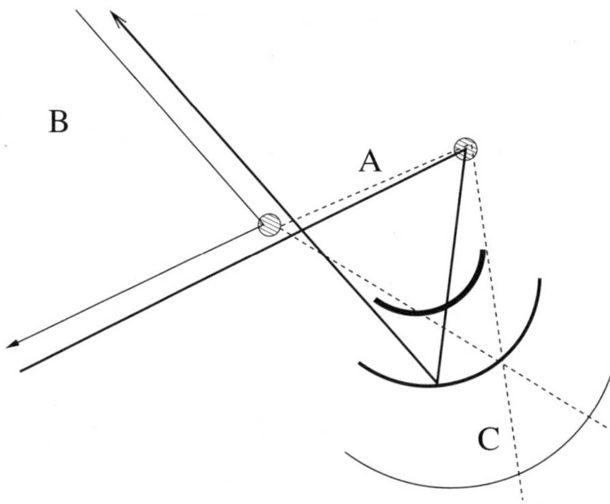

Fig. 5.3. Scattering process involving two impurities and the Friedel oscillation. Scattering at all angles is affected by interference (after [117]).

worked out with diagrammatic techniques by Altshuler and coworkers [107] for the limit of low temperatures $kT \ll \hbar/\tau$ (called the diffusive limit in this context). They consider corrections due to the dominant exchange part of the Coulomb interaction (diffusion channel) in the two-particle scattering, which consists of a singlet and a triplet contribution. They find the correction

$$\Delta\sigma_{\mathrm{I}} = -\frac{e^2}{\hbar}\frac{1}{4\pi^2}\left(2 + \frac{3}{2}\tilde{F}_\sigma\right)\ln\left(\frac{\hbar/\tau}{kT}\right) \quad \text{for } kT \ll \hbar/\tau \tag{5.4}$$

for two-dimensional systems, where τ is the elastic Drude scattering time and $\tilde{F}_\sigma$ is an interaction parameter, which can, for example, be determined by experiment.

5.2.3 A comprehensive theory of interaction corrections based on the Fermi liquid concept

The transition between the diffusive ($kT < \hbar/\tau$) and the ballistic ($kT > \hbar/\tau$) limit has been worked out in Ref. [117], including the modifications caused by the long-range nature of the Coulomb interaction. The main result is that the interaction corrections to the $T = 0$ Drude conductivity can essentially be written as the sum of contributions with a linear T dependence and a logarithmic T dependence plus almost negligible corrections with a very weak T dependence. Although the matter is still under discussion among theorists and a comprehensive theory of interaction corrections has not yet been developed, some agreement has apparently emerged that

- interaction corrections to the conductivity play a very important role in the temperature dependence of the conductivity at very low T within the Fermi-liquid theory;
- the important interaction corrections lead to a superposition of a linear and a logarithmic temperature dependence.

This will be the starting point for the discussion and interpretation of the experimental data shown in chapter 10.

6

Beyond Fermi-liquid theory

Fermi-liquid theory has proven to be extremely useful and successful for the description of electrons in two-dimensional systems. However, it is expected that at large interaction parameters r_s, e.g., at very low electron concentrations, this concept will eventually break down. The most prominent candidate for a strongly correlated ground state is probably the Wigner crystal suggested in 1934 by Wigner for three-dimensional electrons in metals [128, 129]. In two dimensions, a comparison of different Bravais lattices leads to the result that a hexagonal lattice would have minimum energy [130, 131]. A discussion of details of early work can be found in the review on two-dimensional systems by Ando et al. [42]. More recent calculations predict the transition from an electron liquid to a Wigner crystal at $r_s = 37 \pm 5$ in a perfectly pure system [132]. Taking into account the pinning of a Wigner crystal by disorder, Chui predicted a transition at the much lower $r_s = 7.5$ [133]. As far as transport is concerned such a pinned Wigner crystal would be an insulator at zero temperature.

Other possibilities for ground states in two dimensions have been proposed in the past, among them spin-density wave, charge-density wave, and a superconducting ground state, but seem to have become a bit out of fashion in the past twenty years because the quality of the available samples was not sufficient for reaching the necessary $\bar{r}_s$ values. Some details and references can be found in Ref. [42]. As we will show below, this situation may have changed now.

7

Summary of disorder and interaction effects

In an over all picture the conductivity of a two-dimensional electron gas at zero temperature can be characterized by the interplay of disorder and interactions. The former can be described using the dimensionless parameter $k_F l_D$, where k_F is the Fermi wave vector and l_D is the transport mean free path of the electrons at zero temperature. The interactions are quantified by the interaction parameter conventionally defined as $r_s = 1/\sqrt{\pi n_s a_B^{\star 2}}$ in two dimensions, where n_s is the sheet density of electrons (or holes) and $a_B^\star$ is the effective Bohr radius. For two-dimensional systems with no valley degeneracy, r_s coincides with the ratio of the Coulomb interaction between carriers and the Fermi energy, i.e., $r_s = E_C/E_F := \bar{r}_s$. For systems with a valley degeneracy g_v (e.g., $g_v = 2$ for Si-MOSFETS, whereas $g_v = 1$ for p-SiGe), this energy ratio becomes larger by a factor of g_v as compared to the standard definition of r_s, i.e.,

$$\bar{r}_s = \frac{E_C}{E_F} = g_v r_s.$$

In a perturbative sense, the key parameter for a system is given by $\bar{r}_s$ rather than by r_s. In the following we will therefore consistently use $\bar{r}_s$ as the interaction parameter.

In general, weak localization theory is expected to hold in the limit of weak disorder $k_F l_D \gg 1$. The strong localization regime $k_F l_D < 1$ is dominated by hopping transport [100]. On the other hand, Fermi-liquid theory is certainly valid for $\bar{r}_s < 1$, while it is unclear at which $\bar{r}_s$-value the theory actually breaks down. There are indications that at large $\bar{r}_s$ other ground states are energetically more favourable such as the (pinned) Wigner crystal, which is beyond Fermi-liquid theory [132, 133]. Based on these predictions and on the inclusion of correlation effects in the theory of T-dependent screening [127], one would expect correlation effects in two-dimensional systems to become increasingly important with growing $\bar{r}_s$.

The theories of weak localization and interaction effects on the conductivity have been reviewed by numerous authors, for example, by Altshuler and Aronov [107], Lee and Ramakrishnan [101], Kawaji [134], and Vollhardt [135].

The whole theoretical framework of phase-coherent backscattering and interaction corrections to the conductivity developed around 1980 has found numerous experimental support in the 1980s. Logarithmic temperature dependence of the resistance was found in thin metal films [136, 137], silicon MOS devices [138–142], and GaAs/AlGaAs heterostructures [143, 144]. Suppression of the weak localization by a magnetic field was found, for example, for silicon MOS devices [142, 145] and GaAs/AlGaAs heterostructures [144, 146]. Interaction effects and the weak localization effect have been separated using perpendicular magnetic fields [147]. In all these experiments the interaction parameter could not be made too high (typically below the order of 1) without bringing the samples into the strongly localized regime. In recent years, improved sample quality allowed such measurements to be extended to much higher $\bar{r}_s$-values without going through the strong localization threshold. Such experiments will be the subject of the following chapters.

8

Experiments on strongly interacting two-dimensional systems and the metal–insulator transition

In 1994 the belief that all two-dimensional systems are insulating at $T = 0$ was challenged by the interpretation of experiments on high-mobility Si-MOSFETs [148]. In these samples resistivities continuously decreasing with decreasing temperature have been observed for certain electron densities and interpreted as an indication for a metallic ground state at $T = 0$. Lowering the electron density leads to an increasing resistivity with decreasing temperature as expected in the strongly localized (hopping) transport regime. It was suggested that the transition between the metallic and the insulating range of densities might be an interaction-driven quantum phase transition [148–150] (see [151, 152] for a simple account of quantum phase transitions, [153] for a review) from then on called the "metal–insulator transition in two dimensions at zero magnetic field." Such a metal–insulator transition would require a serious modification of the one-parameter scaling theory of localization in the sense that the scaling function $\beta(g)$ should have at least one zero. A possible scenario is indicated in Fig. 4.1 as a dashed line. Another scenario would replace the one-parameter scaling function $\beta(g)$ by a scaling function with more parameters.

Later on, more experiments on Si-MOSFETs [154–156], on GaAs hole [157–162] and electron gases [163], AlAs electron gases [164], and SiGe hole gases [165–168] were shown to exhibit similar features as observed in the Si-MOSFET systems (see Refs. [169, 170] for reviews). Many experiments tried to investigate the detailed properties of the metallic phase and the metal–insulator transition. An overview is given in Table 8.1. A scaling analysis was performed on many measurements of the temperature dependence of the resistivity allowing the interpretation of the data in terms of a quantum phase transition. Investigations in a perpendicular magnetic field aimed at weak localization effects. Experiments in parallel and tilted magnetic fields gave indications that the metallic behavior disappears as soon as the system becomes spin-porlarized. Compressibility studies are the only experiments probing a thermodynamic property in these systems.

The characteristic experimental feature of what has been called the metal–insulator transition in two-dimensional systems is the change in the slope from $\mathrm{d}\varrho_{xx}/\mathrm{d}T > 0$ (metallic behavior) at high carrier density to $\mathrm{d}\varrho_{xx}/\mathrm{d}T < 0$ (insulating behavior) at low carrier density. While the insulating behavior is expected

Material system	Si-MOSFETs	p-SiGe	p-GaAs
T-dependence at $B = 0$ Scaling	[148, 149, 171] [175–177] [154, 180, 181] [156]	[165, 166, 172] [167]	[157, 173, 174] [158, 178, 179] [161, 162, 182] [160]
Non-linear transport	[155, 183, 184]	[167, 172]	[158]
Back-gate dependence	[177, 185]		
Parallel magnetic field	[186–188] [190–192] [193]	[167]	[173, 182, 189]
Perpendicular magnetic field	[154, 194, 195] [200–202]	[168, 196, 197] [203]	[178, 179, 198, 199] [159–162]
Conductance fluctuations	[204]		
Tilted field	[205, 206]		
Compressibility			[207–209]

Table 8.1. Experiments on the metallic behavior of strongly interacting two-dimensional systems: an overview.

from weak localization theory and the gradual transition to strong localization at decreasing density marks the onset of hopping conductivity, the metallic temperature dependence came as a surprise. Most experimentalists were used to the observation that any metallic temperature dependence would at the lowest temperatures eventually be overpowered by the weak-localization correction to the conductivity.

As an example of the signatures of the "metal–insulator transition," Fig. 8.1 shows the resistance of a p-SiGe quantum well as a function of temperature and hole density. The topmost curve corresponds to a hole density $p = 1.0 \times 10^{11}\,\mathrm{cm}^{-2}$, the lowest curve to $p = 2.6 \times 10^{11}\,\mathrm{cm}^{-2}$. At low electron densities above the dashed line in the figure, the resistivity increases monotonically with decreasing temperature, which is the expected insulating behavior. At electron densities below the dashed line, the resistivity shows a slight increase when the temperature is lowered in the range of higher temperatures but then turns down to a monotonous and significant decrease at lower temperatures. This decrease continued beyond the temperature range shown in this figure down to the lowest achievable temperatures in our dilution refrigerator of about 100 mK. The dashed line in the figure corresponds to a density $p_\mathrm{c} = 1.46 \times 10^{11}\,\mathrm{cm}^{-2}$. It can be viewed as the separatrix between metallic and insulating behavior of the conductivity.

These experimental findings are in remarkable agreement with those in other two-dimensional systems, such as the Si-MOSFETs and p-GaAs/AlGaAs structures [169]. In Fig. 8.2 we have compiled characteristic experimental data from various material systems following Ref. [163] for n-GaAs/AlGaAs heterostructures with InAs quantum dots in the two-dimensional electron gas, Ref. [210] for low-density n-GaAs/AlGaAs heterostructures, our data from Ref. [167] for p-SiGe quantum wells, Ref. [148] for Si-MOSFETs, and Refs. [158, 182] for the p-GaAs/AlGaAs heterostructures. The vertical axis describes the degree of disorder quantified by

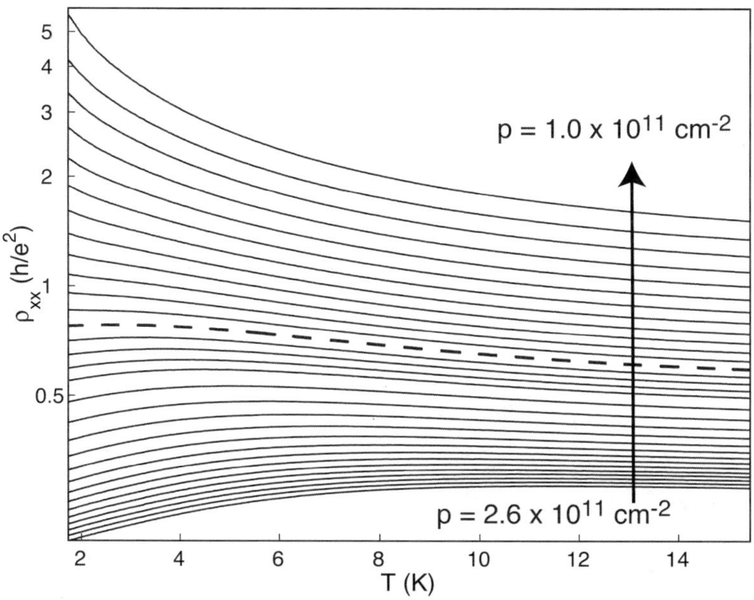

Fig. 8.1. Temperature dependence of the resistance in a p-SiGe quantum well at various densities. The dashed line is the separatrix between metallic and insulating behavior occurring at a density $p_c = 1.46 \times 10^{11}$ cm^{-2}.

$\varrho_{xx}/(h/e^2) = 1/(k_F l_D)$. The numbers for this parameter were taken from the experimental data at the lowest available temperature. On the horizontal axis we plot the interaction parameter $\bar{r}_s$ calculated from the corresponding carrier densities. A factor of 2 was taken into account for the Si-MOSFET system, which is the valley degeneracy g_v on Si(100). Experiments on the metal–insulator transition have been typically performed as a function of gate voltage U_G. With U_G the electron density n_s and the mean free path l_D are changed simultaneously. Traces of resistivity versus gate voltage at the lowest temperatures have been translated into traces $(\bar{r}_s(U_G)$,$1/[k_F l_D(U_G)])$ in Fig. 8.2. All experiments, perhaps with the exception of that reported by Ribeiro and coworkers [163], have been performed at elevated $\bar{r}_s$, where the applicability of the Fermi-liquid description cannot be taken for granted. The points where the change of the temperature dependence from metallic to insulating behavior occur are typically close to $k_F l_D \approx 1$, i.e., at the transition from weak to strong localization. At first sight it seems to be surprising that all the depicted systems with vastly differing interaction parameters exhibit a strikingly similar temperature dependence at various densities and a crossover from metallic to insulating behavior. We emphasize that it is not entirely clear to date whether all the experimental data can be described within the same theoretical framework. In addition, the zero temperature limit can never be reached in an experiment and therefore the question

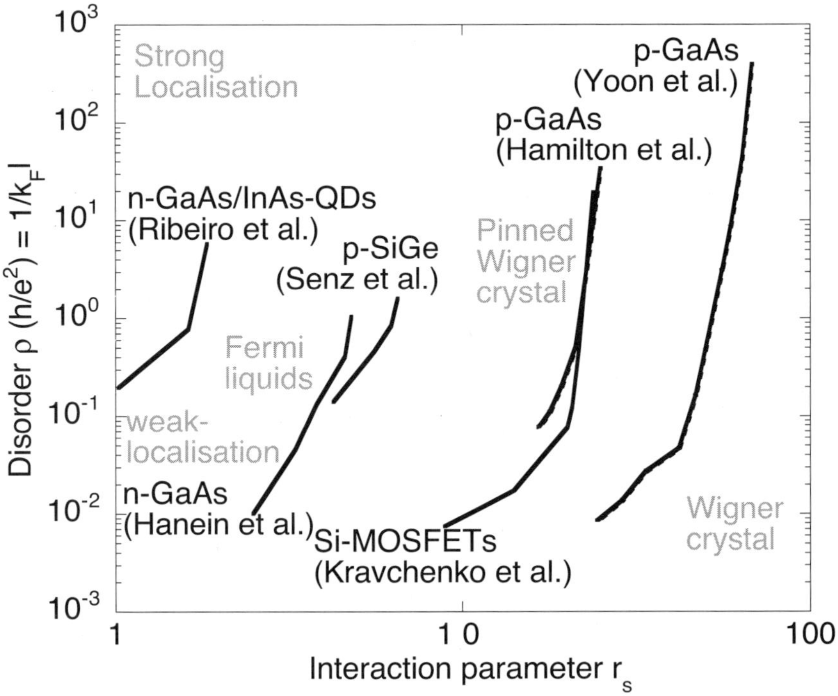

Fig. 8.2. Comparison of systems

of the existence of a metallic state at $T = 0$ cannot directly be answered and any statement relies on an extrapolation of finite temperature data to the $T = 0$ limit.

9

Theoretical work related to the metal–insulator transition

As a consequence of these findings, the existence of a metal–insulator transition (MIT) in 2D systems at zero magnetic field has been intensely and controversially discussed on the theoretical side. No consensus has emerged to date either about the explanation of the metallic behavior or about the question of the existence of a true metal–insulator transition. The approaches can be coarsely classified into those building on the traditional Fermi-liquid concepts or even on non-interacting electrons and those exploring the possibilities of ground states beyond the Fermi-liquid description. In the following we try to give an overview of the enormous amount of theoretical literature that appeared without discussing details and without discussing their relation to the experiments mentioned above. The interested reader can find some additional theoretical discussion in Refs. [169, 211].

Among the traditional models are those of Altshuler, Maslov, and coworkers [212, 213] who suggested the presence of charged hole traps in the oxide layer of the Si-MOSFETs to be responsible for the observed behavior. In a later paper the same authors raised the issue of heating effects and their influence on the measurements [214].

A model building on non-interacting electrons is a percolation description suggested by Y. Meir [215–217]. This author regards the two-dimensional system as separating at low densities into puddles of electrons connected by quantum-point-contact-like tunneling links.

The description put forward by S. Das Sarma and coworkers treats the system in the metallic phase as an ordinary Fermi liquid in the framework of the Boltzmann transport theory including temperature-dependent screening within the random phase approximation (RPA) [123, 124, 218]. The transition is explained as an experimentally given carrier freeze-out, e.g., due to impurity binding [219]. Dolgopolov and Gold modeled the magnetoresistance in parallel magnetic fields in a similar fashion [220]. A model by Kastrinakis including interactions and spin-orbit interactions also remains within the Fermi-liquid model [221]. Interband scattering between spin-split heavy hole bands was suggested in Ref. [160].

However, as pointed out by Abrahams [169], the interaction parameter $\bar{r}_s$ is much larger than one in the experimental systems and theoretically this is "an unambiguous example of the *strong-coupling many-body problem* for which theoretical models are still poorly developed" [169].

V. Dobrosavljevic showed that for interacting electrons the existence of a metal–insulator transition does not disagree with general scaling principles [222]. It was suggested in this paper that the metallic phase is very likely not a Fermi liquid (see also [223]). A scaling approach incorporating disorder *and* interactions has been put forward by Castellani and coworkers [224] using perturbative renormalization group methods. The model was developed further by Si and Varma [225]. Sheng and Weng presented a universal scaling form for the longitudinal resistance [226].

A model by Chakravarty and coworkers describes the metal–insulator transition as the melting of a Wigner glass (insulating state) into a non-Fermi-liquid metallic state [227]. The similarity of two-dimensional charge carrier gases and the low-temperature physics of ^{3}He was pointed out by Spivak [228] who suggested a transition between a Wigner crystal and a Fermi liquid. Near the transition the system becomes inhomogeneous and crystalline regions exist in parallel to liquid regions.

More exotic models suggested superconductivity to account for the observed metallic behavior [229–231]. He and Xi suggested a model for the Si-MOSFET system treating charges in the oxide as being mobile [232]. In this model there is a parameter region in which a metallic electron-hole liquid and an exciton gas coexist.

Numerical models of interacting electrons in disordered media have been presented by a number of authors [150, 233–236] showing indications of metallic states at elevated $\bar{r}_s$ values.

From the above listing of theoretical approaches it becomes clear that the experimental results found enormous interest in the theoretical community. However, it also becomes clear that to date we are far from a well-established description of strongly interacting charge carriers in disordered two-dimensional systems.

10

Metallic behavior in p-SiGe quantum wells

In the following we discuss the metallic behavior observed in p-SiGe quantum wells [167]. Magnetotransport data taken across the MIT will be analyzed. Interference corrections to the conductivity are extracted from weak-localization studies at low magnetic fields and interaction corrections to the conductivity in the diffusive regime are obtained from the temperature dependence of the Hall resistivity, similar to Ref. [159]. In addition, following the suggestion in Ref. [123], we compare the remaining (Drude) part of the conductivity with the theory for temperature dependent screening suitable for the system [122]. Using this program we arrive at the following conclusions: (I) In the metallic regime the 2D-hole gas in our SiGe samples behaves like an ordinary Fermi liquid and exhibits localizing interference and interaction corrections to the conductivity that can be well described by conventional theory. We find empirically that the concept of a Fermi liquid describes the data taken at large $r_s \approx 6$. (II) The metallic temperature dependence of the resistivity can be quantitatively described by the theory of temperature-dependent screening, which can be incorporated in the Drude part of the resistivity.

10.1 Samples and structures

The samples used in this study were grown by MBE (molecular beam epitaxy). A cross-sectional transmission electron microscopy image of the sample is shown in Fig. 10.1. A 20 nm $Si_{0.85}Ge_{0.15}$ quantum well where the 2D-hole gas is formed is sandwiched between undoped Si layers. The structures are remotely doped with boron at a distance of 18 nm above the quantum well and gated with a Ti/Al Schottky gate. Gating p-SiGe quantum wells has been a technological challenge and the tunability of the hole gas distinguishes our samples from those investigated by Coleridge in Refs. [166, 203] in which no gate was available. The SiGe quantum well is compressively strained normal to the growth direction, which leads, together with the confinement, to a splitting between heavy-hole (HH) and light-hole (LH) bands of the order of 25 meV [237]. Figure 10.2 shows a schematic drawing of the valence band edge in the structure around the quantum well. The 2D-hole gas resides in the

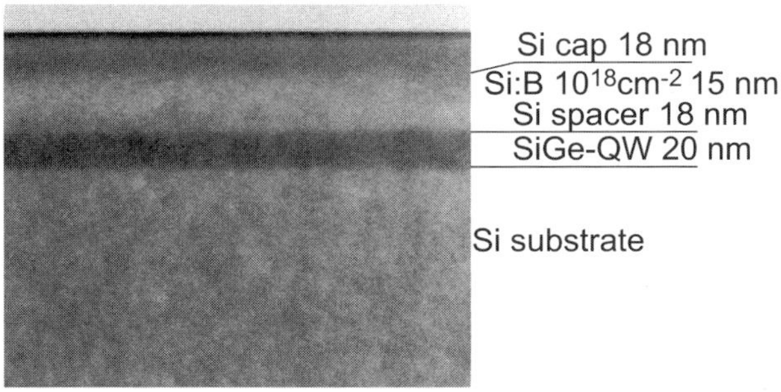

Fig. 10.1. Cross-sectional transmission microscopy image of the SiGe quantum well.

lowest HH subband ($m_J = \pm 3/2$) with effective mass $m^\star = 0.25 m_0$ as determined by measurements of the temperature dependence of Shubnikov–de Haas oscillations. The in-plane LH mass is predicted to be larger than the HH mass [238].

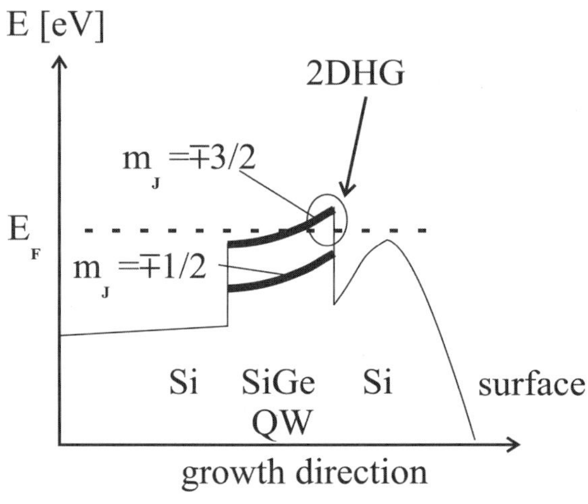

Fig. 10.2. Schematic of the valence band edge in the SiGe samples.

Conventional Hall bar structures of 10 and 30 µm width were fabricated with the length between voltage probes between 7 and 26 µm. Measurements were carried out at temperatures between 190 mK and 15 K in a dilution refrigerator and a ^{4}He-cryostat using standard four-terminal AC lock-in and DC techniques.

The hole density in ungated areas of the device was 4.3×10^{11} cm^{-2}. With the gate the hole density could be tuned between 1.0 and 2.6×10^{11} cm^{-2}, i.e., the

Fermi energy was always smaller than 2 meV. The hole mobility in these structures increases with carrier concentration from 1000 cm^2/Vs at the lowest to 7800 cm^2/Vs at the highest density similar to other p-SiGe structures [238]. The ratio of the Drude scattering time τ_{e} and the quantum lifetime τ_{q} extracted from Shubnikov–de Haas oscillations is close to 1 indicating that transport is dominated by short-range scattering potentials. It was confirmed in other studies that large angle scattering at interface charges dominates the mobility in such structures [237, 238].

10.2 Scaling analysis, quantum phase transition, and heating effects

Coleridge et al. [166] and our group [167, 168] have shown that p-SiGe exhibits the characteristic features of the MIT. As the hole density is increased, the temperature dependence of the zero-magnetic-field resistivity changes from $\mathrm{d}\varrho/\mathrm{d}T < 0$ to $\mathrm{d}\varrho/\mathrm{d}T > 0$ (see Fig. 8.1) [167]. At the critical density p_{c}, we found $r_{\mathrm{s}} = 1/(a_{\mathrm{B}}^{\star}\sqrt{\pi p_{\mathrm{c}}}) \approx 6$, where $a_{\mathrm{B}}^{\star} \approx 2.5\,\mathrm{nm}$ is the effective Bohr radius in SiGe. The critical resistance was $\varrho_{\mathrm{c}} \approx h/e^2$. As discussed above, the r_{s} value is a measure of the importance of interaction effects and ϱ_{xx} is a measure of the degree of disorder in the system. Like in other systems temperature and electric field scaling could be performed [167] allowing the interpretation in terms of a quantum phase transition.

However, great care has to be taken in low-temperature transport experiments and even more in electric field scaling experiments that the electron gas is not significantly heated by the transport current. Weak coupling between electrons and the crystal lattice at low temperatures can lead to this overheating effect [214, 239–242]. Interpreting the electric field scaling results in our samples as a result of carrier heating, we obtain the relation $T_{\mathrm{e}} \propto \mathcal{E}^{1/2}$ between the electron temperature T_{e} and the electric field $\mathcal{E}$. A careful analysis of hot carrier effects in our samples was performed in collaboration with R. Leturcq and D. L'Hote at Saclay [243, 244]. They used a second Hall bar on the same chip as an additional thermometer making sure that the lattice remained in equilibrium with the bath even when the first Hall bar was strongly heated by the applied current. The effective electron temperature was measured by comparing the resistivity $\varrho(T = T_{\min}, \mathcal{E})$ at a given applied electric field $\mathcal{E}$ to the resistivity as a function of temperature, $\varrho(T, \mathcal{E} \to 0)$, i.e., the electron temperature $T_{\mathrm{e}}(\mathcal{E})$ was found from the equation $\varrho(T_{\mathrm{e}}, \mathcal{E} \to 0) = \varrho(T = T_{\min}, \mathcal{E})$. From the same experiment, the dissipated power per electron was determined from the relation $P_{\mathcal{E}} = UI/(pWL) = \mathcal{E}^2/(p\varrho)$, where I is the applied current, U is the voltage drop measured between contacts of separation L along the Hall bar of width W. Both steps taken together allow the determination of $P_{\mathcal{E}}(T_{\mathrm{e}})$, which can then be compared to well-known models of power dissipation. The self-consistency of this analysis will become evident in the following.

Figure 10.3 shows $P_{\mathcal{E}}(T_{\mathrm{e}})$ for two different carrier densities. It turned out that the electric field dependence of the resistivity can be attributed to heating effects. The power dissipated in the system is the sum of deformation potential scattering and piezoelectric coupling which has to be considered in coherently strained SiGe

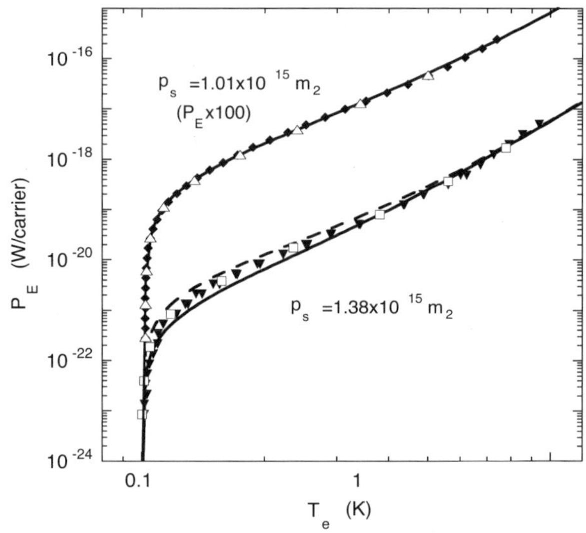

Fig. 10.3. Power loss per carrier vs. hole temperature for two different hole densities.

samples [240, 245], i.e.,

$$P_{\mathcal{E}}(T_{\rm e}) = c_1(T_{\rm e}^{\alpha} - T_{\rm l}^{\alpha}) + c_2(T_{\rm e}^{\alpha'} - T_{\rm l}^{\alpha'}), \tag{10.1}$$

where c_1 and c_2 are material specific constants and $\alpha = 5$ (resp. 7) for deformation potential scattering with weak (resp. strong) screening and $\alpha' = 3$ (resp. 5) for piezoelectric coupling, again with weak (resp. strong) screening. The $T_{\rm l}$-dependent terms in the above equation can be regarded as being constant in the experiment. In the double-logarithmic plot they are responsible for the strong increase of the dissipated power at the lowest electron temperatures. They can be eliminated by adding an offset $P_0(T_{\rm l})$ to $P_{\mathcal{E}}$, which is theoretically given by $c_1 T_{\rm l}^{\alpha} + c_2 T_{\rm l}^{\alpha'}$ but experimentally found by the condition that resulting curves for different $T_{\rm l}$ must collapse onto a single curve. The result of this procedure is shown in Fig. 10.4. At low electron temperatures, i.e., low electric field, it was found that piezoelectric coupling with weak screening dominates the dissipation, while at high electric fields probably deformation potential scattering with weak screening takes over. Piezoelectric coupling with strong screening (also $\alpha = 5$) seems to be unlikely, because, with increasing electron temperature, screening is expected to become even less important. Cooling through the electrical connections of the sample was found to be negligible for the lowest carrier densities, in agreement with the relation [214]

$$P = \left(\frac{2\pi}{e}\right)^2 \frac{kT_{\rm l}(kT_{\rm e} - kT_{\rm l})}{R},$$

where R is the sample resistance. However, for higher densities this effect has to be taken into account as shown by the dashed line in Fig. 10.3.

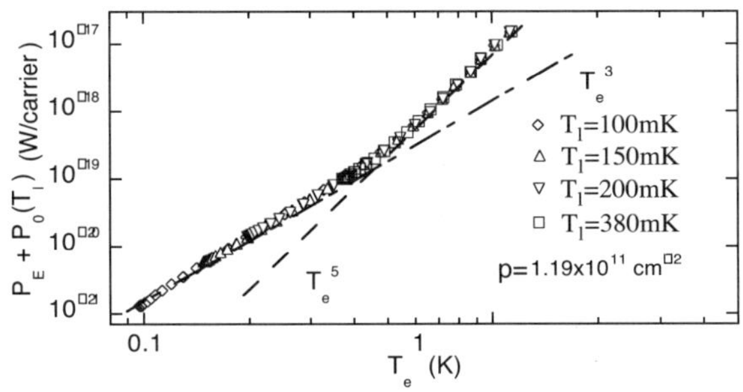

Fig. 10.4. The sum of the power per hole and a constant $P_0(T_l)$ as a function of hole temperature T_e. All the curves obtained at different T_l collapse onto the same master curve. The dashed and dash-dotted lines give the slopes expected for the power laws T_e^3 and T_e^5.

In summary, from the experimental findings presented above we have to conclude that the electric field scaling analysis is not sufficient evidence for the existence of a quantum phase transition in our samples. Heating of holes by the electric field turned out to mask the pure electric field effects.

10.3 Magnetoresistance measurements

It was found in Ref. [168] that in the metallic regime weak localization (WL) reduces the metallic behavior without destroying it, whereas localizing interference corrections dominate the zero field resistivity in the insulating regime. In the following we analyze magnetotransport data in the metallic regime in detail by first extracting the contributions of WL and interaction corrections to the temperature dependence of the conductivity; then the effect of temperature-dependent screening is quantitatively discussed.

Figure 10.5 shows a series of magnetoresistance measurements for various carrier densities and temperatures. The carrier density decreases from top to bottom and the hole gas shows the reversal of the temperature dependence at zero magnetic field from $\mathrm{d}\varrho_{xx}/\mathrm{d}T < 0$ (insulating behavior) for low densities (Fig. 10.5d and e) to $\mathrm{d}\varrho_{xx}/\mathrm{d}T > 0$ (metallic behavior) at high densities (Fig. 10.5a and b). A negative magnetoresistance with a temperature dependence compatible with weak or strong localization is observed for all carrier densities. When the temperature is increased, this peak weakens and becomes broader as expected for the phase-coherent backscattering effect. This peak appears to sit on a very broad background resistance that shows its own distinct temperature dependence. For high carrier densities the background resistance increases with increasing temperature, i.e., it shows metallic behavior. In contrast, for low carrier densities the background resistance decreases with temperature consistent with insulating behavior. The data therefore suggest an interpretation in terms of a sum of at least two effects, one describing the MIT of the background resistance and a "conventional" one being the interference correction of the conductivity in a disordered system.

In order to investigate the possibility of deviations from the Fermi-liquid picture we perform a careful analysis of the magnetoresistance data in the metallic regime (Fig. 10.5a and b) *within* the Fermi-liquid description and look for *inconsistencies*. We therefore analyze the zero-field conductivity $\sigma(T)$ in the spirit of Refs. [107, 117]:

$$\sigma(T) = \sigma_{\mathrm{D}}(T) + \delta\sigma_{\mathrm{WL}}(T) + \delta\sigma_{\mathrm{I}}(T), \tag{10.2}$$

where $\sigma_{\mathrm{D}}(T)$ is the Drude conductivity, $\delta\sigma_{\mathrm{WL}}(T)$ and $\delta\sigma_{\mathrm{I}}(T)$ are the WL and the interaction contributions, respectively. Strictly speaking, eq. (10.2) is based on the validity of the Fermi-liquid description. In addition, the expression for the interaction correction $\delta\sigma_{\mathrm{I}}$ [cf., eq. (5.4)] is based on the assumption $r_{\mathrm{s}} < 1$. Nevertheless, we empirically apply this concept to our system, since phenomenologically the magnetoconductivity exhibits all the features also found in samples with low r_{s}.

10.4 Weak-localization correction

In the well-established theory of WL [246], spin-orbit relaxation mechanisms like the Elliott–Yafet mechanism, the Dyakanov–Perel mechanism or the Rashba effect have been taken into account perturbatively [246–251], which is appropriate for conduction band electrons. WL in 2D p-type systems such as p-GaAs has only recently been studied in detail [252]. In hole gases the valence band is strongly influenced by spin-orbit interaction and strain. The spin relaxation time is of the same order as the momentum relaxation time and can therefore no longer be treated perturbatively [253]. It is therefore not *a priori* clear that the theories in Refs. [246–251] can be applied to our system.

Here we apply the theory developed in Ref. [253] for p-type 2D hole gases that takes the valence band structure into account and includes the effect of HH-LH mixing. Figure 10.6 shows theoretical curves fitted to the measured conductivity according to

$$\begin{aligned}\delta\sigma_{\mathrm{WL}}(B,T) \quad - \delta\sigma_{\mathrm{WL}}(0,T) = \frac{e^2}{2\pi^2\hbar} \times \\ \left[f_2\left(\frac{B}{B_\varphi + B_{||}}\right) + \frac{1}{2}f_2\left(\frac{B}{B_\varphi + B_\perp}\right) - \frac{1}{2}f_2\left(\frac{B}{B_\varphi}\right)\right]. \end{aligned} \tag{10.3}$$

Here $f_2(x) = \ln x + \psi(1/2 + 1/x)$, ψ is the digamma function, and $B_i = \hbar/(4De\tau_i)$ $(i = ||, \perp, \varphi)$, with the diffusion constant D and the phase-coherence time τ_φ. The other relaxation times are given in Ref. [253] to be $1/\tau_{||} = 1/\tau_{\mathrm{e}} \cdot (k_{\mathrm{F}}a/\pi)^4\, I_{||}$ and $1/\tau_\perp = 1/\tau_{\mathrm{e}} \cdot (k_{\mathrm{F}}a/\pi)^6\, I_\perp$, where τ_{e} is the transport relaxation time, a is the width of the quantum well, and k_{F} is the Fermi wavevector. The quantities $I_{||/\perp}$ depend on the ratio of the LH and HH mass and have to be computed numerically. In our samples the holes are effectively confined by a triangular well due to the asymmetric

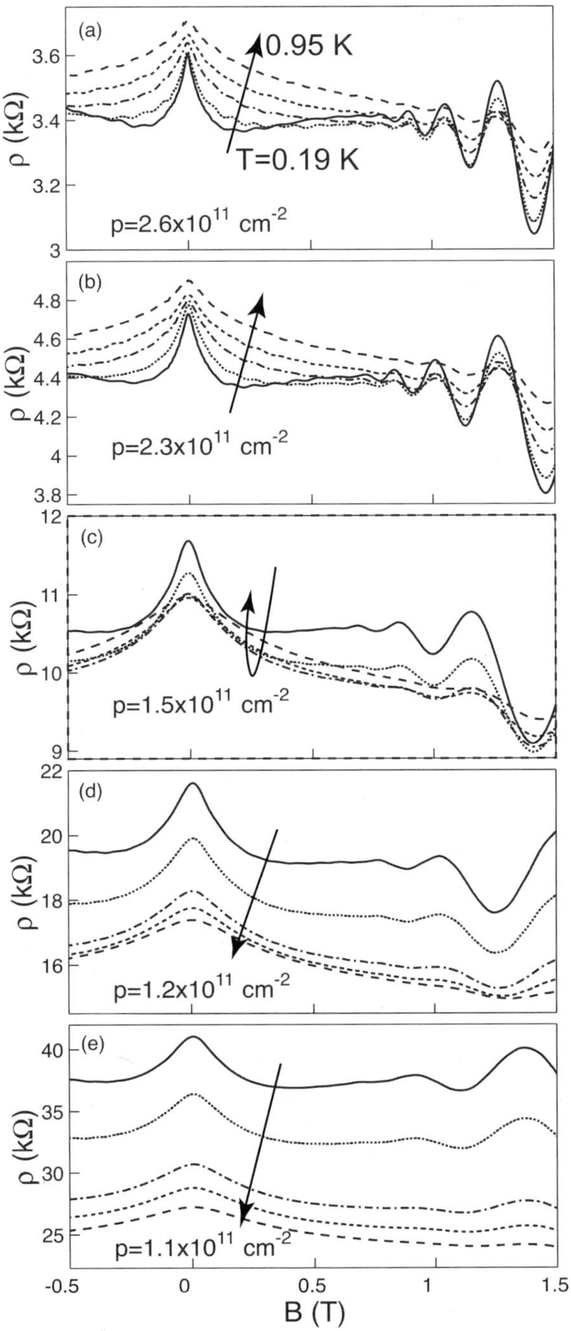

Fig. 10.5. Magnetoresistance at various densities and temperatures in SiGe quantum wells. Arrows indicate increasing temperature.

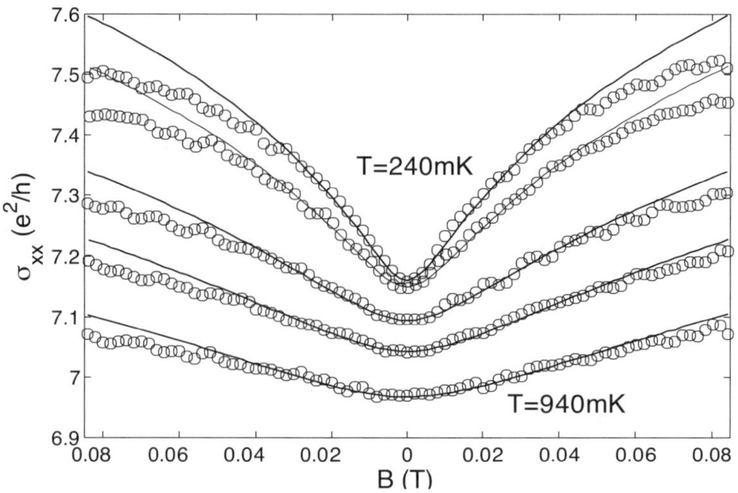

Fig. 10.6. Magnetoconductivity for the density $p = 2.6 \times 10^{11}$ cm^{-2} and the corresponding fits according to eq. (10.3).

T (K)	τ_φ (ps)	$\tau_{||}$ (µs)	$\tau_\perp$ (µs)
0.24	27.77	0.15	0.61
0.33	20.76	0.13	0.05
0.56	12.38	0.13	0.53
0.73	9.00	0.12	0.80
0.94	6.74	0.12	0.86

Table 10.1. Fitting parameters obtained from a fit of eq. (10.3) to the measured magnetoconductivity (see Fig. 10.6).

doping and we estimate $k_F a/\pi \approx 0.2$. We use B_φ, $B_\perp$, and $B_{||}$ as fitting parameters. Table 10.1 shows the resulting parameters.

It turns out that $\tau_{||}, \tau_\perp \gg \tau_\varphi$ or $B_{||}, B_\perp \ll B_\varphi$, i.e., HH–LH mixing seems to be insignificant in our sample. This result is in contrast to the measurements on p-GaAs, where stronger HH–LH mixing leads to appreciable values for $B_\perp$ and $B_{||}$ [252]. In our case eq. (10.3) reduces to the result for negligible spin-orbit scattering [246] also used in Ref. [168]. The result is plausible if we consider the small Fermi energy in our quantum well of less than 2 meV compared to the splitting of about 20 meV between the lowest HH- and LH-subbands.

Coleridge analyzes magnetoresistance data obtained on p-SiGe samples using the expression of Hikami and coworkers [246] and an additional amplitude factor α [203], which is allowed to vary with temperature. The raw data from these experiments are comparable to our results. It has been shown that an amplitude factor changes the absolute values for the extracted dephasing rates τ_φ [168]. A value $\alpha < 1$ leads to larger values of τ_φ for the same set of data. A temperature-dependent α may have an influence on the extracted temperature dependence of τ_φ. Numerical investigations indicate that the physical meaning of $\alpha < 1$ may be attributed to a finite ratio of the elastic scattering rate and the phase breaking rate [254]. The values of $\alpha = 0.6 - 0.7$ determined by Coleridge in Ref. [203] are in agreement with the simulations in Ref. [254] and the value $\alpha = 0.61$ in Ref. [168].

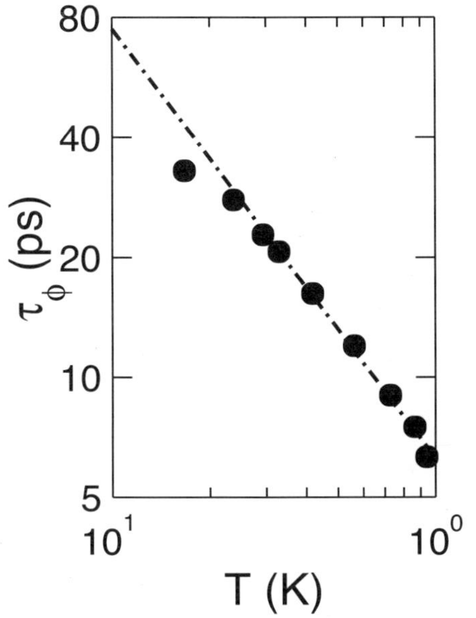

Fig. 10.7. Temperature dependence of the phase coherence time obtained from a fit of the weak-localization peak in the magnetoconductivity at a density $p = 2.6 \times 10^{11}\ \mathrm{cm}^{-2}$. The dash-dotted line is a fit with $\tau_\varphi \propto T^{-\gamma}$, with $\gamma = 1.09 \pm 0.2$. For this fit, the point at the lowest temperatures was left out.

Figure 10.7 shows a plot of $\tau_\varphi(T)$ extracted with an amplitude factor $\alpha = 1$ together with a fit according to the power law $1/\tau_\varphi \propto T^\gamma$. We find a value of $\gamma = 1.09 \pm 0.2$ in good agreement with the expectations from weak-localization theory [cf., eq. (5.2)] [107, 108]:

$$\frac{\hbar}{\tau_\varphi} = kT \frac{e^2/h}{G} \ln \frac{kT}{\hbar/\tau_\varphi}.$$

A similar linear behavior has been obtained for $\alpha = 0.61$ with τ_φ values up to 30% larger than the values shown here for $\alpha = 1$.

A detailed comparison of the behavior at different hole densities exhibits deviations from the above theoretical expression for τ_φ. Although the value of γ turns out to be independent of density, indicating that hole–hole interactions are the dominant dephasing mechanism, the absolute magnitude of the dephasing rate is much higher than expected from the above formula. In Fig. 10.8 we compare the dephasing rate $\hbar/\tau_\varphi$ normalized to kT, which should be a quantity that only depends on the resistivity (conductivity), according to theoretical expectation. It can be seen that there is a significant difference between measurement and theory. The dephasing rate in the experiment is strongly enhanced compared to theory. This will of course weaken the weak-localization correction in the magnetoresistance compared to the theoretical expectation and thereby weaken the insulating trend of the resistivity at zero magnetic field. The enhancement of the dephasing rate over the expectation is common to many experimental systems and its origin remains an interesting topic of future experimental and theoretical research. Taking into account an additional temperature-independent dephasing mechanism in our analysis could not account for the observed behavior.

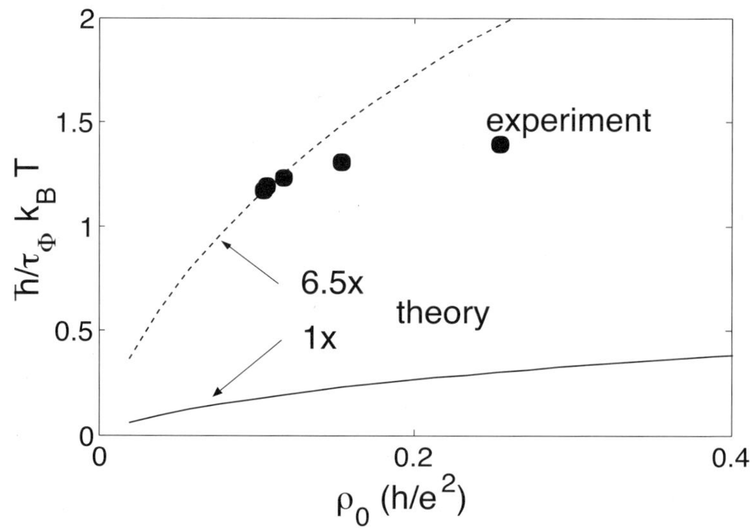

Fig. 10.8. The normalized parameter $(\hbar/\tau_\varphi)/kT$ vs. resistivity in units of h/e^2. The solid line is the prediction of eq. (5.2).

We briefly return to the findings presented in section 10.2 on the electric field dependence of the conductance and heating of the hole gas. The results obtained there can be supported by measurements of the dephasing rate of the holes as a function of temperature and electric field. Figure 10.9 shows the corresponding data. It can be seen that the weak-localization correction becomes weaker with increasing electric field, similar to the effect of temperature. Figure 10.10 shows the extracted values for τ_φ as a function of temperature and electric field, respectively. If we assume that at the lowest electric fields $\tau_\varphi(T)$ is given by the lattice temperature T_l and that in the electric field dependent measurements $\tau_\varphi(\mathcal{E})$ is a measure of the electronic temperature T_e, we find the relation $T_\mathrm{e} \propto \mathcal{E}^\nu$ with $\nu = 0.44$ between electronic temperature and electric field. Assuming only a single dissipation mechanism one obtains with eq. (10.1) and $P_\mathcal{E} = \varrho\mathcal{E}^2$ the relation $T_\mathrm{e} = (T_\mathrm{l}^\alpha + \varrho/c_1 \times \mathcal{E}^2)^{1/\alpha}$, which simplifies to $T_\mathrm{e} \propto \mathcal{E}^{2/\alpha}$ for significant heating above T_l. From our weak-localization analysis we therefore estimate $2/\alpha = \nu = 0.44$. This result is in reasonable agreement to the heating results for large electric fields $2/\alpha = 0.4$ obtained by the Saclay group (see section 10.2) and to earlier electric field scaling results [167].

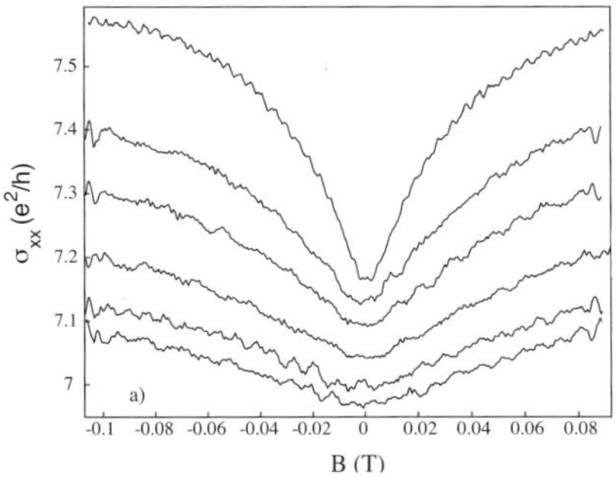

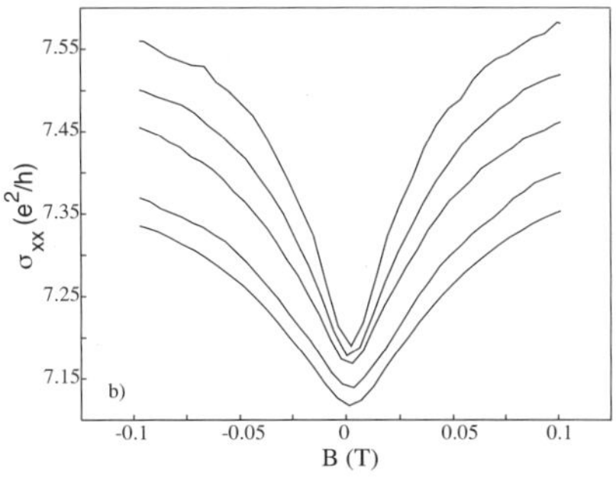

Fig. 10.9. Weak localization peaks measured for different temperatures between 0.1 K and 1 K (a) and for different electric fields between 0.1 V/m and 1 V/m (b). The weak-localization peak becomes less pronounced for increasing temperature or electric field.

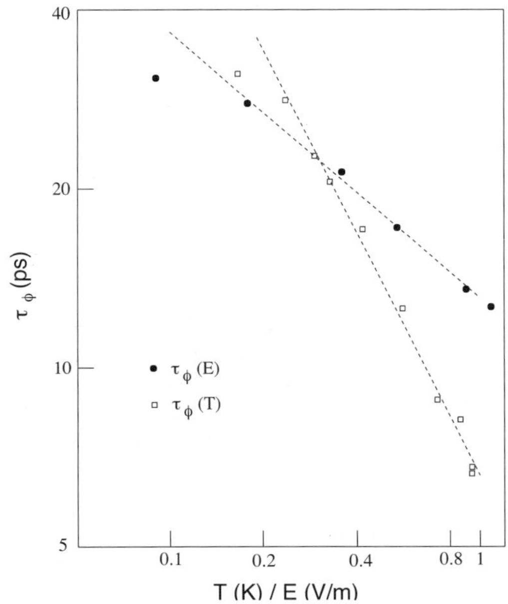

Fig. 10.10. Phase coherence time τ_φ as a function of temperature (open squares) and electric field (filled circles). The dashed lines correspond to fits according to $\tau_\varphi(T) \propto T^\gamma$ with $\gamma = -1.03$ and $\tau_\varphi(E) \propto E^\delta$ with $\delta = -0.45$, respectively.

10.5 Interaction corrections to the conductivity: multiple impurity scattering

Interaction corrections to the conductivity in the diffusive regime can be extracted from the temperature dependence of the Hall resistance at small magnetic fields [143]. It is predicted that the WL correction does not affect the Hall resistance, while interaction corrections obey [107]

$$\delta\sigma_{\mathrm{I}} = -\sigma_{xx} \times \frac{\delta R_{\mathrm{H}}}{2R_{\mathrm{H}}} = \frac{e^2}{2\pi^2\hbar}\left(1 - \frac{3}{4}F^{\star}\right)\ln\left(\frac{kT}{\hbar/\tau_{\mathrm{e}}}\right), \tag{10.4}$$

where $R_{\mathrm{H}} = \mathrm{d}\varrho_{xy}/\mathrm{d}B$ is the Hall coefficient [see eq. (1.2)]. Therefore, these interaction corrections to the conductivity can be extracted from the temperature dependence of the Hall resistance at small magnetic fields. Measurements were taken from negative to positive magnetic fields in order to be able to eliminate admixtures of the longitudinal resistivity $\varrho_{xx}(B)$ to ϱ_{xy}. Such admixtures can be identified through their even symmetry in B. Figure 10.11 shows $\delta\sigma_{\mathrm{I}}(T)$ determined in this way for two different densities in the metallic regime. Measurements of the electron density

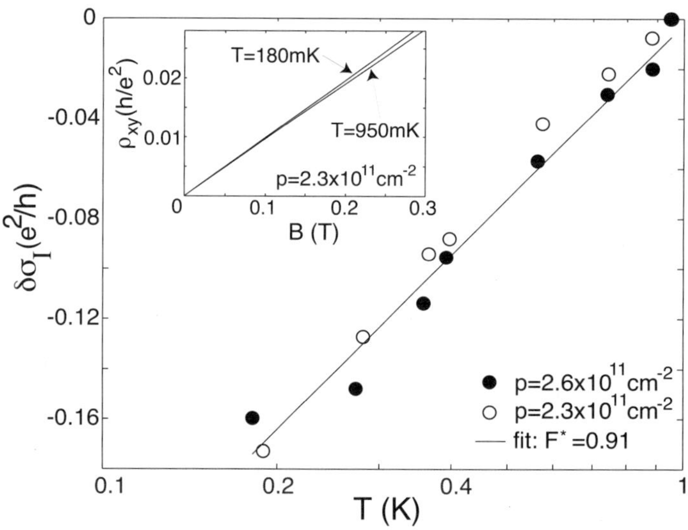

Fig. 10.11. Interaction corrections to the conductivity as determined from the temperature dependence of the Hall resistance at low fields. The inset shows the Hall resistances at two temperatures for an illustration of the method.

via the Shubnikov–de Haas effect at the same temperatures make sure that the electron density in the sample remained unchanged as a function of temperature. This is demonstrated in Fig. 10.12 where we have plotted the hole densities as determined from the Hall slope compared to those obtained from Shubnikov–de Haas measurements. The clear deviation of the Hall density from the Shubnikov–de Haas density can be attributed to the interaction correction to the conductivity.

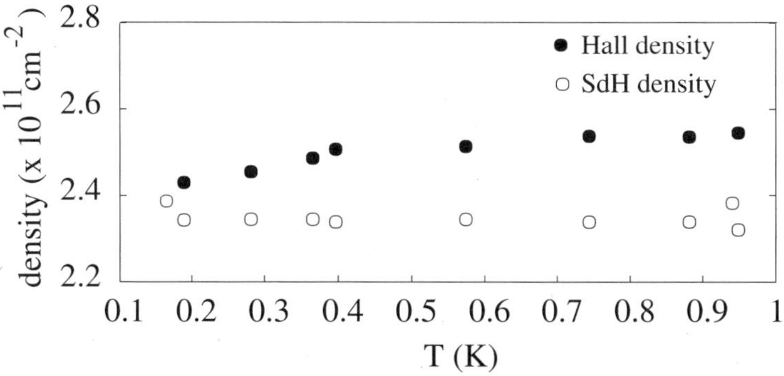

Fig. 10.12. Hall and Shubnikov–de Haas densities at fixed gate voltage as a function of temperature. The Shubnikov–de Haas density remains constant with increasing temperature while the Hall density increases reflecting the influence of the interaction corrections.

We find that $\delta\sigma_{\mathrm{I}}$ is negative and determine $F^{\star} = 0.91$ in agreement with Ref. [255]. This means that interaction corrections to the conductivity give another contribution of insulating type to the total conductivity in the metallic phase. This result, which agrees with Ref. [159] for p-GaAs, is of special importance, since interaction corrections have been suggested to lead to metallic behavior [169].

Coleridge and coworkers performed a similar analysis for their p-SiGe samples [203]. They found values of $F^{\star} \approx 1.33$ for a hole density $p = 5.7 \times 10^{11}\,\mathrm{cm}^{-2}$, which is deep in the metallic regime, and $F^{\star} \approx 1.1$ for $p = 1.2 \times 10^{11}\,\mathrm{cm}^{-2}$, close to the transition to insulating behavior, in rough agreement with our result. A difference between the samples in this paper and our samples is the presence of a front gate in our case. It may be that screening effects due to the gate metal reduce the interaction effects in our case slightly. Their values are, however, not in agreement with much larger values $F^{\star} = 2.45$ and $F^{\star} = 1.95$ determined in the same paper for the same samples from the residual temperature dependence of ϱ_{xx} after subtraction of the weak-localization correction.

On the theoretical side, Zala and coworkers have shown recently that the factor of 2 appearing in the denominator in front of the Hall coefficient R_{H} on the right-hand side of eq. (10.4) is correct only for the $T \to 0$ limit and can be strongly reduced at higher temperatures $kT/(\hbar/\tau) > 1$ [256]. We have neglected this effect in our analysis and estimate it to be small in our measurements for the lowest temperatures where $kT/(\hbar/\tau) \approx 0.007$.

10.6 Interaction corrections of the Drude conductivity due to T-dependent screening

Now that we have experimentally determined $\delta\sigma_{\mathrm{WL}}(T)$ and $\delta\sigma_{\mathrm{I}}(T)$, we are in the position to extract the bare $\sigma_{\mathrm{D}}(T)$ according to eq. (10.2) by subtracting the two corrections from the measured zero-field conductivity (Fig. 10.13). The obtained Drude conductivity shows a linear metallic temperature dependence in accordance with eq. (5.3) describing temperature-dependent screening, i.e., scattering of electrons at single impurities and the Friedel oscillations of their screening cloud.

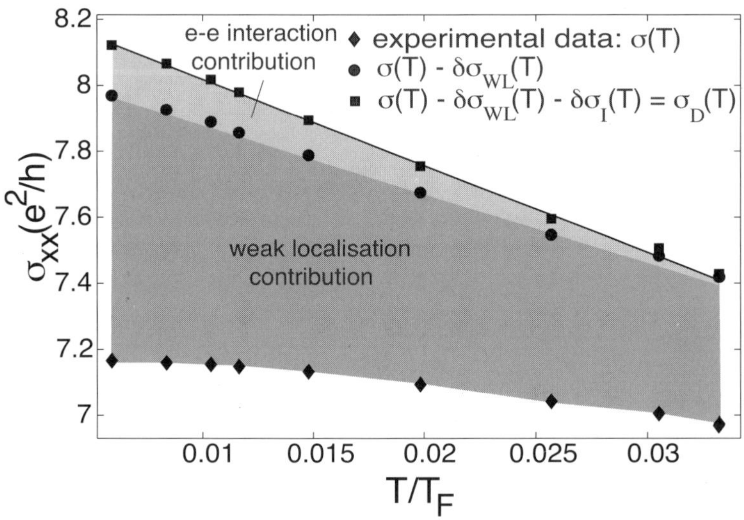

Fig. 10.13. Contributions to the conductance at zero magnetic field for a hole density $p = 2.6 \times 10^{11}\,\mathrm{cm}^{-2}$. From the measured curve (filled diamonds) we have subtracted the weak-localization contribution $\delta\sigma_{\mathrm{WL}}(T)$ determined from the magnetoresistance measurements and the interaction correction $\delta\sigma_{\mathrm{I}}(T)$ determined from the T-dependent Hall slope. The remaining conductivity $\sigma_{\mathrm{D}}(T)$ has a linear temperature dependence.

In the following we will briefly sketch the ingredients leading to the linear metallic temperature dependence. To this end we follow the conventional Drude description of conductivity in metals including screening effects via Lindhard's dielectric function in the random-phase approximation (RPA). In this description the temperature-dependent Drude conductivity is

$$\sigma_{\mathrm{D}}(T) = \frac{pe^2\tau_{\mathrm{e}}(T)}{m^\star},$$

where p is the sheet density of the holes and the average Drude scattering time $\tau_{\mathrm{e}}(T)$ has to be calculated according to [110, 123, 257]

$$\tau_{\mathrm{e}}(T) = \frac{\int \mathrm{d}E\, E\tau_{\mathrm{e}}(E,T)(-\mathrm{d}f/\mathrm{d}E)}{\int \mathrm{d}E\, E(-\mathrm{d}f/\mathrm{d}E)},$$

with the Fermi distribution function $f(E, T, \mu_{\mathrm{ch}})$ and the energy-dependent scattering rate [123]

$$\frac{\hbar}{\tau_{\mathrm{e}}(E,T)} = 2\pi N_{\mathrm{i}} \int \frac{\mathrm{d}^2 k'}{(2\pi)^2} \left|\frac{V(q)}{\varepsilon(q,T)}\right|^2 (1-\cos\theta)\delta\left(E_k - E_{k'}\right).$$

Here N_{i} is the density of ionized impurities, $V(q)$ is the matrix element for scattering by a wavevector $q = k_{\mathrm{F}}\sqrt{2(1-\cos\theta)}$, and $\varepsilon(q,T)$ is Lindhard's dielectric function. It is the explicit temperature dependence of this screening function that leads to the metallic temperature dependence. The origin of the explicit temperature dependence is the polarization function $\Pi(q, T, \mu_{\mathrm{ch}})$ entering the expression [258]

$$\varepsilon(q,T) = 1 + \frac{1}{q a_{\mathrm{B}}^{\star}} \Pi(q, T, \mu_{\mathrm{ch}}) v(q),$$

where $v(q)$ is the form factor associated with the finite thickness of the two-dimensional gas of carriers. The polarization function was for two-dimensional systems calculated by Stern [118] and is within the RPA given by

$$\Pi(q, T, \mu_{\mathrm{ch}}) = \frac{2}{A} \sum_{\boldsymbol{k}} \frac{f(E_{\boldsymbol{k}}) - f(E_{\boldsymbol{k}+\boldsymbol{q}})}{E_{\boldsymbol{k}} - E_{\boldsymbol{k}+\boldsymbol{q}} - \mathrm{i}0^+},$$

where A is a normalization area. At zero temperature the sum over wave vectors can be performed analytically and the polarization function is given by [258]

$$\Pi(q, T=0, \mu_{\mathrm{ch}} = E_{\mathrm{F}}) = \frac{m^\star}{\pi\hbar^2} \begin{cases} 1 & \text{for } q < k_{\mathrm{F}} \\ 1 - \sqrt{1-(2k_{\mathrm{F}}/q)^2} & \text{for } q > k_{\mathrm{F}} \end{cases}.$$

The finite temperature polarization function can then be calculated following Maldague [259]

$$\Pi(q, T, \mu_{\mathrm{ch}}) = \int_0^\infty \frac{\Pi(q, T=0, \mu')\mathrm{d}\mu'}{4kT\cosh^2[(\mu_{\mathrm{ch}} - \mu')/2kT]},$$

which essentially averages the zero-temperature polarization function around $q = 2k_{\mathrm{F}}$ and therefore removes the Kohn singularity at this point [see Fig. 5.2 for a plot of $\Pi(q, T, \mu_{\mathrm{ch}})$ at different temperatures]. This zero-temperature singularity has been demonstrated to be the origin of Friedel oscillations around an impurity [118]. Zala and coworkers have shown that the same result can be obtained with diagrammatic techniques in the single impurity limit [117] as discussed in section 5.2.

The above formalism is suitable for describing the conductivity of a two-dimensional system including interactions within the RPA, if the scattering matrix elements $V(q)$ are specified. Different scattering mechanisms will lead to different expressions for $V(q)$, which can be found in Refs. [119, 125, 260–262] for remote ionized impurity scattering, in Refs. [125, 261, 262] for homogeneous background doping, in Refs.

[119, 125, 260–262] for interface roughness scattering, in Refs. [119, 125, 262] for alloy-disorder scattering and in Refs. [125, 258] for acoustic phonon, piezoelectric, and optical phonon scattering.

Detailed model calculations for our structures within the RPA were performed in collaboration with E.H. Hwang and S. Das Sarma at the University of Maryland. Figure 10.14 shows their results which include Coulomb scattering at interface charges, acoustic phonon scattering, and the collision broadening of the polarization function [121] fitted to our high-temperature experimental data. Very nice agreement is

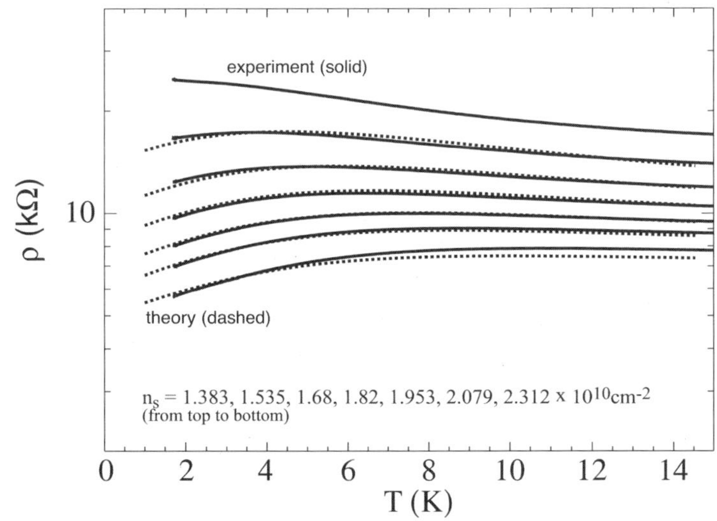

Fig. 10.14. Comparison between experiments (solid lines) and calculations (dashed lines) performed by E.H. Hwang and S. Das Sarma. Coulomb scattering at interface charges and acoustic phonon scattering were included. The effect of collision broadening on the screening function [121] has been taken into account.

obtained for the whole range of densities and temperatures.

For dominant large-angle scattering (i.e., scattering for $q \approx 2k_\mathrm{F}$), it has been shown analytically that [122, 127]

$$\sigma_\mathrm{D}(T) = \sigma_\mathrm{D}(0)\left[1 - c(p)\frac{T}{T_\mathrm{F}}\right] + \mathcal{O}\left[\left(\frac{T}{T_\mathrm{F}}\right)^{3/2}\right]. \tag{10.5}$$

Theory predicts values for the constant $c(p)$, which depend on the scattering mechanism and on the hole density p. If we apply eq. (10.5) to $\sigma_\mathrm{D}(T)$ in Fig. 10.13 we determine $c(p) = 3.1$ in reasonable agreement with a predicted value of $c(p) \approx 2.8$ [122] for charged interface impurity scattering at low p. The theoretical value has been shown to increase if correlation effects are included in the calculations [127].

Similar agreement has been found for all densities in the metallic regime. This together with the results of the model calculation presented above demonstrates that temperature-dependent screening can indeed explain the metallic temperature dependence of the resistivity in our p-SiGe system *without invoking a novel metallic phase*. It also implies that at sufficiently low temperatures insulating behavior with $\mathrm{d}\varrho/\mathrm{d}T < 0$ is expected to be recovered. Similar results in p-SiGe were obtained for much higher electron densities in Ref. [255].

In the p-SiGe samples of the NRC-group a similar analysis was performed [203]. A linear temperature dependence was again found down to moderately low temperatures of the order of a percent of the Fermi temperature. The coefficient $c(p)$ in eq. (10.5) was found to be larger than expected from the initial calculation by Gold and Dolgopolov [122]. The revised theory presented in Ref. [127] may remedy this discrepancy. Very small deviations from linear temperature dependence were identified at the lowest temperatures in this paper.

10.7 Reentrant insulating behavior

For higher carrier densities p we expect $\sigma_{\mathrm{D}}(T)$ to exhibit a weaker temperature dependence, because $c(p)$ decreases [122]. In Fig. 10.15 we therefore show the magnetic field and the temperature dependence measured at two densities. The lower surface shows the curves measured at $p = 2.6 \times 10^{11}\,\mathrm{cm}^{-2}$. At large $|B|$ the metallic temperature dependence can be seen. At $B = 0$ the WL peak tends to counteract this metallic behavior without overpowering it in the range of temperatures shown. The upper surface was measured at a density of $4.3 \times 10^{11}\,\mathrm{cm}^{-2}$ on an ungated part of the same sample. The temperature dependence at large $|B|$ is weaker than for the other surface [$c(p) = 0.8$], in agreement with temperature-dependent screening. At $B = 0$, however, the WL correction is strong enough to restore the insulating temperature dependence of the conductivity. Such a reentrant insulating behavior is consistent with the observations made in Refs. [176, 182].

10.8 Parallel magnetic field

The behavior of the magnetoresistance of p-SiGe in parallel magnetic fields is of particular interest compared to other systems like Si-MOSFETs and p-GaAs. The reason is that in the p-SiGe system the holes occupy the degenerate $(J, m_J) = (3/2, \pm 3/2)$ (J is the total angular momentum) heavy-hole subbands. Owing to strong spin-orbit coupling $\boldsymbol{J}$ is aligned normal to the quantum well. This orientation anisotropy is further stabilized by strain, which increases the heavy-hole light-hole separation compared to the unstrained case. The total angular momentum therefore has an easy axis normal to the plane of the quantum well. Application of an external magnetic field perpendicular to the well therefore leads to strong spin-splitting, whereas the orientation of the total angular momentum $\boldsymbol{J}$ is insensitive to a parallel magnetic field. As

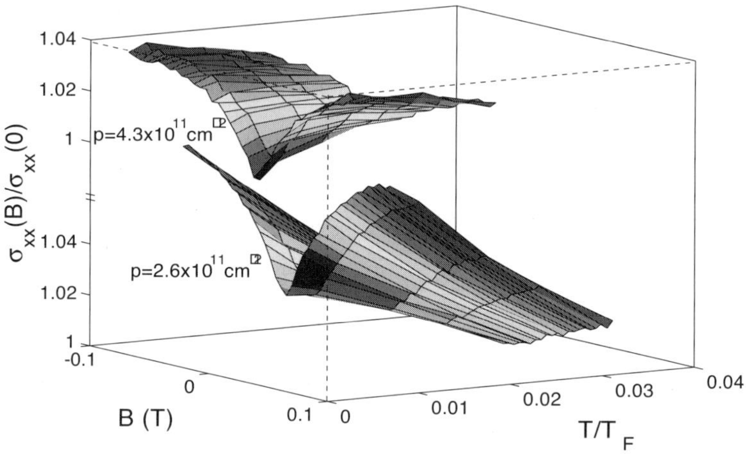

Fig. 10.15. Temperature and magnetic field dependence of the conductivity for two different densities. The upper curve at high density shows reentrant insulating behavior.

a result, the two-dimensional hole gas cannot be polarized with an in-plane field, in contrast to the other experimental systems.

Figure 10.16 shows the dependence of the resistivity on temperature with an in-plane field of $B = 6\,\mathrm{T}$ applied and for densities above and below the critical density at which the temperature dependence changes from metallic to insulating at zero magnetic field. Except for an overall increase of the resistivity, the behavior remains essentially the same. This result supports the hypothesis that spin polarization of the two-dimensional electron gas in a parallel magnetic field is responsible for the large positive magnetoresistance observed in Si-MOSFETs [169].

10.9 Discussion of the results and conclusions

In the following, we address the range of validity of the presented analysis. The theories of linear screening, WL, and interaction corrections are expected to hold as long as the disorder in the system is low enough, i.e., as long as $k_{\mathrm{F}}l > 1$. Therefore, the analysis is limited to the metallic regime. Non-linear screening models have to be invoked at the MIT, where our system undergoes the transition from WL ($k_{\mathrm{F}}l > 1$) to strong localization ($k_{\mathrm{F}}l < 1$). We have empirically applied concepts developed for weakly interacting systems with $r_{\mathrm{s}} \leq 1$. This attempt was successful and a Fermi-liquid description of the 2D hole gas appears to be consistent with the experiment even at our large interaction parameter $r_s \approx 6$. This fact suggests that the Fermi-liquid concept and the perturbative screening theory are still valid in two-dimensional systems even when the interaction parameter is considerably larger than one. A deeper theoretical understanding and interpretation of this finding remains an open issue.

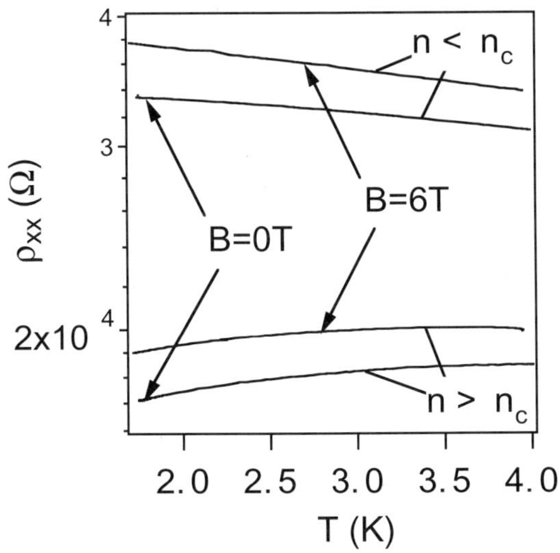

Fig. 10.16. Resistivity versus temperature for two different carrier densities for $B_{||} = 0\,\mathrm{T}$ and $6\,\mathrm{T}$, respectively.

How relevant is this analysis for other systems in which the MIT was observed? The analysis was simplified by the fact that large-angle scattering is dominant in p-SiGe. For this special case, the analytical results of Ref. [122] apply. Candidates for a similar analysis are therefore low-mobility Si-MOSFETs (for which this theory was originally developed) or the n-GaAs system with InAs quantum dots near the 2D electron gas [163]. Kravchenko et al. stated in Ref. [148] explicitly that temperature-dependent screening can not account for the metallic behavior in their high-mobility Si-MOSFET samples. However, at the time the extended theory of temperature-dependent screening including correlation effects [127], which predicts an enhancement of the metallic behavior, was not available and it would certainly be interesting to compare. Performing a similar analysis on the n-type GaAs samples of Ref. [163] we find that the temperature dependence in the metallic range is much too large compared to the theory used above. For p-type GaAs samples exhibiting the MIT temperature-dependent screening was suggested in Refs. [159, 161] as the relevant mechanism. Under the condition of small-angle scattering this effect is much weaker than in our case. The absence of a strong magnetoresistance in p-SiGe in parallel magnetic field is certainly of some value for the research on strongly interacting two-dimensional systems in Si-MOSFETs or p-GaAs heterostructures with much weaker spin-orbit coupling. It supports the observation in these systems that restoring insulating behavior at large $B_{||}$ comes along with spin-polarization of the electron gas.

In conclusion, we have analyzed the magnetoresistance and the Hall resistance in p-SiGe samples in terms of interference and interaction corrections to the conductivity and found the measurements to be consistent with ordinary Fermi-liquid behavior, RPA screening, and diffusive interaction corrections at large interaction parameters $r_\mathrm{s} \approx 6$. Both corrections tend to localize the system as the temperature is lowered.

The temperature dependence of the background Drude conductivity has been found to be linear in temperature. Its behavior is in good agreement with the theory of temperature-dependent screening for systems in which large-angle scattering dominates. The analysis applies to densities where the system is in the metallic regime and cannot easily be extended to the transition region, where $k_{\mathrm{F}}l \approx 1$. Although these observations cannot exclude the possibility of a novel metallic ground state in our or in other systems unambiguously, we are inclined to discard this exciting possible interpretation for p-SiGe hole gases on the basis of our experiments and analysis. The theoretical understanding of experimental consistency with Fermi-liquid behavior at these large r_{s}-values remains a challenging topic for further research.

Part III

Electron Transport through Quantum Dots and Quantum Rings

11

Introduction to electron transport through quantum dots

In section 2.3.3 we have phenomenologically introduced the experimental observations made in very small conducting islands weakly coupled to leads, i.e., the Coulomb-blockade effect and the associated transmission resonances (see Fig. 2.8). In this chapter we give a brief discussion of the basic physical concepts needed in order to interpret the experimental data presented later. For more extensive and comprehensive overviews over the many experiments performed and theories developed in recent years we refer to the many books [32, 87–90] and reviews [81, 91–95] published on this topic.

11.1 Resonant tunneling and the quantization of the particle number on weakly coupled islands

Electron transport through quantum dot structures is closely related to the quantum mechanical phenomenon of resonant tunneling discussed in many textbooks (see e.g., [27, 31]). In order to illustrate the properties of a resonant tunneling structure which are of importance for quantum dots, we consider a one-dimensional system in which non-interacting quantum particles scatter at a double barrier consisting of two δ-potentials separated by a region with length L of free propagation. The model schematically depicted in Fig. 11.1 is a simplified version of the model that will be used for the description of quantum scattering in wires in chapter 17 (see also Appendix C). The transfer matrix T_k of an individual δ-scatterer is given by

$$\begin{pmatrix} b_1 \\ b_2 \end{pmatrix} = \underbrace{\frac{1}{2k}\begin{pmatrix} 2k - \mathrm{i}\gamma & -\mathrm{i}\gamma \\ \mathrm{i}\gamma & 2k + \mathrm{i}\gamma \end{pmatrix}}_{\mathsf{T}_k} \begin{pmatrix} a_1 \\ a_2 \end{pmatrix},$$

where $k = \sqrt{2mE/\hbar^2}$ and $\gamma = -2mV_\delta/\hbar^2$, V_δ is the strength of the δ-scattering potential and the a_i and b_i are the amplitudes of the plane wave states left and right of the barrier, respectively. Propagation between the scatterers is described by the equation

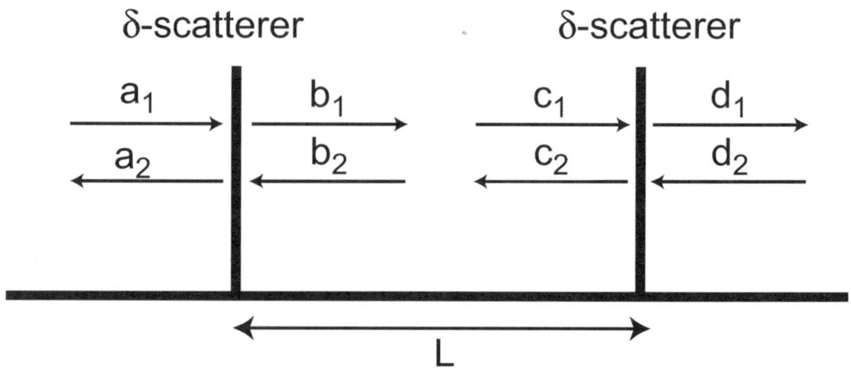

Fig. 11.1. One-dimensional scattering potential comprising two δ-scatterers separated by the distance L. The $a_1, a_2, b_1, b_2, \ldots$ are the amplitudes of right- and left-going plane wave states forming the solution of the scattering problem.

$$\begin{pmatrix} c_1 \\ c_2 \end{pmatrix} = \underbrace{\begin{pmatrix} e^{\mathrm{i}kL} & 0 \\ 0 & e^{-\mathrm{i}kL} \end{pmatrix}}_{\mathsf{P}_k} \begin{pmatrix} b_1 \\ b_2 \end{pmatrix},$$

with the Propagator P_k. With the two matrices T_k and P_k, we can write the solution of the transmission problem as $\boldsymbol{d} = \mathsf{T}_k\mathsf{P}_k\mathsf{T}_k\boldsymbol{a} := \mathsf{M}_k\boldsymbol{a}$, where M_k is the total transmission matrix. The quantity $t = M_{k;1,1} - M_{k;1,2}M_{k;2,1}/M_{k;2,2}$ is the transmission coefficient for a particle incident from the left with wave vector k and $T = |t|^2$ is the corresponding total transmission. The transmission for $\gamma L = 10$ is shown in Fig. 11.2a. It exhibits the well-known transmission resonances at certain values of kL, i.e., at certain energies, where the transmission approaches 1. The resonances become broader with increasing energy and as a consequence the background transmission between the resonances increases steadily with energy.

Within the same model we can calculate the wave-functions $\psi_k(x)$ in the region between the two δ-scatterers. The quantity $\mathcal{D}_k = 2\int_0^L \mathrm{d}x\, |\psi_k(x)|^2/(2\pi)$ is the density of states in this "island." The factor of 2 in front of the integral accounts for the fact that scattering states from the left and the right lead contribute. The density of states for $\gamma L = 1$ is depicted in Fig. 11.2b. It shows peaks at the same values of kL (i.e., at the same energies) as the transmission. However, as the energy increases, the density of states peaks become wider and smaller, in contrast to the transmission resonances. These peaks are reminiscent of the discrete level spectrum of a system with opaque barriers. With increasing strength of the delta scatterers γ, the density of states peaks become sharper (not shown).

Filling the quantum states of this one-dimensional non-interacting system with fermions up to a certain Fermi wavevector k_F leads to a certain number of particles N on the island, which is given by $N = \int_0^{k_\mathrm{F}} \mathcal{D}_k \mathrm{d}k$. This number N is shown in Fig. 11.2c as a function of $k_\mathrm{F}L$. Although N is a monotonic and continuous function

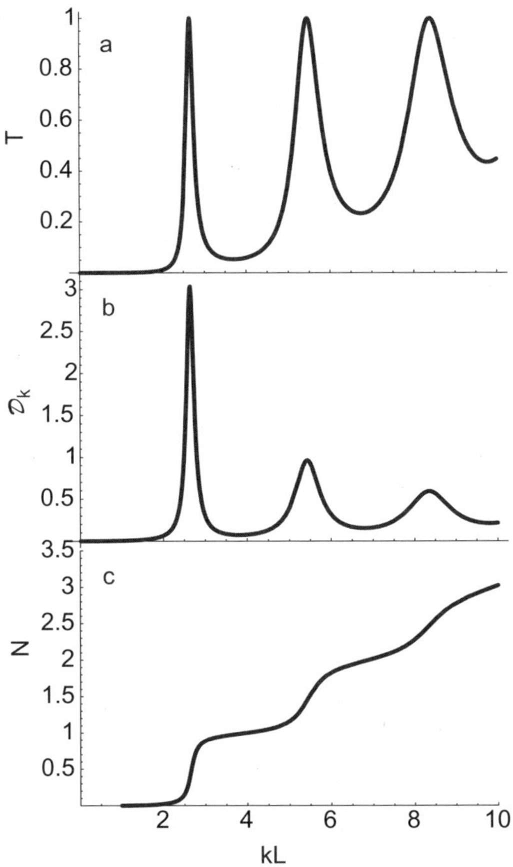

Fig. 11.2. Transmission T (a), density of states $\mathcal{D}_k$ (b) and number of particles N (c) calculated for the double δ-barrier problem with parameter $\gamma = 10$.

of energy, it exhibits steps at integer N, which are more pronounced at lower energies (lower kL) and wash out at higher energies. The step-like character of this function will be the more pronounced the larger γ is, i.e., the weaker the tunneling coupling of the island to the leads. This one-dimensional model of an island demonstrates that the number of particles on a weakly coupled island is an integer number, i.e., it is quantized.

More involved models of resonant tunneling with islands of higher dimensionality or more realistic tunneling barriers will all show the same conceptual behavior. Such models are typically hard to solve because one needs to solve Schrödinger's equation in three dimensions for the given potential. In addition, the tunneling coupling can in most cases only be taken into account in the weak coupling approximation. Most common is the transfer hamiltonian approach developed by Bardeen [40, 41].

11.2 Quantum dot states: from a general hamiltonian to the constant-interaction model

The quantization of N occurs also in systems of interacting particles that are weakly coupled to leads. However, the interactions lead to a serious modification of the density of states $\mathcal{D}_k$ compared to the non-interacting case. This is the basic ingredient for the Coulomb-blockade effect in electron transport through a quantum dot.

The theoretical description of the Coulomb-blockade effect typically involves two separate steps: the states of the quantum dot are described with some hamiltonian neglecting the coupling to source and drain contacts via tunneling barriers. In a separate step, the tunneling coupling of the dot states to the leads is introduced and the electrical current is calculated.

In this section we will give an overview of how a relatively simple but very intuitive model hamiltonian for a quantum dot can be found starting from very general considerations. This model is called the "constant-interaction model." Although it neglects many details that are sometimes of interest in experiments, it captures the basic effects leading to the Coulomb-blockade phenomenon in electron transport.

We consider a typical system containing dielectric media, metallic gate electrodes, and fixed ionic charges (e.g., dopants or surface charges). The gate electrodes are used for defining a lateral confinement potential for electrons, the vertical confinement is assumed to be provided by a certain layered sequence of materials like in a heterostructure or in a quantum well. A very general hamiltonian for such a quantum dot system with N electrons is given by [263]

$$
\begin{aligned}
H_N = \sum_{n=1}^{N} \Bigg\{ & \frac{\boldsymbol{p}_n^2}{2m^\star} - e\int_V \mathrm{d}V \rho_{\mathrm{ion}}(\boldsymbol{r}) G(\boldsymbol{r}_n, \boldsymbol{r}) + \frac{e^2}{2} G(\boldsymbol{r}_n, \boldsymbol{r}_n) - e\sum_i \phi_i \alpha_i(\boldsymbol{r}_n) \\
& + e^2 \sum_{m=1}^{n-1} G(\boldsymbol{r}_m, \boldsymbol{r}_n) \Bigg\}. \qquad (11.1)
\end{aligned}
$$

This is an effective mass hamiltonian assuming a parabolic conduction band dispersion at the Γ-point (first term, kinetic energy). It contains a number of single-particle contributions to the potential energy and a term describing electron–electron interactions. The function $G(\boldsymbol{r}, \boldsymbol{r}')$ is the electrostatic Green's function of the system fulfilling Poisson's equation

$$\nabla \left[\varepsilon(\boldsymbol{r})\varepsilon_0 \nabla G(\boldsymbol{r}, \boldsymbol{r}_1)\right] = -\delta(\boldsymbol{r} - \boldsymbol{r}_1)$$

with boundary condition

$$G(\boldsymbol{r}, \boldsymbol{r}_1)|_{\boldsymbol{r} \in S_i} = 0,$$

i.e., the function is zero on the metallic electrodes. It therefore contains image charge effects and screening by the gate electrodes. The last term in the hamiltonian (11.1) describes the electron–electron interactions. The term $e^2 G(\boldsymbol{r}_n, \boldsymbol{r}_n)/2$ contains the image charge potential for an individual electron and a self-energy term that has to be renormalized away by appropriate methods. The second term on the right contains the fixed ionic charges in the system, $\rho_{\mathrm{ion}}(\boldsymbol{r})$, and represents their effective electrostatic potential. The functions $\alpha_i(\boldsymbol{r})$ are the characteristic potentials of the gate electrodes, which are indexed by i. They are connected to the Green's function by the surface integral

$$\alpha_i(\boldsymbol{r}) = -\int\int_{S_i} \mathrm{d}s_{1i} \left[\varepsilon(\boldsymbol{r}_1)\varepsilon_0 \nabla_1 G(\boldsymbol{r}_1, \boldsymbol{r})\right] \cdot \boldsymbol{n}_i,$$

where $\varepsilon(\boldsymbol{r})$ is the relative dielectric constant of the system and $\boldsymbol{n}_i$ is vector normal to the surface element ds_{1i} pointing into the gate electrode. The $\alpha_i(\boldsymbol{r})$ represent the electrostatic potential created by a unit potential on gate i with all other gates grounded. The term in the hamiltonian (11.1) containing these functions represents the potential created by the voltages on the gates i with electrostatic potential ϕ_i. In general, differences of applied voltages (differences of electrochemical potentials) $U_{ij} = U_i - U_j$ on the gates translate into differences of electrostatic potentials via

$$U_{ij} = \frac{\mu_{\mathrm{elch}}^{(i)} - \mu_{\mathrm{elch}}^{(j)}}{-e} = \phi_i - \phi_j - \frac{\mu_{\mathrm{ch}}^{(i)} - \mu_{\mathrm{ch}}^{(j)}}{e}.$$

The electrostatic potentials ϕ_i are apart from the offset $\mu_{\mathrm{ch}}^{(i)}$ identical to the applied voltages.

A number of methods exist to tackle the many-body eigenvalue problem with the hamiltonian (11.1), among them exact diagonalization [264–269], the Hartree approximation [270], Hartree–Fock methods [271–275], density functional theory [276–278] or quantum Monte Carlo simulations [279, 280]. In the following we sketch the assumptions and simplifications made for arriving at the simplest approximate model, the so-called constant-interaction model.

Hartree–Fock approximation

In a first simplifying step we neglect some parts of the interaction term, and assume the wave-functions to be Slater determinants of single-particle wave functions. This approximation is known as the Hartree–Fock approximation for a many-electron system requiring a self-consistent solution of the Hartree–Fock equation [281]

$$\left\{ \frac{\boldsymbol{p}^2}{2m^\star} - e \int_V \mathrm{d}V_1 \left[\rho_{\mathrm{ion}}(\boldsymbol{r}_1) + \rho_{\mathrm{e}}(\boldsymbol{r}_1)\right] G(\boldsymbol{r}, \boldsymbol{r}_1) + \frac{e^2}{2} G(\boldsymbol{r}, \boldsymbol{r}) \right.$$
$$\left. - e \sum_i \phi_i \alpha_i(\boldsymbol{r}) \right\} \psi_n(\boldsymbol{r}) + e \int \mathrm{d}^3 r' \, G(\boldsymbol{r}, \boldsymbol{r}') \rho_{\mathrm{xc}}^{(s_n)}(\boldsymbol{r}, \boldsymbol{r}') \psi_n(\boldsymbol{r}') = E_n \psi_n(\boldsymbol{r}) \tag{11.2}$$

with the electron density

$$\rho_{\mathrm{e}}(\boldsymbol{r}) = -e \sum_{n_{\mathrm{occ}}} |\psi_n(\boldsymbol{r})|^2$$

and the non-local exchange density

$$\rho_{\mathrm{xc}}^{(s_n)}(\boldsymbol{r}, \boldsymbol{r}') = -e \sum_{m_{\mathrm{occ}}} \psi_m^\star(\boldsymbol{r}') \psi_m(\boldsymbol{r}) \delta_{s_n, s_m}.$$

The wave functions $\psi_n(\boldsymbol{r})$ are the self-consistently determined eigenfunctions with energies E_n. The exchange density $\rho_{\mathrm{xc}}^{(s_n)}$ depends on the quantum number s_n of the z-component of the spin in level n. The number of electrons in the system is determined from the charge neutrality condition.

Neglecting self-consistent effects

Let us assume that we have solved the Hartree–Fock problem self-consistently for a given set of gate potentials $\{\phi_i^{(0)}\}$ and we have found a large number N_0 electrons (say, $N_0 > 100$) to reside on the dot. We now consider the effect of changing the gate voltates slightly by $\{\Delta\phi_i\}$. This changes the gate-voltage-dependent term in the Hartree–Fock equation (11.2) and there will be a first-order shift of energy levels. However, there will also be a first-order modification of the self-consistent wave functions leading to a first-order change in the electron density $\rho_{\mathrm{e}}(\boldsymbol{r})$ and the exchange density $\rho_{\mathrm{xc}}^{(s_n)}$, which in turn causes an additional induced potential perturbation. In addition we may have to adapt the electron number in order to satisfy charge neutrality. These intricate changes in the system may be regarded as a screening effect in response to the gate voltage change.

A tremendous simplification arises if we neglect the self-consistent density changes in $\rho_{\mathrm{e}}(\boldsymbol{r})$ and $\rho_{\mathrm{xc}}^{(s_n)}$. We denote the electron densities under these conditions $\rho_{\mathrm{e}}^{(0)}(\boldsymbol{r})$ and $\rho_{\mathrm{xc0}}^{(s_n)}$. This approximation essentially means neglecting the polarizability of the quantum dot. Under this assumption the change in gate potentials will in first-order lead to a linear shift of the energy levels E_n of the quantum dot according to

$$\Delta E_n = -e \sum_i \Delta\phi_i \left\langle n \left| \alpha_i(\boldsymbol{r}) \right| n \right\rangle = -e \sum_i \Delta\phi_i \alpha_i.$$

The expectation value α_i of the characteristic potential $\alpha_i(\boldsymbol{r})$ of gate i is called the lever arm of this gate. Although the lever arm may depend on the state index n, it turns out that in many-electron dots the lever arm is to a good approximation the same for relevant states near the Fermi level.

Constant interaction and constant exchange

The change in gate voltages may eventually lead to a change in the number of electrons. Given that we consider a dot with a large number of electrons N_0, the self-consistent changes in the local Hartree potential and the non-local Hartree–Fock potential will again be neglected.

Nevertheless, the energy levels E_n will be shifted by the Hartree potential created by the additional charge. Assuming that the electron density distribution does not change a lot with electron number, we can define an electron number independent "shape" function for the electron density $\beta(\boldsymbol{r}) = \rho_{\mathrm{e}}^{(0)}(\boldsymbol{r})/N_0$. The Hartree potential will then have the expectation value

$$E_{\mathrm{H}}(N_0) = N_0 \left\langle n \middle| -e \int_V \mathrm{d}V_1 \beta(\boldsymbol{r}_1) G(\boldsymbol{r}_1, \boldsymbol{r}) \middle| n \right\rangle := N_0 V_{\mathrm{c}}.$$

Adding an electron to the quantum dot into an empty state n lifts — under the assumptions introduced above — each energy level by a constant amount V_{c}, if the matrix element is about the same for all states.

The exchange energy will also lead to a shift of energy levels E_n. In contrast to the Hartree shift, the exchange-induced shift will depend on the spin of the electron in level E_n. The non-local exchange potential can be approximated if we take into account that $G(\boldsymbol{r}, \boldsymbol{r}')$ is dominated by small values of $|\boldsymbol{r} - \boldsymbol{r}'|$. Approximating $G(\boldsymbol{r}, \boldsymbol{r}') = V_{\mathrm{xc}} \delta(\boldsymbol{r} - \boldsymbol{r}')$, we obtain a kind of "local density approximation" for the exchange term leading to the local spin density

$$\rho_{\mathrm{xc,loc}}^{s_n}(\boldsymbol{r}) = -e \sum_{m_{\mathrm{occ}}} |\psi_m(\boldsymbol{r})|^2 \, \delta_{s_n, s_m}.$$

The expectation value of the exchange interaction can in this approximation be treated similar to the Hartree interaction term leading to

$$E_{\mathrm{xc}}^{\uparrow/\downarrow}(N_0) = -N_0^{\uparrow/\downarrow} V_{\mathrm{xc}},$$

where $N_0^{\uparrow}$ ($N_0^{\downarrow}$) is the number of electrons with spin $\uparrow$ ($\downarrow$).

Without the exchange term the obtained approximations lead to the constant-interaction model [282]. Exchange can be included in terms of a "constant exchange" term. We can summarize the energy required for adding an electron with spin $\uparrow$ ($\downarrow$) to the quantum dot by

$$\mu_{N+1}^{\uparrow/\downarrow} = E_{N+1} + V_{\mathrm{c}}(N+1) - V_{\mathrm{xc}}(N_0^{\uparrow/\downarrow} + 1) - e \sum_i \Delta\phi_i \alpha_i. \tag{11.3}$$

The exchange contribution always lowers this energy, i.e., $V_{\mathrm{xc}} > 0$. For the determination of the spin of an added electron, it is decisive if $\mu_{N+1}^{\uparrow} > \mu_{N+1}^{\downarrow}$ or vice versa. Within this model, the added electron will have spin up, if the difference

$$\mu_{N+1}^{\uparrow} - \mu_{N+1}^{\downarrow} = E_{N+1}^{\uparrow} - E_{N+1}^{\downarrow} - V_{\mathrm{xc}}(N_0^{\uparrow} - N_0^{\downarrow}) < 0.$$

Here, $E_{N+1}^{\uparrow/\downarrow}$ are the single-particle energy levels in which the up (down) electron is filled and $N_0^{\uparrow} - N_0^{\downarrow}$ is the occupation difference of spin up and spin down states already occupied in the dot. It becomes immediately clear that the exchange term favors parallel alignment of spins (in atoms this leads to Hund's rules). If the energy splitting $E_{N+1}^{\uparrow} - E_{N+1}^{\downarrow}$ is larger than the gain in exchange energy, the spin polarization of the dot will be lowered or kept minimized (extreme case: alternating up/down filling), otherwise the spin polarization will be increased (extreme case: fully spin-polarized ferromagnetic ground state).

Constant interaction from a capacitance model

The constant-interaction model (without the exchange interaction term) is analogous to a capacitance model in which the quantum dot is regarded as a floating metallic island with integer charge. In this model the interaction of electrons on the island among each other and with surrounding metallic electrodes is described by an electrostatic capacitance matrix in which the capacitances are considered to be independent of the voltages applied to the electrodes.

We consider $n+1$ metallic objects $0 \ldots n$. The objects $1 \ldots n$ (gates) are connected to voltage sources keeping them at constant potential. Island 0 (dot) carries charge Q_0, but is kept floating. We want determine the charging energy required to bring N electrons onto the dot. This is a standard problem of classical electrostatics closely related to the electrostatics problem discussed in chapter 15 (see also Appendix A).

The charges on the metallic islands are given by the relation [283, 284]

$$Q_i = \sum_{j=0}^{n} C_{ij}\phi_j + Q_i^{(0)}.$$

The C_{ii} are called the capacitances of island i and they are always positive. The C_{ij} with $i \neq j$ are called capacitance coefficients. They are always negative. The capacitance matrix can be formally represented in terms of the system's Green's function containing all the geometrical information about the system (see Appendix A). Owing to overall charge neutrality, the capacitance coefficients obey the condition $\sum_{i=0}^{n} C_{ij} = 0$. When all the potentials ϕ_j are changed by a constant amount, the induced charges Q_i do not change and therefore $\sum_{j=0}^{n} C_{ij} = 0$. The charge $Q_i^{(0)}$ is the charge on electrode i when all the potentials ϕ_i are set to zero. It is due to fixed charges in the system such as dopant ions or residual impurities.

The charge Q_0, in particular, is

$$Q_0 = \sum_{j=0}^{n} C_{0j}\phi_j + Q_0^{(0)} = C_{00}\phi_0 + \sum_{j=1}^{n} C_{0j}\phi_j + Q_0^{(0)}.$$

For C_{00} we can write $C_{00} = -\sum_{j=1}^{n} C_{0j} := C_{\Sigma}$. Because we regard the charge Q_0 as being given, we solve this equation for ϕ_0 in order to obtain the electrostatic potential of the dot determined by the voltages on all the gate electrodes:

$$\phi_0(Q_0) = \frac{1}{C_\Sigma}\left(Q_0 - Q_0^{(0)} - \sum_{j=1}^{n} C_{0j}\phi_j\right).$$

If $Q_0 = -eN$, the electrostatic energy $E_{\mathrm{elstat}}(N)$ needed to change the number of electrons from 0 to N at fixed potentials ϕ_j ($j = 1\ldots n$) is given by the integral

$$\begin{aligned}E_{\mathrm{elstat}}(N) &= \int_0^{-eN} \mathrm{d}Q_0\phi_0(Q_0) = \frac{1}{C_\Sigma}\int_0^{-eN} \mathrm{d}Q_0\left(Q_0 - Q_0^{(0)} - \sum_{j=1}^{n} C_{0j}\phi_j\right)\\ &= \frac{e^2N^2}{2C_\Sigma} + eN\left(\frac{Q_0^{(0)}}{C_\Sigma} + \sum_{j=1}^{n}\frac{C_{0j}}{C_\Sigma}\phi_j\right).\end{aligned}$$

The increase of the total electrostatic energy of the whole *system* is larger than $E_{\mathrm{elstat}}(N)$ by the amount

$$W_{\mathrm{gates}}(N) = \sum_{i=1}^{n}\int \mathrm{d}Q_i\phi_i = \sum_{i=1}^{n}\phi_i\int \mathrm{d}Q_i = \sum_{i=1}^{n}\phi_i Q_i.$$

When the charge $-eN$ is brought onto the dot at constant voltages ϕ_i this is work done by the voltage sources connected to the gates.

If we wish to add a single electron to the dot, we have to pay the charging energy

$$E_{\mathrm{c}} = E_{\mathrm{elstat}}(N+1) - E_{\mathrm{elstat}}(N) = \frac{e^2}{C_\Sigma}\left[N + \frac{1}{2}\right] + e\left(\frac{Q_0^{(0)}}{C_\Sigma} + \sum_{j=1}^{n}\frac{C_{0j}}{C_\Sigma}\phi_j\right).$$

In general, the charging energy increases with the number of electrons N that are already on the dot. This is (apart from a constant energy offset) already identical to the interaction term and the gate-induced energy change in the constant-interaction model [see eq. (11.3)], if we identify the charging energy $V_{\mathrm{c}} = e^2/C_\Sigma$ and the lever arm $\alpha_i = -C_{0i}/C_\Sigma > 0$.

We are interested in systems where the dot is sufficiently small that quantization effects of electronic states on the dot become important. In this case we may describe the total energy $E(N)$ of the dot by the sum of its single particle energies E_n plus the electrostatic energy $E_{\mathrm{elstat}}(N)$:

$$E(N) = \sum_{n=1}^{N} E_n + E_{\mathrm{elstat}}(N) = \sum_{n=1}^{N} E_n + \frac{1}{2}N^2V_{\mathrm{c}} + eN\left(\frac{Q_0^{(0)}}{C_\Sigma} - \sum_{j=1}^{n}\alpha_j\phi_j\right).$$

The electrochemical potential μ_N is by definition the energy required to add the Nth electron to the system. It is given by

$$\mu_N = E(N) - E(N-1) = E_N + V_{\mathrm{c}}\left(N - \frac{1}{2}\right) + e\left(\frac{Q_0^{(0)}}{C_\Sigma} - \sum_{j=1}^{n}\alpha_j\phi_j\right),$$

which is (up to a constant) the expression for the constant-interaction model (11.3) without the exchange contribution. It can be regarded as the sum of the chemical potential $\mu_{\mathrm{ch}} = E_N$ and the electrostatic potential

$$\phi(N) = V_{\mathrm{c}} \left(N - \frac{1}{2} \right) + e \left(\frac{Q_0^{(0)}}{C_\Sigma} - \sum_{j=1}^{n} \alpha_j \phi_j \right) .$$

At fixed electron number the electrostatic potential of the dot is a linear function of the gate voltages. It jumps discontinuously when an electron is added to the dot. This has, for example, been measured in Ref. [285].

11.3 Transport through quantum dots

11.3.1 Coulomb-blockade oscillations

In a typical quantum dot arrangement (cf. inset of Fig. 2.8) there are, in addition to the dot, at least three metallic islands: the source and drain contacts and a plunger-gate. Tunneling junctions connect the dot to source and drain while the coupling of the gate to the dot is purely capacitive. This situation is schematically depicted in Fig. 11.3(a). In the following we denote source-related quantities with subscript 1, drain-related voltages with subscript 2, and gate-related quantities with subscript 3. We consider the case of vanishing bias, e.g., $\mu_1 \equiv \mu_{\mathrm{S}} \approx \mu_{\mathrm{D}} \equiv \mu_2$ and therefore $\phi_1 = \phi_2 = 0$. (Dot and source-drain contacts are usually made of the same materials and no differences in chemical potentials exist.) In order to keep the discussion as

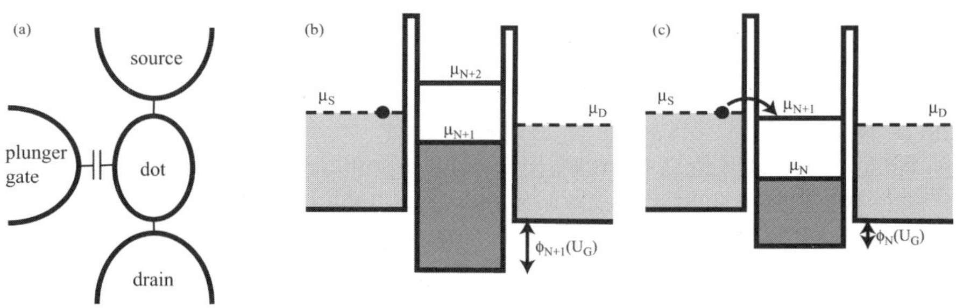

Fig. 11.3. (a) Schematic drawing of the system under consideration. (b) The dot is in the blockade. No electron can be transferred from source into the dot or from the dot into the drain contact. (c) The dot is in resonance with source and drain. An electron can tunnel from source into the dot and then out into drain (or vice versa; a small bias causes a net current in a preferred direction).

simple as possible we assume zero temperature at this point. Figure 11.3(b) shows the energetic relations for a typical plunger-gate voltage. The dot is filled with $N+1$

electrons. Since $\mu_{N+2} > \mu_{\rm S}, \mu_{\rm D}$, no electrons can tunnel into the dot, the current is blocked. This situation is called the Coulomb-blockade.

Figure 11.3(c) shows the situation with a very special plunger-gate voltage aligning the electrochemical potential in the dot μ_{N+1} with those in source and drain. In this case, an electron can be exchanged between the leads and the dot and a small bias voltage will cause a net current flow in one direction.

In general, the dot will make a transition from N to $N-1$ electrons if $\mu_{\rm S} \leq \mu_N$. It will make a transition from N to $N+1$ electrons if $\mu_{\rm S} \geq \mu_{N+1}$. This means that conductance peaks like those in Fig. 2.8 arise under the condition (we neglect the exchange contributions in the following for simplicity)

$$\mu_{\rm S} = \mu_N = E_N + V_{\rm c}\left(N - \frac{1}{2}\right) - e\left(\sum_{j=3}^{n} \alpha_j \phi_j - \frac{Q_0^{(0)}}{C_\Sigma}\right),$$

which can be solved for the gate potential $\phi_{\rm G} \equiv \phi_3$ to give

$$\phi_{\rm G}^{(N)} = \frac{1}{e\alpha_{\rm G}}\left[E_N + V_{\rm c}\left(N - \frac{1}{2}\right) - e\sum_{j=4}^{n} \alpha_j \phi_j + \frac{eQ_0^{(0)}}{C_\Sigma} - \mu_{\rm S}\right]. \tag{11.4}$$

This equation is the basis for a very straightforward interpretation of Coulomb-blockade measurements as a function of an external parameter like the magnetic field. If we assume that the capacitance matrix of our system and the chemical potentials in source and drain are not significantly modified by the field, we can see that the position of a conductance peak in plunger-gate voltage, $\phi_{\rm G}^{(N)}(B)$ (the voltage and the potential are the same up to an additive constant) is proportional to the magnetic-field-dependent single-particle energy level $E_N(B)$. Under certain conditions it is therefore possible to reconstruct the single-particle energy spectrum of a quantum dot in a magnetic field, at least over a limited range of energies.

Two consecutive conductance peaks have the spacing

$$\Delta\phi_{\rm G}^{(N+1)} = \phi_{\rm G}^{(N+1)} - \phi_{\rm G}^{(N)} = \frac{1}{e\alpha_{\rm G}}\left[E_{N+1} - E_N + V_{\rm c}\right] = \frac{1}{e\alpha_{\rm G}}\left[\Delta_{N+1} + V_{\rm c}\right],$$

where Δ_{N+1} is the single-particle level spacing. The term "constant-interaction model" implies that it is assumed that $V_{\rm c}$ does not change with electron number, at least in a certain energy interval of interest. This assumption, although not strictly valid in semiconductor quantum dots, turns out to be a very good approximation in many-electron dots. In the most typical cases where the typical single-particle level spacing in a dot is only a fraction of the charging energy, the conductance peaks are almost equidistant in plunger-gate voltage. Within the constant-interaction model, fluctuations in peak spacing occur due to fluctuations in the single-particle level spacing.

11.3.2 Coulomb-blockade diamonds

We now consider the case where the conductance is measured as a function of the plunger-gate voltage $U_{\mathrm{G}} \equiv \mu_{\mathrm{ch,G}} - e\phi_3$ and of the source-drain voltage. For simplicity, we still assume zero temperature.

Source grounded

We assume that μ_{S} is kept constant, $\mu_{\mathrm{D}} = \mu_{\mathrm{S}} - eU_{\mathrm{SD}}$. For the dot to be in a stable region with N electrons, we have the requirements

$$\begin{aligned} \mu_N &< \mu_{\mathrm{S}} - eU_{\mathrm{SD}} \\ \mu_{N+1} &> \mu_{\mathrm{S}}, \end{aligned}$$

if $U_{\mathrm{SD}} > 0$. If $U_{\mathrm{SD}} < 0$, we have

$$\begin{aligned} \mu_N &< \mu_{\mathrm{S}} \\ \mu_{N+1} &> \mu_{\mathrm{S}} - eU_{\mathrm{SD}}. \end{aligned}$$

The first two conditions give the two border lines for $U_{\mathrm{SD}} > 0$:

$$U_{\mathrm{G}} = \frac{1}{e\alpha_{\mathrm{G}}}\left[E_N + V_{\mathrm{c}}\left(N - \frac{1}{2}\right) - \mu_{\mathrm{S}} + e(1-\alpha_{\mathrm{D}})U_{\mathrm{SD}} - e\sum_{j=4}^{n}\alpha_j\phi_j + \frac{eQ_0^{(0)}}{C_\Sigma}\right],$$

$$U_{\mathrm{G}} = \frac{1}{e\alpha_{\mathrm{G}}}\left[E_{N+1} + V_{\mathrm{c}}\left(N + \frac{1}{2}\right) - \mu_{\mathrm{S}} - e\alpha_{\mathrm{D}}U_{\mathrm{SD}} - e\sum_{j=4}^{n}\alpha_j\phi_j + \frac{eQ_0^{(0)}}{C_\Sigma}\right].$$

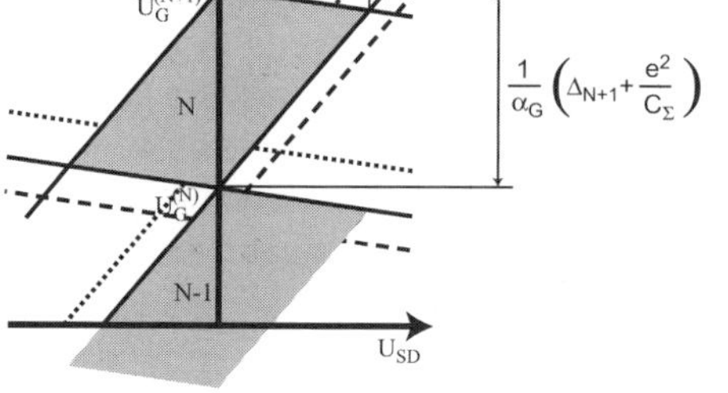

Fig. 11.4. Coulomb-blockade diamonds. The current is blocked in the diamond shaped areas shaded in gray. In these areas the number of electrons in the dot, N, is constant. Conductance peaks occur on the U_{G}-axis at points where neighboring diamonds touch.

They cross the U_{SD} axis at the positions of two neighboring conductance peaks. They cross each other at

$$eU_{\mathrm{SD}} = \Delta_{N+1} + V_{\mathrm{c}}.$$

The difference of their slopes is

$$\frac{1-\alpha_{\mathrm{D}}}{\alpha_{\mathrm{G}}} - \left(-\frac{\alpha_{\mathrm{D}}}{\alpha_{\mathrm{G}}}\right) = \frac{1}{\alpha_{\mathrm{G}}}.$$

This situation is schematically illustrated in Fig. 11.4. A measurement of these so-called Coulomb-blockade diamonds (for symmetrically biased dots to be discussed in the following) is shown in Figs. 12.3 and 13.2.

Symmetrically biased dot

We assume that $\mu_{\mathrm{S}} = \mu_{\mathrm{S}}^{(0)} + eU_{\mathrm{SD}}/2$ and $\mu_{\mathrm{D}} = \mu_{\mathrm{S}}^{(0)} - eU_{\mathrm{SD}}/2$ and have the requirements for the stable region with N electrons ($U_{\mathrm{SD}} > 0$):

$$\mu_N < \mu_{\mathrm{S}}^{(0)} - eU_{\mathrm{SD}}/2,$$
$$\mu_{N+1} > \mu_{\mathrm{S}}^{(0)} + eU_{\mathrm{SD}}/2.$$

They lead to the border line equations

$$U_{\mathrm{G}} = \frac{1}{e\alpha_{\mathrm{G}}}\left[E_N + V_{\mathrm{c}}\left(N - \frac{1}{2}\right) - \mu_{\mathrm{S}}^{(0)} + e(1-\alpha_{\mathrm{S}}+\alpha_{\mathrm{D}})U_{\mathrm{SD}}/2 \right.$$
$$\left. - e\sum_{j=4}^{n}\alpha_j\phi_j + \frac{eQ_0^{(0)}}{C_\Sigma}\right],$$
$$U_{\mathrm{G}} = \frac{1}{e\alpha_{\mathrm{G}}}\left[E_{N+1} + V_{\mathrm{c}}\left(N + \frac{1}{2}\right) - \mu_{\mathrm{S}}^{(0)} - e(1+\alpha_{\mathrm{S}}-\alpha_{\mathrm{D}})U_{\mathrm{SD}}/2 \right.$$
$$\left. - e\sum_{j=4}^{n}\alpha_j\phi_j + \frac{eQ_0^{(0)}}{C_\Sigma}\right].$$

If $\alpha_{\mathrm{S}} = \alpha_{\mathrm{D}}$ (symmetric dot), the border lines have exactly the opposite slope of $\pm 1/(2\alpha_{\mathrm{G}})$. Again we find the crossing of the two lines at $eU_{\mathrm{SD}} = \Delta_{N+1} + V_{\mathrm{c}}$ and the difference of the slopes is again $1/\alpha_{\mathrm{G}}$.

Excited states

Tunneling through excited states of the dot leads to additional structure in the Coulomb-blockade diamonds. For example, tunneling transport can be achieved starting from the dot in the $(N-1)$-electron ground state into an excited N-electron intermediate state back into the excited $(N-1)$-electron ground state. Another scenario is tunneling from an $(N-1)$-electron excited state via the N-electron ground state into the same $(N-1)$-electron excited state. At zero temperature such processes are suppressed within the Coulomb-blockade diamonds in Fig. 11.4. At finite bias voltages outside the Coulomb-blockade diamonds, such processes become possible. Experimentally, they can be observed in differential conductance measurements, i.e., measurements of $\mathrm{d}I_{\mathrm{SD}}/\mathrm{d}U_{\mathrm{SD}}$, which reveal the excitation spectrum of the quantum dots. In such measurements, excited states manifest themselves in additional peaks running in parallel to the Coulomb-blockade diamond boundaries (see additional lines in Fig. 11.4).

11.3.3 Conductance peak line shape at finite temperatures

A theory for the temperature dependent line shape of conductance peaks has been given by Beenakker and van Houten [30] based on the constant-interaction model. In this model the tunneling processes through the dot are described as two sequential incoherent elastic tunneling events, the tunneling-in and the tunneling-out process. It is assumed that the tunneling coupling of dot states to the leads Γ is smaller than the single-particle level spacing Δ and smaller than temperature kT. We illustrate the model with the example of a dot with four energy levels E_0 to E_3. Each level can be occupied by one electron due to Pauli's exclusion principle. The state of the dot is given by the occupation numbers of the individual levels, e.g., the dot state $(0,1,1,0)$ denotes a state with two electrons in the dot occupying the levels E_1 and E_2. We can now calculate the probability that the dot is in a certain state if the temperature and the external gate voltages are given:

$$P_{\mathrm{eq}}(\{n_i\}) = Z^{-1} \exp\left[-\frac{1}{kT}\left(\sum_{i=0}^{\mathcal{N}} E_i n_i + E_{\mathrm{elstat}}(N) - NE_{\mathrm{F}}\right)\right].$$

Here, Z is the partition function

$$Z = \sum_{\{n_i\}} \exp\left[-\frac{1}{kT}\left(\sum_{i=0}^{\mathcal{N}} E_i n_i + E_{\mathrm{elstat}}(N) - NE_{\mathrm{F}}\right)\right],$$

$\mathcal{N}$ is the maximum level index, E_{F} is the Fermi energy in the leads, and $E_{\mathrm{elstat}}(N)$ is the electrostatic energy of the dot filled with $N = \sum_i n_i$ electrons within the constant-interaction model. An example for $P_{\mathrm{eq}}(\{n_i\})$ for some occupation states is shown in Fig. 11.5a. It has already become clear in the discussion of the previous sections that gate-voltage intervals exist, in which a certain electron number N is dominant. The borders of these intervals are smeared out by temperature. Several states can have significant probability of occupation for a fixed electron number (see, e.g., the two states for $N = 1$ in the figure). Narrow regions of gate voltages exist, where states with N and $N+1$ electrons have a significant probability simultaneously. These are the regions where conduction processes through the dot can take place, because if an electron tunnels through the dot, the dot undergoes the occupation sequence $N \to N+1 \to N$ (electron tunneling) or $N+1 \to N \to N+1$ (hole

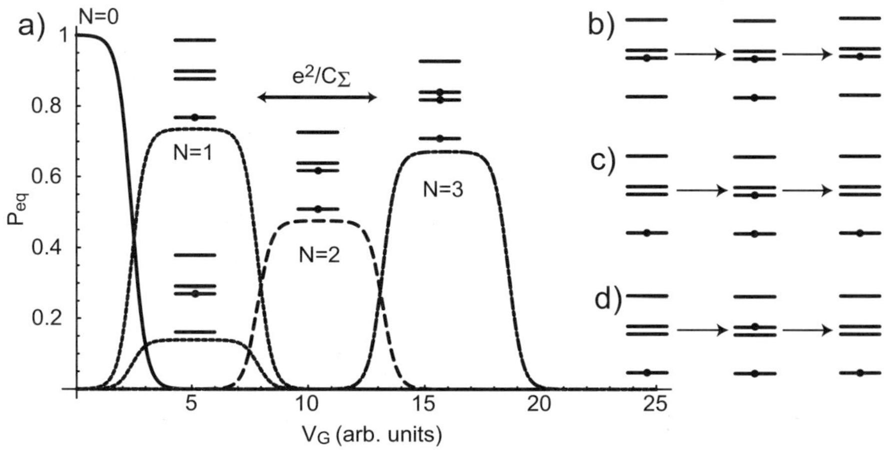

Fig. 11.5. a) Occupation probability for some states of a four-level quantum dot. The states are indicated in the occupation diagrams next to the curves. b)–d) show three different sequences of dot states during a sequential tunneling event. All three contribute to the total conductance at a gate voltage where the $N = 1$ and $N = 2$ probabilities overlap.

tunneling). Because we only consider elastic tunneling processes, the dot state after the sequence is exactly the initial state. The tunneling electron (hole) will therefore occupy a certain energy level E_p during its stay in the dot, which will eventually be empty again. In this picture it is fair to say that transport takes place through level E_p. However, more than one level E_p can contribute to the total conductance seen in a particular conductance peak. Examples are illustrated in Fig. 11.5b)–d).

For a quantitative description of the tunneling conductance, we need the probability that N electrons occupy the dot *and* level E_p is occupied

$$F_{\mathrm{eq}}(E_p, N) = \sum_{\{n_i\}} P_{\mathrm{eq}}(\{n_i\})\delta_{N,\sum_i n_i}\delta_{n_p,1}$$

and the probability that the state in the emitter (collector) at the same energy is empty $1 - f(E_p + E_{\mathrm{elstat}}(N) - E_{\mathrm{elstat}}(N-1) - E_{\mathrm{F}})$. The contribution of the level E_p to the conductance in an N-electron dot is then given by [30]

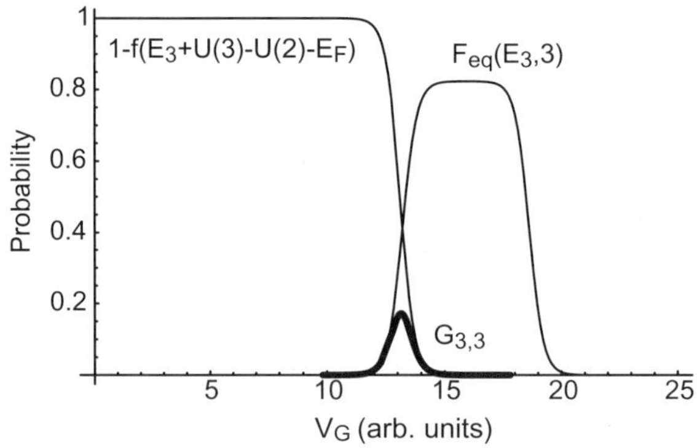

Fig. 11.6. Contribution $G_{p,N}$ to a conductance peak (thick line). It comes about by the product of the occupation statistics of the dot $F_{eq}(E_p, N)$ and the occupation factor in the leads. The prefactor containing the dot-lead coupling constants has been neglected here.

$$
\begin{aligned}
G_{p,N} = & \frac{e^2}{kT} \frac{\Gamma_p^{\mathrm{l}} \Gamma_p^{\mathrm{r}}}{\Gamma_p^{\mathrm{l}} + \Gamma_p^{\mathrm{r}}} \times \\
& F_{\mathrm{eq}}(E_p, N) \left[1 - f(E_p + E_{\mathrm{elstat}}(N) - E_{\mathrm{elstat}}(N-1) - E_{\mathrm{F}})\right] .
\end{aligned}
$$

The shape of the conductance contribution is determined by the thermally broadened statistical functions. An example is shown in Fig. 11.6. The peak amplitude of $G_{p,N}$ is determined by the tunneling coupling of state E_p to the left (Γ_p^{l}) and the right (Γ_p^{r}) lead.

The total tunneling conductance is then the sum over all the contributions of the levels E_p and all the dot occupation numbers N [30]:

$$
G = \sum_{p=0}^{\mathcal{N}} \sum_{N=1}^{\mathcal{N}} G_{p,N} .
$$

At sufficiently low temperatures only a single N gives a significant contribution to a particular conductance peak. However, depending on the energetic separations between individual single-particle levels compared to kT, one or several E_p-levels can contribute to a conductance peak. Figure 11.7 shows, how one can go from single-level transport (c) to multilevel transport (a) by increasing the temperature.

In the single-level transport regime only a single level E_p and a single electron number N contributes to the total conductance, and it can be shown [30] that in this case the shape of a conductance peak can be represented as

$$
G = \frac{e^2}{\hbar} \frac{\Gamma_p^{\mathrm{r}} \Gamma_p^{\mathrm{l}}}{\Gamma_p^{\mathrm{r}} + \Gamma_p^{\mathrm{l}}} \frac{1}{4kT} \cosh^{-2}\left(\frac{\delta}{2kT}\right) ,
$$

where $\delta = e\alpha_{\mathrm{G}}(U_{\mathrm{G}} - U_{\mathrm{G}}^{\mathrm{res}})$ and $U_{\mathrm{G}}^{\mathrm{res}}$ is the gate voltage at maximum conductance.

The temperature dependence of the conductance peaks can be different for the single-level transport regime and the multi-level transport regime, especially if one takes into account that neighboring orbital levels can have vastly different tunneling

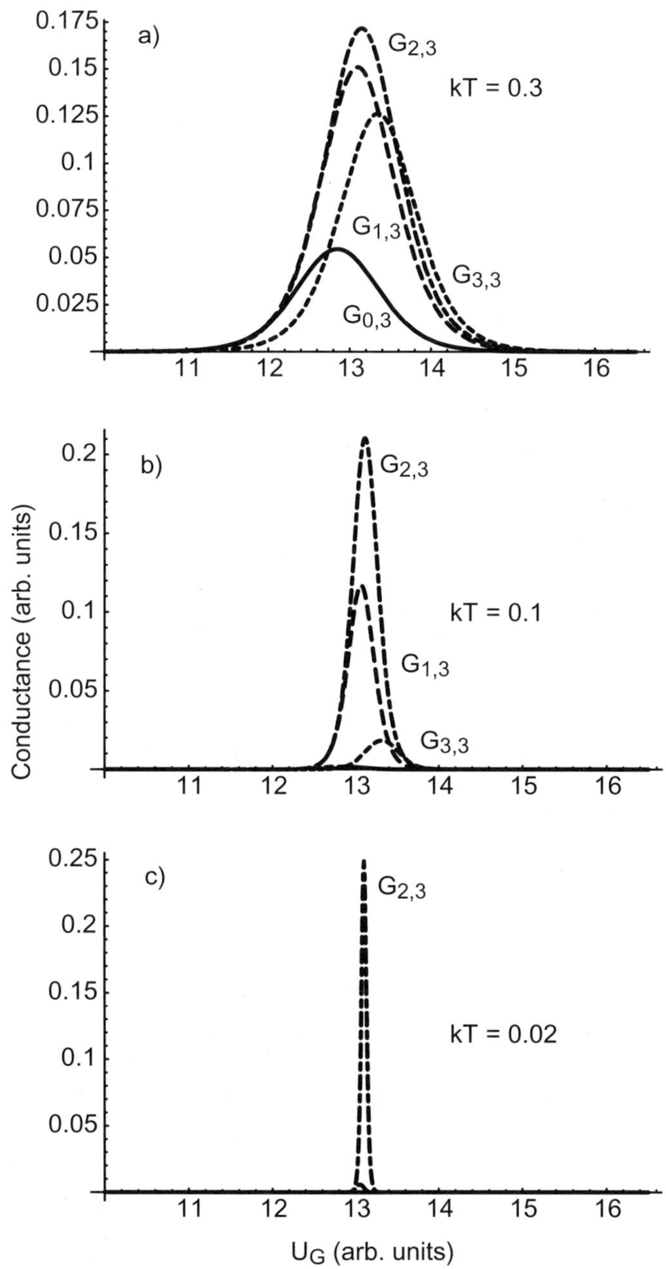

Fig. 11.7. All the contributions $G_{p,3}$ to a conductance peak in a four level dot. The single-particle energies are $E_0 = 0$, $E_1 = 0.5$, $E_2 = 0.6$ and $E_3 = 1$. Figures a) to c) show the contributions at different temperatures. While in c) only a single level contributes to the conductance, in b) there are already two and in a) all four levels contribute.

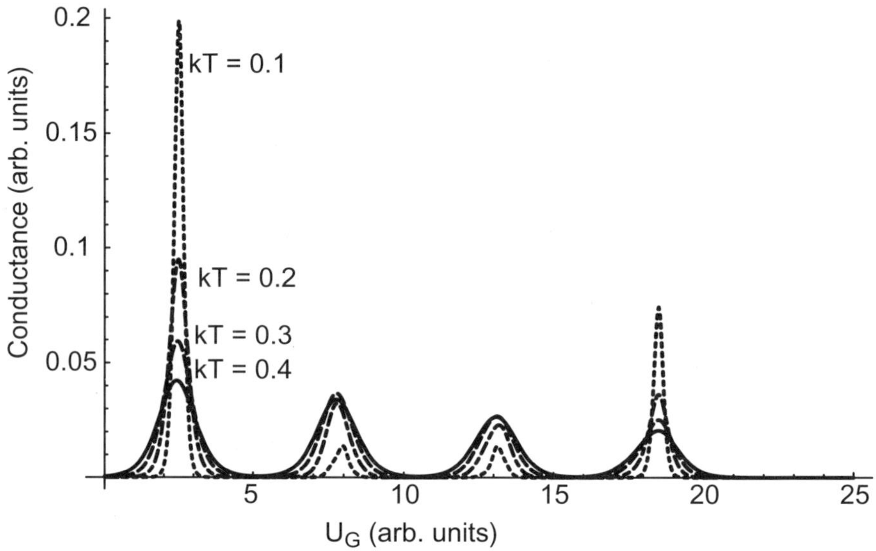

Fig. 11.8. Temperature dependence of the conductance peaks of the four-level system. Energies are $E_0 = 0$, $E_1 = 0.5$, $E_2 = 0.6$, $E_3 = 1$. Different coupling constants $\bar{\Gamma}_p = \Gamma^{\rm l}_p \Gamma^{\rm r}_p / (\Gamma^{\rm l}_p + \Gamma^{\rm r}_p)$ were chosen for the different single-particle levels, namely, $\bar{\Gamma}_0 = 0.08$, $\bar{\Gamma}_1 = 0.03$, $\bar{\Gamma}_2 = 0.03$, and $\bar{\Gamma}_3 = 0.03$.

coupling to the leads. Figure 11.8 shows an example for our four-level model system. The peaks at $U_{\rm G} = 2.5$ and at $U_{\rm G} = 18.5$ show the conventional temperature dependence found in the single-level transport regime. With increasing temperature the peak height decreases and the width increases. However, the energy levels and tunneling rates were chosen such that the two conductance peaks in the middle are in the multi-level transport regime. Increasing the temperature leads to an *increase* in the peak height because the strongly coupling E_0 and E_3 levels contribute more strongly due to thermal activation. In addition to this inverted temperature dependence, these peaks also show an asymmetric line shape at the lowest temperatures displayed.

11.4 Beyond the constant-interaction model

The constant-interaction model for quantum dots has been very successful for the description of many-electron quantum dots in limited ranges of plunger-gate voltages. As the number of electrons is decreased or the gate-voltage ranges are extended, modifications of this description are necessary. Experimentally, it is found that the capacitances of many-electron dots depend on the plunger-gate voltages. In some cases the constant-interaction picture can still be used over extended gate-voltage ranges if linear changes of the lever arms α_j with voltage are assumed. However, even in such cases, interaction effects fluctuate from level to level. Peak spacings have to be analyzed by statistical methods and random matrix theory is frequently applied (see Refs. [93, 94] for reviews). Self-consistent Hartree calculations of quantum dots [270] and Hartree–Fock calculations [271–275] have been applied going beyond the constant-interaction model. Other models [276–278] are based on density functional theory and the Kohn–Sham equations [286]. Also quantum Monte Carlo simulations have been performed [279, 280]. For very small electron numbers, even exact diagonalization calculations could be performed [264–269], which include all the interaction effects beyond the Hartree approximation, i.e., exchange and correlation effects. These calculations can be made only for dots coupled very weakly to source and drain contacts.

If a quantum dot becomes gradually more strongly coupled to source and drain contacts, other corrections to the above theories have to be taken into account. For example, the assumption of sequential resonant tunneling as the dominant process breaks down and tunneling processes of higher order have to be considered. An example are co-tunneling processes, which are correlated tunneling processes of two (or more) electrons [287, 288]. Another example is the formation of correlated states between electrons on the dot and in the leads such as a Kondo state predicted for quantum dots theoretically at the end of the 1980s [84–86] and found experimentally about 10 years later [82, 289–291].

12

Energy spectra of quantum rings

12.1 Introduction to quantum rings

Probably the most famous ring structure occurring in nature is the benzene molecule. Its structure has been suggested by Kekulé[1] as early as 1865. In physics, ring structures became especially important in connection with the Aharonov–Bohm effect [59] and related phenomena [66, 292].

In the context of the discussion of flux quantization in superconducting samples with the topology of a ring, Byers and Yang [293] and later Bloch [294] determined the characteristics of the energy spectra of ring-shaped samples pierced by a magnetic flux Φ. They predicted that the physical properties of such rings should be periodic in Φ with period of one flux quantum $\Phi_0 = h/e$ in normal conductors. A special example of this general statement is the prediction of an oscillatory magnetization of phase-coherent rings due to persistent currents [295].

The ring shape also plays a role for the very general concept of Berry's topological phase [296]. In this context the Aharonov–Bohm phase can be regarded as a special realization of a Berry-phase effect [297]. Other suggestions have been made for the observation of Berry's phase in a transport experiment on a ring where electronic spins rotate coherently due to a radial magnetic field or due to spin-orbit coupling [298–301].

On the experimental side, the Aharonov–Bohm effect has been observed in metal rings [65, 302] and in semiconducting rings [61, 303]. The existence of experimental evidence for Berry's phase different from the Aharonov–Bohm effect in mesoscopic structures remains an open issue to date, but a few experiments have been reported [304–307].

Recently, the phase coherence of transport through a quantum dot embedded in one arm of an open ring has been demonstrated [308, 309]. In this experiment the ring geometry was employed for the measurement of the amplitude and phase of the transmission coefficient through the quantum dot and the results still await a detailed understanding.

---

[1] August Kekulé von Stradonitz, ⋆ 1829, † 1896

Persistent currents were experimentally detected by measuring the magnetic response of 10^7 mesoscopic copper rings [310] measured with a SQUID magnetometer. Later, persistent currents in a single isolated gold loop were measured [311]. The only measurements of persistent currents in a ring on a GaAs/AlGaAs heterostructure are due to Mailly [312] using a SQUID fabricated on chip.

The energy spectrum of self-assembled closed rings has recently been analyzed by optical experiments [313]. A transition of the ground state of the ring from angular momentum $\ell = 0$ to $\ell = -1$ was observed when one flux quantum threaded the area of the ring. When the rings are successively filled with electrons, a shell structure can be detected [314] similar to the shell structure in real atoms and in artificial semiconductor quantum dots [80].

Here magnetotransport experiments on a ring-shaped semiconductor quantum dot in the Coulomb-blockade regime [92] will be described [315]. The measurements allow one to extract the discrete energy levels of a realistic ring, which are found to agree well with theoretical expectations. Such an agreement between a model spectrum predicted by theory and an experimentally determined spectrum in quantum dots was so far limited to few-electron systems [80], but is here extended to a many-electron system. In addition, the interaction effects important for the ring structure are discussed and analyzed in detail. It is shown that the experimentally observed charging energy can be quantitatively understood within the Hartree approximation including a strong screening contribution due to the top gate. The strong screening effect is the reason for the frequent occurrence of spin pairs in the ring. A semiclassical interpretation of the results indicates that electron motion is governed by regular rather than chaotic motion, an unexplored regime in many-electron quantum dots.

12.2 Samples and structures

Quantum-ring samples have been fabricated on AlGaAs/GaAs heterostructures containing a two-dimensional electron gas (2DEG) with density $5 \times 10^{11}\,\mathrm{cm}^{-2}$ and mobility $90\,\mathrm{m}^2$/Vs at $T = 4.2\,\mathrm{K}$ only 34 nm below the sample surface. The surface of the heterostructure has been locally oxidized by applying a voltage between the conductive tip of an atomic force microscope (AFM) and the 2DEG [316]. The electron gas is depleted below the oxidized regions, which was used in other studies for defining high-quality quantum dots [317]. Figure 12.1 shows the principle of this patterning technique. The tip of a scanning force microscope is brought close to the surface of a GaAs heterostructure with a shallow 2DEG at room temperature under well-controlled humidity conditions. A water film will then be present on the sample surface. When the tip is moved along a line with a large voltage applied between the tip and the sample, an oxidation process takes place along this line. The oxidation has the same effect on the 2DEG as a well-controlled local shallow etch. The electron gas under the line is depleted and self-aligned conducting regions are created. The details of the fabrication process, which is crucial for the high-electronic quality of the quantum ring, are described in Ref. [318].

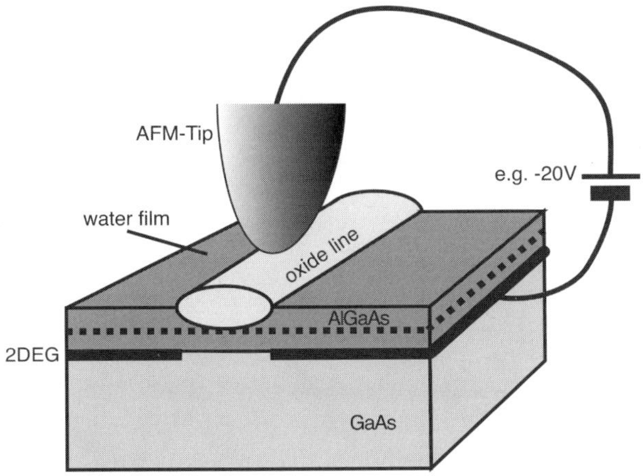

Fig. 12.1. Principle of AFM-Lithography: the sample with a shallow 2DEG is kept at room temperature under well-controlled humidity conditions. An SFM tip with a voltage between -10 and -20 V applied relative to the sample writes oxide lines into the surface under which the electron gas is depleted.

Figure 12.2(a) shows an SFM image (taken with an unbiased tip directly after the oxidation process) of the oxide lines defining the quantum ring. The width of the quantum point contacts connecting the ring to source (drain) is controlled by voltages applied to the lateral gate electrodes qpc1a and b (qpc2a and b). The number of electrons in the ring can be tuned via the lateral plunger gates pg1 and 2. Shape deformations due to applied in-plane gate voltages are known to be relatively weak [317, 318]. The schematic in Fig. 12.2(b) shows the dimensions of the quantum ring. After the oxidation step, the sample is covered with a metallic top-gate electrode. We will show below that with the combination of in-plane and top-gate electrodes the quantum ring can be tuned into the Coulomb-blockade regime with the single-particle level spacing being much larger than the thermal energy kT.

From the ring geometry and the electron density of the two-dimensional electron gas surrounding the ring, we obtain an upper limit for the number of electrons in the dot of about 270. An upper limit for the number of radial subbands in the dot can be estimated from the width $\Delta r \approx 65\,\mathrm{nm}$ of the ring and the Fermi wavelength of the electrons in the host material $\lambda_{\mathrm{F}} \approx 35\,\mathrm{nm}$ to be 3. However, it is known that the electron density in strongly confined quantum dots can be much smaller than the two-dimensional electron density of the host material and λ_{F} can be much larger. This means, the number of electrons can be greatly reduced below the upper limit quoted above and the number of radial modes is very likely to be smaller than 3.

12.3 Magnetotransport measurements on a quantum ring

Figure 12.3(b) presents a logarithmic color plot of the current through the quantum ring as a function of plunger-gate voltage and magnetic field B (applied normal to the 2DEG plane). This measurement was performed at a constant DC source-drain voltage $U_{\mathrm{SD}} = 20\,\mu\mathrm{V}$ and at a temperature of 100 mK in a dilution refrigerator. The bandwidth of the current measurement was limited to about 1 Hz.

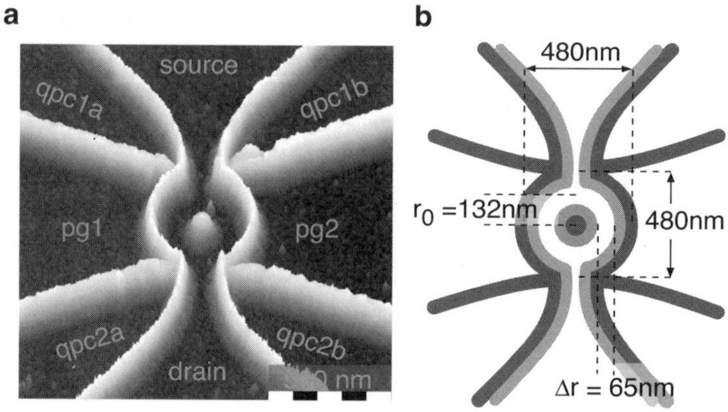

Fig. 12.2. Sample layout. (a) Micrograph of the quantum ring taken with the unbiased SFM tip after writing the structure. The oxide lines (bright regions) deplete the 2DEG 34 nm below the surface separating the sample into several conductive (dark) regions. The current is passed from source to drain. The so-called in-plane gates (qpc1a, qpc1b, qpc2a, qpc2b, pg1, and pg2) are used to tune the point contacts and two arms of the ring. (b) Schematic sketch of the ring. The dark curves represent the oxide lines. From transmission measurements of the point contacts at source and drain we estimate the depletion length to be about 50 nm, which results in an estimated channel width of $\Delta r \approx 65$ nm. The average radius of the ring is $r_0 = 132$ nm.

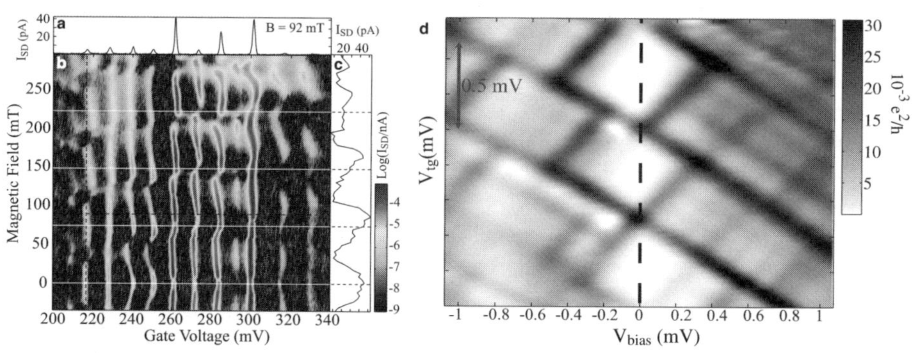

Fig. 12.3. *Please see the insert for a color reproduction of this figure.* The addition spectrum. (a) Measurement of Coulomb-blockade resonances at fixed magnetic field. The current is measured as a function of a voltage applied to both plunger gates (pg1 and 2) simultaneously. (b) The evolution of such sweeps with magnetic field results in the addition spectrum shown in color. The regions of high current (yellow/red) mark configurations in which a bound state in the ring aligns with the Fermi level in source and drain. The Aharonov–Bohm period expected from the ring geometry is indicated by the thin white horizontal lines. (c) Magnetic field sweep for constant gate voltage $U_{\mathrm{PG}} = 218$ mV (dashed line in the color plot). While this peak shows a maximum in amplitude for $B = 0$, other peaks ($U_{\mathrm{PG}} = 270$ mV) display a minimum in amplitude. (d) Typical Coulomb-blockade diamonds measured at zero magnetic field.

In Fig. 12.3(a) the Coulomb-blockade oscillations have been extracted along the horizontal dashed line in Fig. 12.3(b), i.e., at constant $B = 92\,\mathrm{mT}$. A series of current peaks can be seen that vary strongly in height from peak to peak. The highest peaks correspond to a conductance of a few percent of the conductance quantum e^2/h. The width of most peaks is of the order of $4kT$, indicating thermal smearing of conductance resonances. In between the peaks the current is typically unmeasurably small with some occasional regions in the U_{PG}-B plane where it rises slightly above noise. All these observations indicate that the ring is in the Coulomb-blockade regime.

Extracting the conductance as a function of magnetic field along the dashed vertical line of constant gate voltage leads to the trace in Fig. 12.3(c). It shows oscillatory behavior with a period of $\Delta B = 75\,\mathrm{mT}$, i.e., one flux quantum h/e per area πr_0^2. This is exactly the Aharonov–Bohm period of an open ring with the same radius. In fact, by opening the point contacts (not shown) we find the well-known Aharonov–Bohm oscillations in the conductance with the same ΔB [61, 302, 303, 319, 320]. It can be seen that the positions and amplitudes of most peaks in Fig. 12.3(b) show this period as well. This is the manifestation of Aharonov–Bohm-type effects in the quantum ring.

The grayscale plot in Fig. 12.3(d) shows the conductance of the ring in the U_{SD}-U_{TG} plane measured at zero magnetic field. Typical Coulomb-blockade diamonds are observed where the conductance is unmeasurably small. The diamonds are separated by the conductance peaks at $U_{\mathrm{SD}} = 0$. At larger U_{SD} outside the white Coulomb-blockade diamonds, conductance peaks of excited dot states can be seen.

12.4 Interpretation within the constant-interaction model

In the following we interpret the data using the constant-interaction model of the Coulomb-blockade effect [92, 282] (see 11.2). From measurements of the Coulomb-blockade diamonds [for a typical example see Fig. 12.3(d)] in the U_{SD}-U_{TG} plane and in the U_{SD}-U_{PG} plane we determine the lever arms of the top- and in-plane gates allowing gate voltages to be translated into energies (see 11.2). The minimum separation of neighboring conductance peaks occurs for degenerate single-particle levels and gives the bare charging energy ΔE_{c}. From the smallest experimentally observed peak spacings, we determine[2] the charging energy $\Delta E_{\mathrm{c}} = e^2/C_{\Sigma} \approx 310\,\mu\mathrm{eV}$, which is much larger than $kT \approx 10\,\mu\mathrm{eV}$. This is another indication that the ring is well in the Coulomb-blockade regime. The observed single-particle level spacings obtained from the conductance peak separations after subtraction of ΔE_{c} are found to be as large as $\Delta \approx 300\,\mu\mathrm{eV}$ (see below). This value agrees with the large separation of excited states from the Coulomb-blockade diamond boundaries in Fig. 12.3(d). We emphasize that such a large single-level spacing is usually not achieved for quantum dots of comparable size. In the ring it is due to the exclusion of electrons from the

[2] In Ref. [315] the charging energy that serves as a reference for all other energy scales was underestimated. The values quoted here are taken from A. Fuhrer, PhD thesis ETH Zurich, 2003.

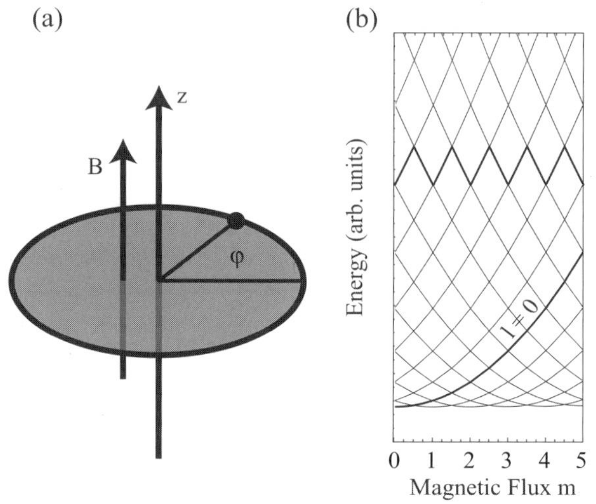

Fig. 12.4. (a) The model of the ideal ring. (b) Theoretical spectrum of a single-mode ring. The parabolas with constant ℓ (bold parabolic line) have a minimum at $\ell = m$. The bold zig-zag line corresponds to a Coulomb peak after the charging energy has been subtracted in the constant-interaction model.

central region accomplished by the presence of the antidot. Because $\Delta \gg kT$, we can state that we are in the single-level transport regime if we neglect occasional level crossings.

Within the constant-interaction model, the positions of the Coulomb-blockade peaks in gate voltage are given by [cf. eq. (11.4)]

$$U_{\mathrm{G}}^{(N)}(B) = \frac{1}{e\alpha_{\mathrm{G}}}\left[\epsilon_N(B) + \Delta E_{\mathrm{c}}\left(N - \frac{1}{2}\right) - \mu_{\mathrm{S}}\right]. \qquad (12.1)$$

Here we have assumed that the lever arm α_{G}, the charging energy ΔE_{c}, and μ_{S}, the electrochemical potential in source and drain contacts, are independent of magnetic field. This assumption is reasonable at small magnetic fields where the orbital wave functions in the ring are not significantly changed compared to $B = 0$. Under this assumption the magnetic field dispersion of individual conductance peaks in the measured addition spectrum in Fig. 12.3(b) directly reflect the dispersion of individual single-particle levels.

In the following sections, we therefore discuss models of increasing complexity showing the characteristic features of the energy spectra of quantum rings.

12.5 One-dimensional ring model

We start the discussion from the energy spectrum of a one-dimensional perfect ring of radius r_0 enclosing $m = B\pi r_0^2/\Phi_0$ flux quanta [295] [see Fig. 12.4(a)]. The eigenvalue problem

$$\left[-\frac{\hbar^2}{2m^\star r_0^2}\left(\frac{\mathrm{d}}{\mathrm{d}\varphi} - \mathrm{i}m\right)^2\right]\psi = E\psi,$$

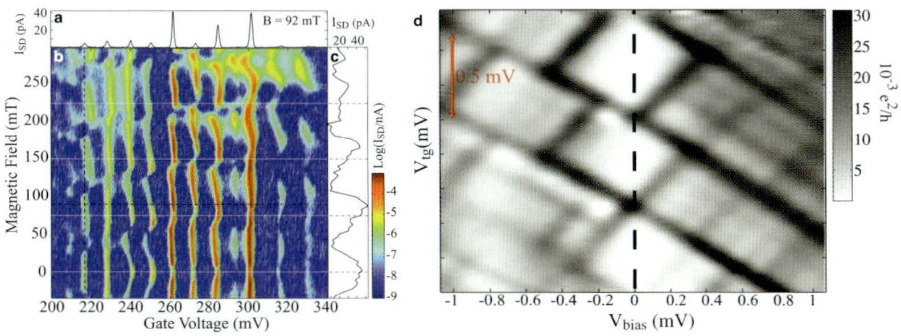

Fig. 12.3. The addition spectrum. For full caption, please see p. 90.

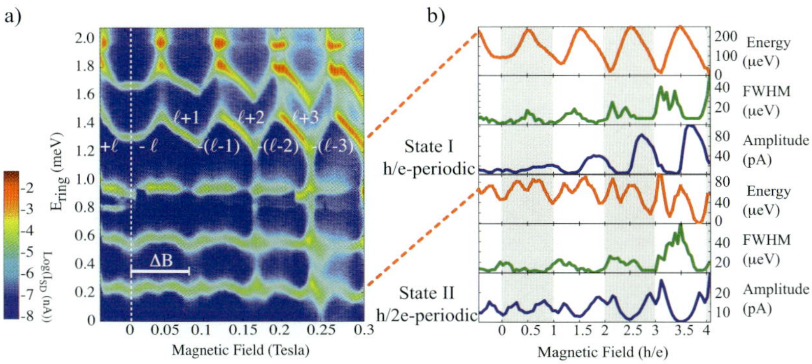

Fig. 12.5. Detail of the addition spectrum showing selected conductance peaks. For full caption, please see p. 94.

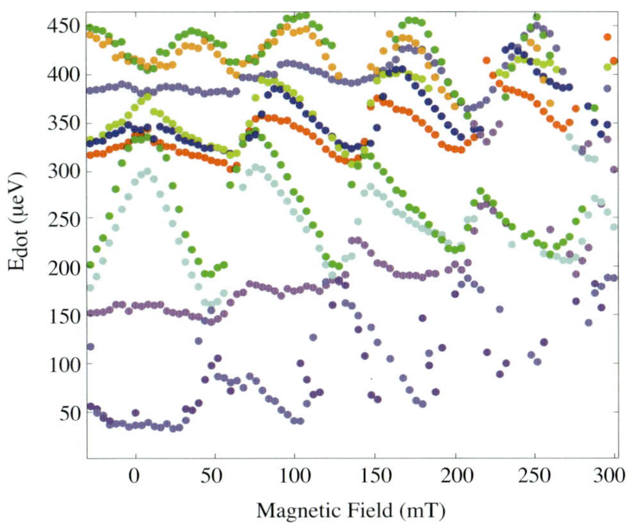

Fig. 12.8. Reconstruction of the energy spectrum of the ring from the data shown in Fig. 12.3. For full caption, please see p. 97.

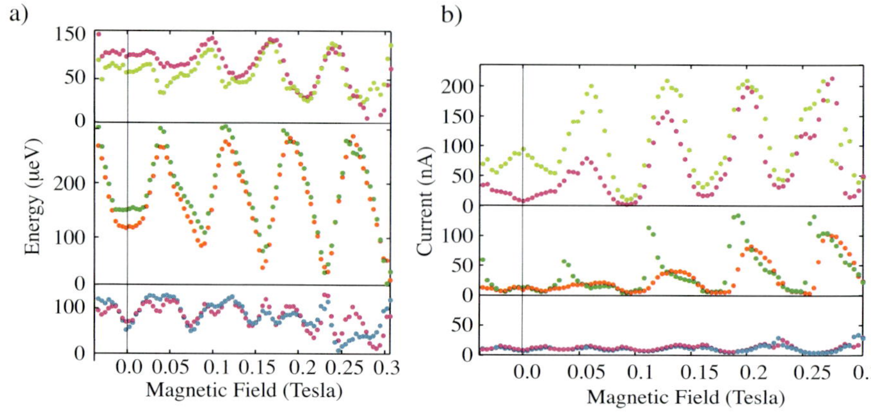

Fig. 12.12. Position and amplitude of conductance peaks for three different spin pairs. Typically position and amplitude are strongly correlated.

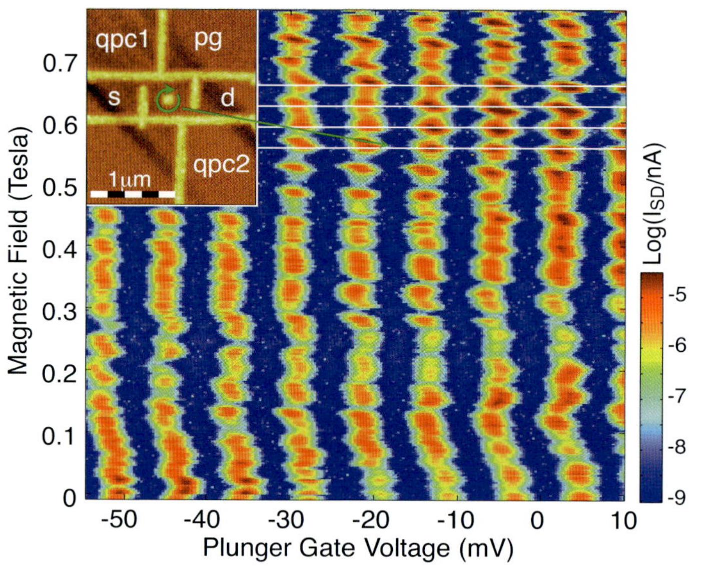

Fig. 12.13. Addition spectrum of a Sinai billard. The inset shows an SFM micrograph of the square-shaped quantum dot with a circular antidot in the center. The color plot is a similar measurement as to the one in Fig. 12.3(b) and also shows the evolution of the conductance as a function of plunger-gate voltage and magnetic field. Around $B = 0$ the Coulomb maxima fluctuate irregularly with changing magnetic field. As the magnetic field is increased to a value where the classical cyclotron diameter orbit matches the antidots circumference (indicated by the green circle and the arrow in the inset), a well-pronounced h/e-periodic behavior (white lines) of the Coulomb maxima in amplitude and position of the Coulomb peaks is recovered, indicating the quenching of the chaotic behavior.

is solved by [66]

$$\psi_{m,\ell}(\varphi) = \frac{1}{\sqrt{2\pi}} e^{\mathrm{i}\ell\varphi} \tag{12.2}$$

leading to the energy spectrum [66]

$$E_{m,\ell} = \frac{\hbar^2}{2m^\star r_0^2}(m-\ell)^2. \tag{12.3}$$

Here φ is the angle coordinate, $m^\star$ is the mass of the particle, and ℓ is the angular momentum quantum number. For a given angular momentum state the energies as a function of magnetic field (or flux m) lie on a parabola with its apex at $m = \ell$, as depicted in Fig. 12.4(b). The characteristic energy scale for the ring is given by $E_\mathrm{r} = \hbar^2/(2m^\star r_0^2) = 33\,\mu\mathrm{eV}$.

According to this simple picture a single Coulomb-blockade peak should oscillate as a function of B along a zig-zag line (bold curve) with the Aharonov–Bohm period ΔB. Comparison with the measurement in Fig. 12.3(b) shows that, indeed, some peaks move along a zig-zag line. For example, the two neighboring peaks in Fig. 12.5(a) labeled "State II" show this behavior. These two conductance peaks correspond to a single orbital wave function that is successively populated by a spin-up and then by a spin-down electron, i.e., we can talk about a spin pair. Comparing the measurement with Fig. 12.4(b) and eq. 12.3 we can assign angular momentum numbers to straight sections of the moving peaks starting from some unknown initial number ℓ. A peak follows a line of constant electron number [bold line in Fig. 12.4(b)]. The current is successively carried by states $(-\ell), (\ell+1), -(\ell-1), (\ell+2), -(\ell-2), \ldots$ when B is increased from zero, i.e., a change in state occurs every half flux quantum.

In Fig. 12.5(b) we show position, width, and amplitude of a "State II" peak as the peak moves as a function of magnetic field. The h/e-period becomes very clear in all three quantities. From the slopes of the peak positions versus magnetic field we can estimate the angular momenta of the states involved. A typical slope is $\pm 500\,\mu\mathrm{eV}$ per flux quantum leading to a typical angular momentum $\ell \approx \pm 8$. We conclude that under these conditions not more than 17 spin-degenerate angular momentum states are available below the Fermi energy in the dot leading to a lower limit of 34 for the number of electrons in the dot. This number is an order of magnitude smaller than the upper limit obtained above from the geometry and the density of the two-dimensional electron gas in the host material. However, because it is very likely that more than one radial subband is occupied in the ring, we expect the actual electron number to be larger than this lower limit. In addition, as we will see below, the presence of a symmetry-breaking potential in the ring can alter the dispersion of individual levels significantly and our estimate of ℓ is rather a lower limit for the maximum angular momentum states involved.

Some other peaks in Fig. 12.3 show barely any B dependence, a behavior that will be discussed later. As a specific example, we show the pair of peaks labeled "State I" in Figs. 12.5(a) and (b). This peak exhibits a $h/2e$-periodicity in magnetic field. We will show below that the occurrence of this period is not related to the

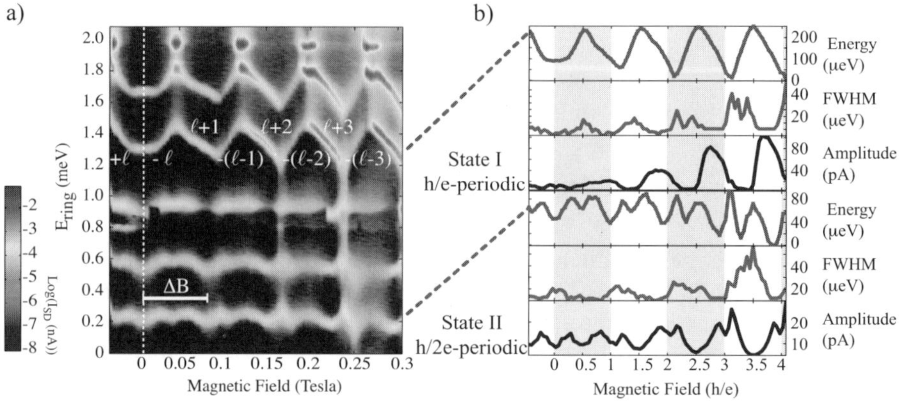

Fig. 12.5. *Please see the insert for a color reproduction of this figure.* (a) Detail of the addition spectrum showing selected conductance peaks. State II resembles the behavior expected from the ideal ring model. State I shows a $h/2e$-period. (b) From top to bottom: conductance peak position, full width half maximum of the peaks, and peak amplitude (for both states).

Altshuler–Aronov–Spivak [321] oscillations but rather a lucky coincidence of level crossings.

The amplitude of Coulomb-blockade peaks is determined by the wave function overlap between the confined states in the ring and the extended states in source and drain [92]. However, the wave functions in eq. (12.2) are independent of magnetic field. The (lateral) overlap [proportional to $|\psi_{m,\ell}(\varphi)|^2 = 1/(2\pi)$] with source and drain is the same for all states. This model does therefore not predict the observed — for some peaks h/e-periodic — modulation of the peak amplitude.

12.6 Ring with finite width

A more realistic but still analytically soluble model due to Tan and Inkson takes the finite extent of the ring in radial direction into account [322] (see Fig. 12.6). This model incorporates the confinement potential in radial (r-) direction

$$V(r) = \frac{\hbar^2 \mathcal{L}^2}{2m^\star r_0^2}\left[\left(\frac{r}{r_0}\right)^2 + \left(\frac{r_0}{r}\right)^2 - 2\right].$$

For parameters $r_0 > 0$ and $\mathcal{L} > 0$ this potential describes a ring geometry with a potential minimum at r_0 with $V(r_0) = 0$ and $V''(r_0) = 4\hbar^2\mathcal{L}^2/(m^\star r_0^4)$. The corresponding Schrödinger equation is solved using the separation Ansatz

$$\psi_{n,\ell}(r,\varphi) = \frac{1}{\sqrt{2\pi}} e^{i\ell\varphi} \chi_{n,\ell}(r).$$

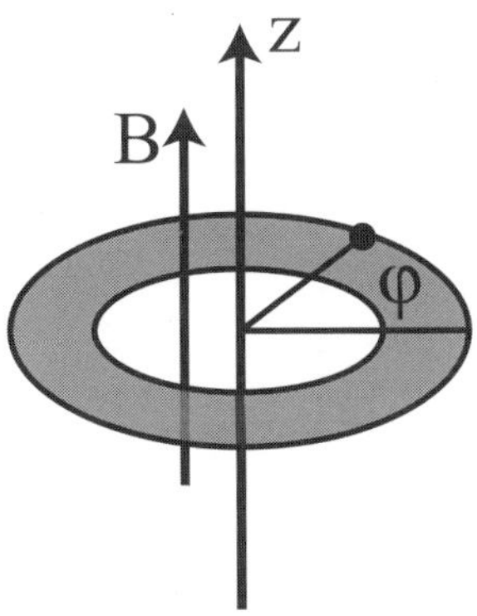

Fig. 12.6. The Tan–Inkson model for a ring takes the finite width of the ring into account.

Different radial channels are indexed by the quantum number n. The energy spectrum is then given by

$$E_{n,\ell,m} = \frac{\hbar^2}{m^\star r_0^2}\left[\sqrt{\mathcal{L}^2+m^2}\left(2n+1+\sqrt{\ell^2+\mathcal{L}^2}\right)-m\ell-\mathcal{L}^2\right], \qquad (12.4)$$

where m is the number of flux quanta h/e within the ring area πr_0^2. Figure 12.7 shows the confinement potential and the energy spectrum of this model. The properties of the ideal ring spectrum in Fig. 12.4(b) are retained for each of the radial subbands [see Fig. 12.7(b)]. For a given radial quantum number n the dispersion as a function of ℓ has a minimum at $\ell = m$. For small magnetic fields ($m < \mathcal{L}$) the dispersion of each state is linear in magnetic field with slope identical to the model of the one-dimensional ring. The prefactor $\sqrt{\mathcal{L}^2+m^2}$ describes the diamagnetic shift of energy levels that is parabolic at small fields and linear at high fields. At zero magnetic field, states with $\ell < \mathcal{L}$ do not differ significantly in energy on the scale of the characteristic energy $E_\mathrm{r} = \hbar^2/(2m^\star r_0^2)$. The energetic separation of radial subbands is constant (at $m = 0$) and given by $4\mathcal{L}E_\mathrm{r}$. In Fig. 12.7(c) (blow-up of part of the spectrum) a small charging energy was added in order to separate lines of constant electron number. In the lowest radial mode the zig-zag behavior of individual peaks is found again. The lowest states of the second radial subband appear as rather flat lines decorated by occasional kinks at positions where states of the lower radial mode cross.

The full wave functions in radial direction are given by

$$\chi_{n,\ell}(r) = \frac{1}{l_\mathrm{eff}}\sqrt{\frac{2n!}{\Gamma(\nu+n+1)}}\left(\frac{r}{l_\mathrm{eff}}\right)^\nu \exp\left(-\frac{r^2}{2l_\mathrm{eff}^2}\right) L_n^{(\nu)}\left(\frac{r^2}{l_\mathrm{eff}^2}\right),$$

with $\nu = \sqrt{\ell^2+\mathcal{L}^2}$, $l_\mathrm{eff} = r_0/(m^2+\mathcal{L}^2)^{1/4}$. For small magnetic fields ($m < \mathcal{L}$) the effective length scale is independent of the field and so is the wave function. We call this the Aharonov–Bohm regime. For intermediate magnetic fields, when $m \sim \mathcal{L}$ the wave functions become squeezed (diamagnetic effect). For large magnetic fields the effective length scale becomes equal to $\sqrt{2}l_\mathrm{c} = \sqrt{2\hbar/eB}$.

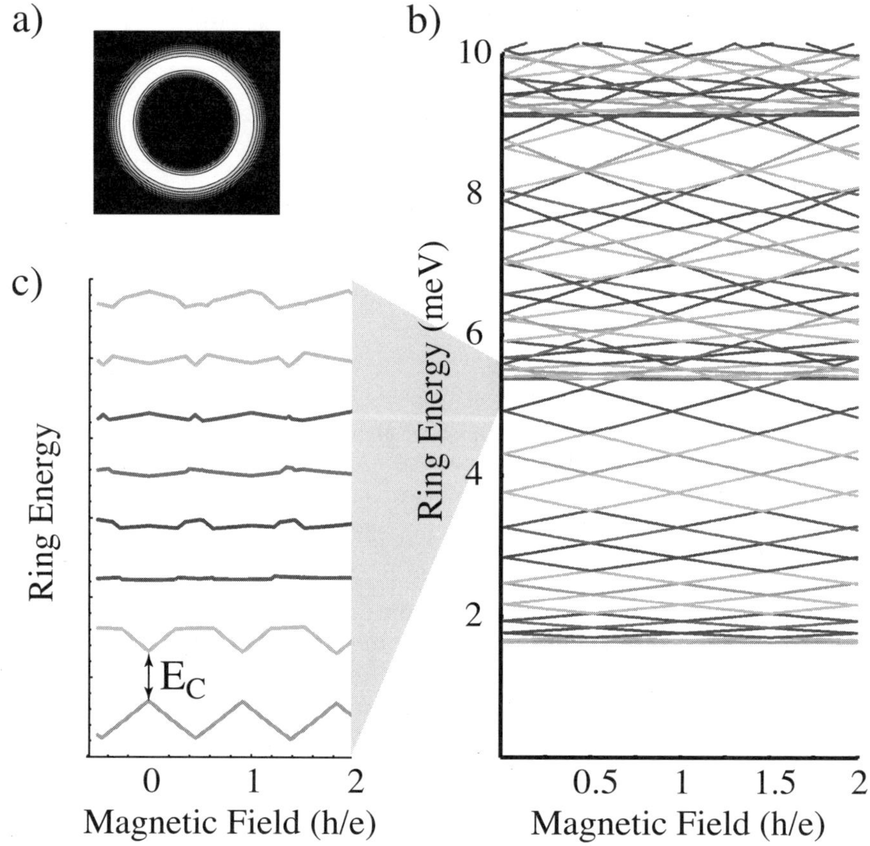

Fig. 12.7. Tan–Inkson model: (a) Contour plot of the confinement potential. (b) Single-particle level spectrum. (c) Level spectrum after addition of a constant charging energy between individual levels.

This model predicts that the exponential decay of the wave functions in radial direction depends on the value of $|\ell|$, but is relatively insensitive to the value of m (for small m). Because all states move in zig-zag lines with an h/e-periodicity in magnetic field, crossings of states with different angular momentum ℓ may lead to a different wave function overlap and therefore to a modulation of the corresponding Coulomb peak amplitude. This is in close agreement with "State II" in Figs. 12.5(a) and (b).

12.7 Experimental single-particle level spectrum

We now turn to the analysis of the experimental conductance peak positions. They are obtained from measurements like the ones shown in Fig. 12.3(b) by converting the gate-voltage axis into an energy scale using the appropriate lever arm as determined from the analysis of the Coulomb-blockade diamonds [323]. A constant charging energy of 310 μeV is subtracted from the position of a Coulomb maximum [92, 317] and the resulting energies are plotted in Fig. 12.8 as a function of magnetic

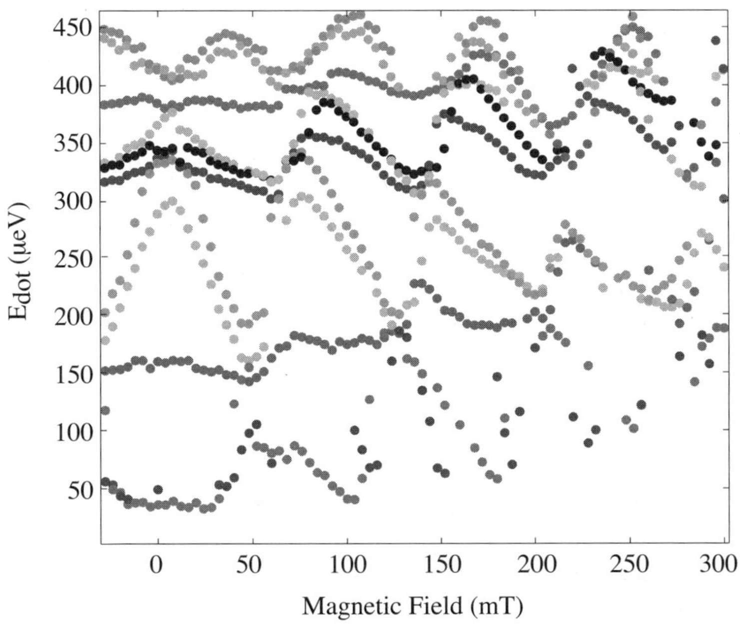

Fig. 12.8. *Please see the insert for a color reproduction of this figure.* Reconstruction of the energy spectrum of the ring from the data shown in Fig. 12.3. The plunger-gate voltage was converted into dot energy using measurements of the Coulomb diamonds and a constant charging energy of 190 μeV was subtracted.

field. Clearly many of the peaks move in pairs (see, e.g., the light green and light-blue points), previously identified as spin pairs [317]. Electrons successively populate these orbital states with spin-up and spin-down electrons. For other resonances spin pairing is not clearly observed. As depicted in Fig. 12.8 the orbital states move up and down in magnetic field with the Aharonov–Bohm period ΔB. Diamond-like patterns can be identified which are characteristic for ring spectra (cf. Fig. 12.6). In this respect our experiments show the long-predicted energy spectrum characteristic for quantum rings [293]. However, the experimental spectrum shows features that are not found in the spectra of ideal rings. For example, there are some states (e.g., magenta) having a very weak magnetic field dispersion. These states could be identified as the lowest-lying states of the second subband in the model of Tan and Inkson, but in this case they should come in big bundles, and we would not expect them to intersect successive diamonds with a separation of about 100 μeV. We will therefore consider additional effects in the following.

12.8 Effects of broken symmetry

The angular uniformity of the probability density in the model of the infinitely thin ring or in the model due to Tan and Inkson stems from the cylindrical symmetry, which, for the real sample, will be broken by the pure presence of source and drain, by dopants, and by the limits of the fabrication procedure. This leads to pinning of the wave function and therefore to a distinct amplitude of the probability density at source and drain. The perturbation will become especially important at the degeneracy points of levels where it can lead to anti-crossing behavior [324]. In the simplest case the probability density changes from a uniform to a sinusoidal angle dependence at the degeneracy points. In this picture the AB-periodic oscillation of the amplitude along a single Coulomb peak can be understood in terms of changing contributions of single-particle levels to the current-carrying state.

Let us look at the states in Fig. 12.8 that have a very small dispersion in the magnetic field. In the framework of the model of Ref. [322] such "flat" states can occur at the onset of the occupation of the next higher radial channel. However, in the experiment we observe such states over wide ranges of gate voltages and an extended model including the effects of the broken symmetry is necessary in order to obtain a qualitative understanding. We emphasize that we do not intend to fit a theoretical spectrum to the data quantitatively because the details of the symmetry-breaking potentials in our sample are not known.

The effects of disorder can be incorporated in a ring model by introducing a symmetry-breaking potential perturbation $V(r,\varphi)$. In order to be more specific, we add this potential term to the analytically solvable model of Tan and Inkson [322] and express the resulting hamiltonian as a matrix using the unperturbed wave functions as basis states. The result is the hamiltonian matrix

$$H_{n,\ell;n',\ell'} = E^{(0)}_{n,\ell}\delta_{n,n'}\delta_{\ell,\ell'} + V_{n,\ell;n',\ell'},$$

where $E^{(0)}_{n,\ell}$ is given by the dispersion in equation 12.4 and the matrix elements of the potential perturbation are given by

$$V_{n,\ell;n',\ell'} = \int \mathrm{d}r\, r\chi_{n,\ell}(r)\chi_{n',\ell'}(r)v_{\ell-\ell'}(r)$$

with

$$v_\ell(r) = \frac{1}{2\pi}\int_0^{2\pi} \mathrm{d}\varphi\;\; V(r,\varphi)e^{-\mathrm{i}\ell\varphi}.$$

For a specific choice of the perturbing potential we use the following plausibility arguments. The symmetry of the ideal ring will be definitely broken by the presence of the the quantum point contacts connecting the dot to source and drain. Electronic states can extend into the contact regions by tunneling and their energy is therefore lowered. Following an idea of Berman et al. [325] we have simulated this effect with the potential perturbation

$$V(r,\varphi) = \frac{\hbar^2 \mathcal{L}^2}{2m^\star r_0^2} \left\{ [1+\epsilon\cos(2\varphi)]^2 - 1 \right\} \left(\frac{r}{r_0} \right)^2 ,$$

where ϵ is a parameter characterizing the size of the perturbation. The matrix elements of this perturbation can be conveniently calculated because the angular and radial integrations separate. The perturbation breaks the symmetry leading to inter- as well as intra-subband coupling. We then truncate the infinite hamiltonian matrix to a finite set of basis states (typically two radial modes are taken into account with angular momentum numbers $|\ell|$ up to 20) and determine its eigenvalues numerically.

In Fig. 12.9 we show a calculation with ring parameters typical for our dot. The spectra in Fig. 12.9(a) and (b) are obtained from the Tan–Inkson model. The spectrum in Fig. 12.9(c) and (d) is obtained from the diagonalization of the truncated hamiltonian matrix described above. The values used for the calculation were $\mathcal{L} = 27.2$, $r_0 = 132\,\mathrm{nm}$, and $\epsilon = 0.06$. The perturbation mixes states of positive and negative angular momenta, which leads in some cases to eigenenergies that barely depend on magnetic field. Such states can intersect the diamonds formed by strongly oscillating levels in close resemblance to the experimental findings.

The model calculation allows a closer inspection of the wave functions associated with certain states. As a general tendency, states with a large magnetic field dispersion have a quite uniform probability density around the ring and are certainly delocalized [Fig. 12.10(top)]. On the other hand, states with vanishing magnetic field dispersion tend to have localized or delocalized but rippled probability densities around the ring as depicted in Fig. 12.10(bottom). It makes clear that one can learn something about the structure of individual wave functions from the energy spectrum of the quantum ring.

This model obviously can only give the general tendency of the experimental spectra. Nevertheless, the dominant deviations from the perfect ring spectrum can be understood as the result of a symmetry-breaking potential perturbation.

12.9 Interaction effects and spin-pairing

So far we have limited our models to a single-particle picture including interaction effects within the constant-interaction model. In this section we want to show that the observed charging energy can be well understood in a Hartree approximation of the quantum ring. In particular, we will also show that screening effects due to the nearby gate electrodes play a very important role in suppressing the Hartree and exchange contributions to the interaction energies.

A quite general treatment of the electrostatics of quantum dots in the Hartree approximation has been given by Hallam and coworkers [263]. The Hartree energy contribution for the addition of an electron in state $\varphi_i(\mathbf{r})$ to an N-electron system is given by [7, 326]

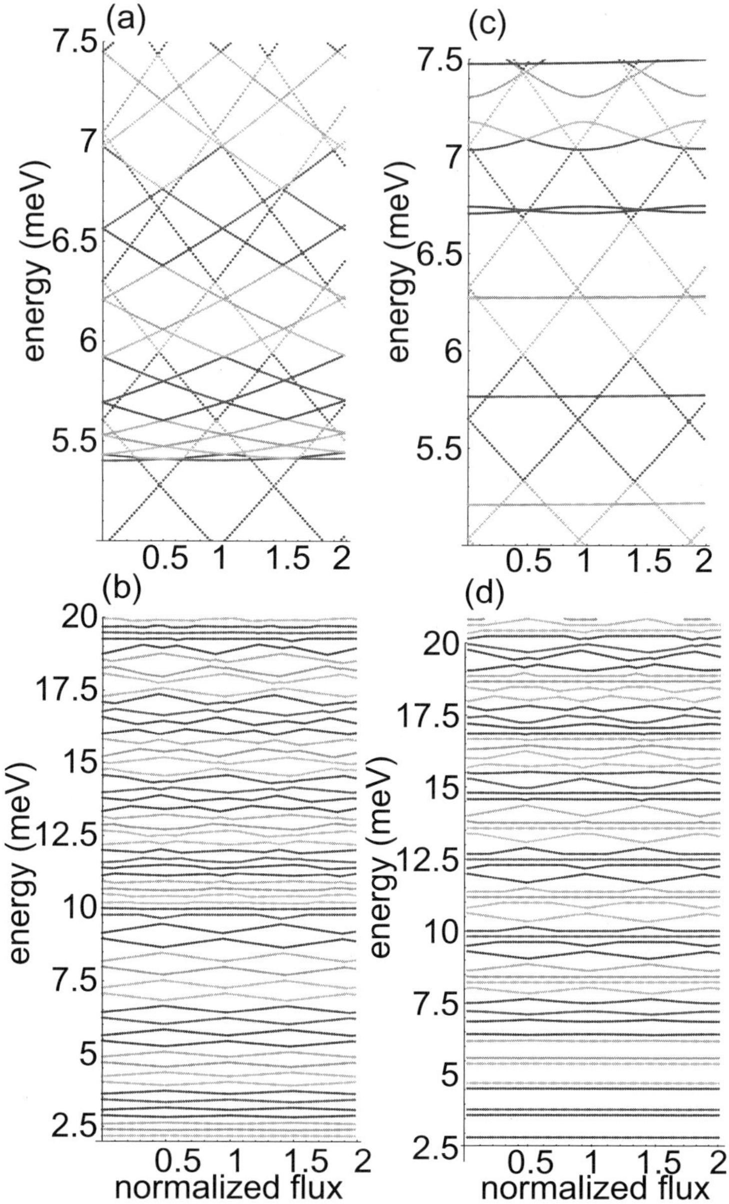

Fig. 12.9. (a) Detail of a spectrum calculated according to the Tan–Inkson model. (b) The same detail with a small charging energy added. (c) Detail of spectrum calculated including a symmetry-breaking potential. (d) The same detail with a small charging energy added.

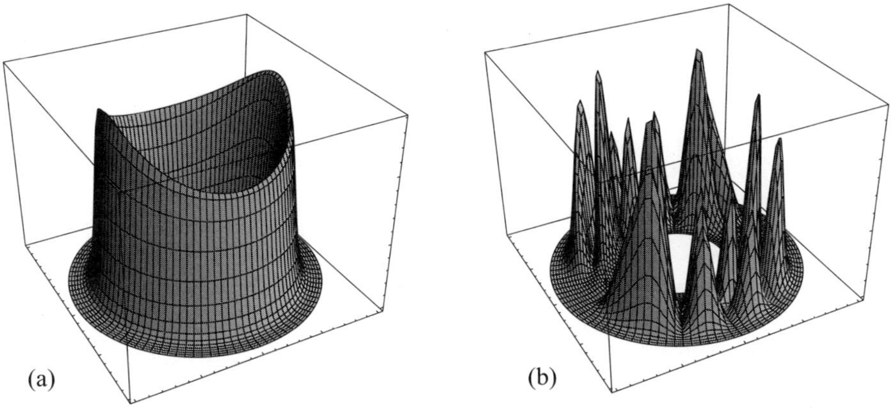

Fig. 12.10. Rippled and flat wave functions

$$E_{\mathrm{H}} = e \int \mathrm{d}^3 r \, |\varphi_i(\boldsymbol{r})|^2 \, V_{\mathrm{H}}(\boldsymbol{r}),$$

where the Hartree potential created by the N electrons already residing on the dot is given by

$$V_{\mathrm{H}}(\boldsymbol{r}) := \int \mathrm{d}^3 r' \rho(\boldsymbol{r}') G(\boldsymbol{r}, \boldsymbol{r}').$$

Here the function $G(\boldsymbol{r}, \boldsymbol{r}')$ is the electrostatic potential at $\boldsymbol{r}$ created by a unit charge at position $\boldsymbol{r}'$. For our purposes we chose the expression

$$G(\boldsymbol{r}, \boldsymbol{r}') = \frac{1}{4\pi\varepsilon\varepsilon_0} \left(\frac{1}{|\boldsymbol{r} - \boldsymbol{r}'|} - \frac{1}{\sqrt{(x-x')^2 + (y-y')^2 + (z+z')^2}} \right).$$

The first term in the bracket is the direct Coulomb interaction, while the second term describes an image charge induced in the top gate. In general, it is not possible to calculate the Hartree energy analytically for arbitrary wave functions. Usually, a three-dimensional numerical self-consistent calculation is necessary for this purpose [276]. Here we restrict ourselves to a simplified scheme for calculating the Hartree energy. We assume that the N electrons on the dot create the homogeneous electron density

$$\rho(\boldsymbol{r}) = \begin{cases} -\frac{eN}{\pi(r_2^2 - r_1^2)} \cdot \delta(z) & \text{for } r_1 < r < r_2 \\ 0 & \text{elsewhere.} \end{cases}$$

We further assume that the electron to be added goes into a state that can be approximated by the homogeneous probability density

$$|\varphi_i(\boldsymbol{r})|^2 = \begin{cases} \frac{1}{\pi(r_2^2 - r_1^2)} \cdot \delta(z) & \text{for } r_1 < r < r_2 \\ 0 & \text{elsewhere.} \end{cases}$$

For this simplified case the Hartree energy can be evaluated analytically (see Appendix B). The result is

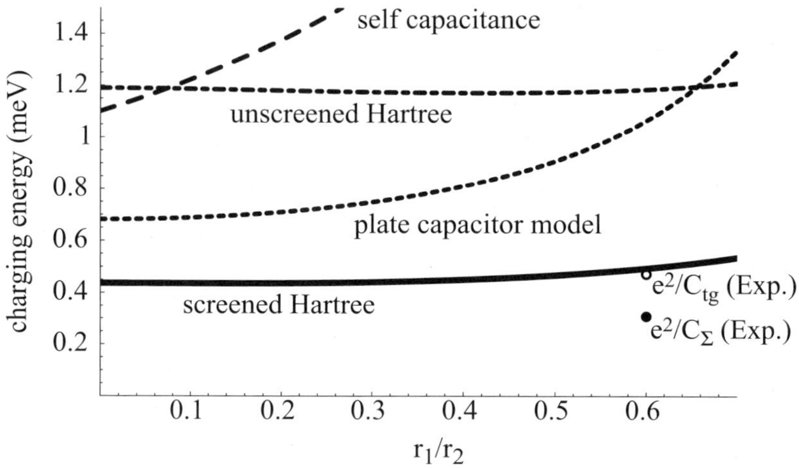

Fig. 12.11. Quenching of the charging energy due to screening of the Hartree contribution of the interaction by the nearby top-gate electrode. A value of $\eta = 0.486$ was used for this calculation. The ratio $r_1/r_2 \approx 0.6$ in our experiment. The dash-dotted curve is the result of the Hartree calculation without the screening effect, while the solid curve contains screening by the top gate. For comparison we show the charging energies obtained from the constant-interaction model using a self-capacitance model (dashed) and a plate-capacitor model (dotted). The open and filled dot are experimentally determined values.

$$E_{\mathrm{H}}(N) = \frac{e^2}{8\varepsilon\varepsilon_0 r_2}\frac{32N}{3\pi^2}\frac{1}{[1-(r_1/r_2)^2]^2}\left[E_{\mathrm{u}}\left(\frac{r_1}{r_2}\right) - E_{\mathrm{s}}\left(\frac{r_1}{r_2},\eta\right)\right],$$

with the contribution of the unscreened Coulomb interaction specified by

$$E_{\mathrm{u}}\left(\frac{r_1}{r_2}\right) = 1 + \left(\frac{r_1}{r_2}\right)^3 - \frac{3\pi}{4}\left(\frac{r_1}{r_2}\right)^2 {}_2F_1\left(-\frac{1}{2},\frac{1}{2};2;\frac{r_1^2}{r_2^2}\right)$$

and the screening reduction given by

$$E_{\mathrm{s}}\left(\frac{r_1}{r_2},\eta\right) = \frac{3}{2}\int_{r_1/r_2}^{1} \mathrm{d}x\, x \int_{r_1/r_2}^{1} \mathrm{d}x'\, x' \frac{K\left(-\frac{4xx'}{(x-x')^2+\eta^2}\right)}{\sqrt{(x-x')^2+\eta^2}},$$

where $\eta = 2d/r_2$ and d is the separation of the plane of the electrons in the dot from the top gate. The double integral for the determination of S has no divergences and can easily be solved numerically.

Figure 12.11 shows a comparison of the charging energies calculated from the above equations with and without the screening effect of the top gate. It can be seen that the presence of the top gate reduces the charging energy to about 40% of its unscreened value. In the ring structure measured in these experiments, the ratio $r_1/r_2 \approx 0.6$. From the figure we read a charging energy of 500 μeV for this value.

This charging energy can be compared to the charging energy $e^2/C_\Sigma = 310\,\mu\text{eV}$, which is considerably smaller than the calculated value. This is not surprising, because in the experiments, other electrodes contribute to C_Σ in addition to the top gate. It is therefore appropriate to compare the calculated charging energy with the energy e^2/C_{tg} in the experiment, where C_{tg} is the dot–top-gate capacitance. From the measurement of the Coulomb-blockade diamonds we determine the lever arm of the top gate to be $\alpha_{\text{tg}} = C_{\text{tg}}/C_\Sigma = 0.6$. From these values we conclude that the experimental $e^2/C_{\text{tg}} = e^2/C_\Sigma/\alpha_{\text{tg}} = 520\,\mu\text{eV}$. This value is in excellent agreement with the theoretical prediction of 550 µeV for the parameters $r_1 = r_0 - \Delta r/2 = 102\,\text{nm}$, $r_2 = r_0 + \Delta r/2 = 162\,\text{nm}$, and $d = 45\,\text{nm}$ (34 nm AlGaAs cap layer and 11 nm stand-off distance of the wave function from the interface). This quantitative result is also relevant for measurements in the Coulomb-blockade regime on other quantum dots fabricated with AFM lithography [317].

In order to check the validity of the simplest capacitive models used for estimating the charging energy, we compare to a naive model of the self-capacitance of a ring with capacitance

$$C_{\text{ring}} = 8\varepsilon\varepsilon_0(r_2 - r_1)$$

and a ring-shaped plate-capacitor model with

$$C_{\text{plate}} = \frac{\varepsilon\varepsilon_0\pi(r_2^2 - r_1^2)}{d}.$$

Figure 12.11 shows that the self-capacitance model strongly overestimates the charging energy of the ring. Even compared to the unscreened Hartree result, this model is only good in the limit $r_1 \to 0$. The plate-capacitor model does account for part of the screening effects. However, for the parameters of our dot, the error of this model compared to the screened Hartree result is more than 100% because it underestimates the effective area of the top gate.

The strong reduction of electron–electron interactions due to screening effects in our ring is also important for interaction effects beyond the Hartree approximation of a uniformly charged ring with hard walls. In particular, screening will also strongly reduce exchange effects in our system. It has been predicted by Blanter and coworkers [327] that subsequent occupation of the same orbital level by spin-up and spin-down electrons should be observed in systems with small interaction effects. In a Coulomb-blockade experiment this will show up as paired neighboring conductance peaks with strongly correlated position and amplitude in a varying magnetic field, so-called spin pairs. This is in strong contrast to the occupation of levels in real atoms or in artificial semiconductor atoms [80], where according to Hund's rules the successive filling of parallel spins is favored. In general, the occurrence of ground-state spins of more than $S = 1/2$ depends in disordered dots on the parameter ξ/Δ, where ξ is the mean exchange energy and Δ is the mean single-particle level spacing [328].

In the experiment it turns out that many spin pairs can be observed in the quantum ring. In Fig. 12.12 we show three examples. On the left, the positions of conductance peaks as a function of magnetic field are shown with the charging energy subtracted.

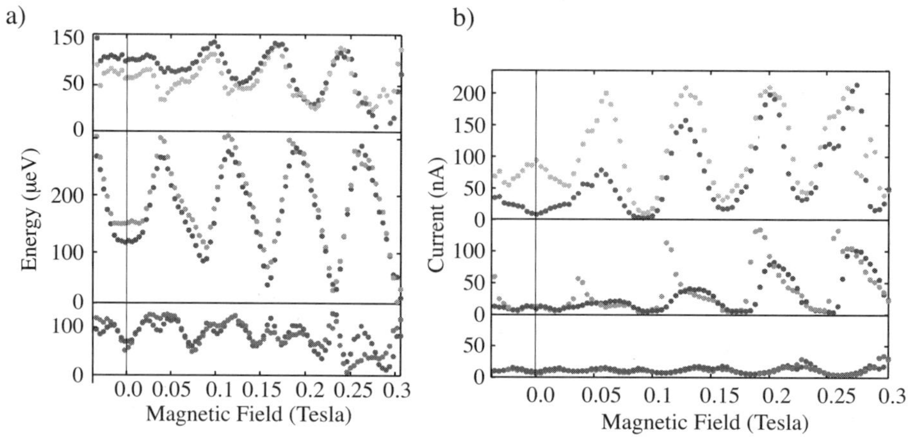

Fig. 12.12. *Please see the insert for a color reproduction of this figure.* Position and amplitude of conductance peaks for three different spin pairs. Typically position and amplitude are strongly correlated.

The curves of spin pairs almost perfectly collapse onto a single curve. On the right, the corresponding amplitudes of the conductance peaks are shown. Although they are not perfectly identical for the spin pairs, strong correlations are evident. Loosely speaking, the screening effect leads to a strong reduction of the interaction parameter r_s, which is the ratio between the Coulomb interaction energy and the Fermi energy in a two-dimensional electron gas. For the ring we estimate a two-dimensional electron density of less than $2 \times 10^{11}\,\mathrm{cm}^{-2}$, resulting in $r_\mathrm{s} = 1/\sqrt{\pi n_\mathrm{s} a_\mathrm{B}^{\star\,2}} = 1.3$. If we assume a reduction factor for the interactions of 0.4 as found for the Hartree interaction due to the presence of the top gate we arrive at an effective $r_\mathrm{s}^\mathrm{eff} \approx 0.5$. According to a theoretical paper by Baranger and coworkers [329], the ratio ξ/Δ can be estimated according to

$$\frac{\xi}{\Delta} \approx \frac{1}{\sqrt{2}\pi} r_\mathrm{s} \ln \frac{1}{r_\mathrm{s}}, \tag{12.5}$$

strictly valid only for $r_\mathrm{s} \ll 1$. Inserting the above $r_\mathrm{s}^\mathrm{eff}$ into this equation we estimate $\xi/\Delta \approx 0.07$, i.e., a relatively small value. With an average single-particle level spacing $\Delta \approx 50\,\mu\mathrm{eV}$, we find $\xi \approx 3.5\,\mu\mathrm{eV}$. According to Ref. [328] the probabilities of finding ground-state spins larger than $S = 1$ in such a system are negligibly small, and the frequent occurrence of spin pairs in the experiment agrees well with these predictions.

12.10 Coulomb-blockade in a Sinai billiard

For quantum dots containing a small number of electrons, the shell structure of the level occupancy can clearly be detected because of the dominating cylindrical symmetry [80]. Quantum dots containing many electrons are usually described in the context of an underlying classically chaotic geometry, because small perturbations of parameter space such as potential shape or magnetic field can induce parametric fluctuations in the energy levels and, consequently, in the Coulomb peak positions. Several theoretical and experimental publications address the question, whether the spectra of such many-electron quantum dots can be adequately described by Random Matrix Theory [93]. Our quantum ring represents a many-electron Coulomb-blockaded system with regular geometry. This together with the small number of radial modes is the reason why we can qualitatively understand the principal features of the observed energy spectrum without the need of a statistical analysis. The source of this striking observation lies in the circular geometry of our ring.

In order to support this view we have fabricated a square-shaped quantum dot with a circular antidot in the center. This system is considered a Sinai billiard and is

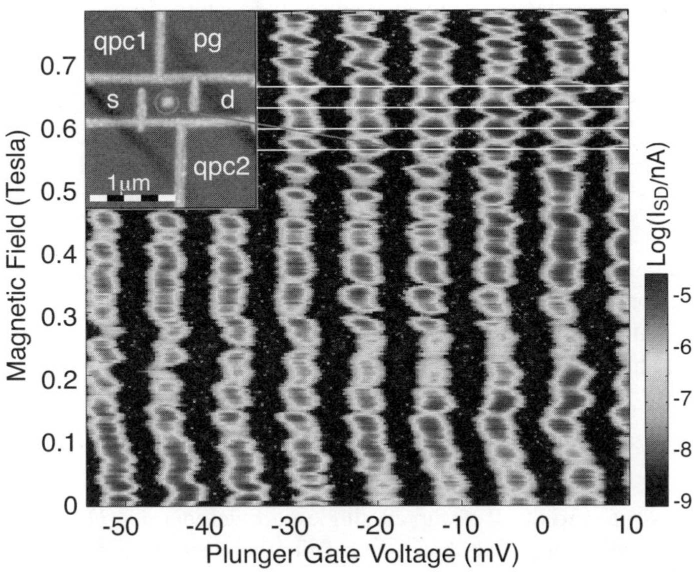

Fig. 12.13. *Please see the insert for a color reproduction of this figure.* Addition spectrum of a Sinai billiard. The inset shows an SFM micrograph of the square-shaped quantum dot with a circular antidot in the center. The color plot is a similar measurement as to the one in Fig. 12.3(b) and also shows the evolution of the conductance as a function of plunger-gate voltage and magnetic field. Around $B = 0$ the Coulomb maxima fluctuate irregularly with changing magnetic field. As the magnetic field is increased to a value where the classical cyclotron diameter orbit matches the antidots circumference (indicated by the green circle and the arrow in the inset), a well-pronounced h/e-periodic behavior (white lines) of the Coulomb maxima in amplitude and position of the Coulomb peaks is recovered, indicating the quenching of the chaotic behavior.

one of the theoretically best studied examples of a classically chaotic system. Figure 12.13 shows the evolution of the conductance as a function of plunger gate and magnetic field of this system presented in a way comparable to Fig. 12.3 for the quantum ring. Around $B = 0$ the Coulomb peak maxima fluctuate irregularly as a function of magnetic field with an average period compatible with an Aharonov–Bohm-type argument. As the magnetic field is increased to a value where the classical cyclotron diameter matches the antidots circumference, a well-pronounced B-periodic behavior of the Coulomb peak maxima in amplitude and position is recovered. Similar to antidot lattices [330], the magnetic field is expected to induce regular parts in the predominately chaotic phase space existing at $B = 0$.

12.11 Relation of the ring spectra to persistent currents

The existence of persistent currents in coherent mesoscopic rings was brought to the attention of the experimentalists by Büttiker et al. [295]. We will briefly sketch the main idea behind this effect below, emphasizing the relation to the single-particle energy spectrum of a ring.

We start by recalling that the magnetic moment of an individual single-particle state i is given by

$$m_i = -\frac{\partial E_i(B)}{\partial B},$$

where $E_i(B)$ is the state's energy and B is the magnetic field. Talking about sharp energy levels here implicitly assumes that the system is closed (or weakly coupled to contacts) and phase-coherent. We restrict ourselves to zero temperature and determine the total magnetic moment of the whole system:

$$M = \sum_i m_i = -\sum_i \frac{\partial E_i}{\partial B} = -\frac{\partial E_{\mathrm{tot}}}{\partial B},$$

where E_{tot} is the sum of all occupied single-particle energies E_i. We know from elementary magnetostatics that there is a relation between the magnetic moment M and a circulating current I enclosing an area A, namely,

$$M = IA.$$

In the particular case of a ring geometry, the area A is given by the enclosed area πr_0^2. If we solve this equation for the current, we obtain the basic expression for the persistent current

$$I = \frac{M}{A} = -\frac{\partial E_{\mathrm{tot}}}{\partial (BA)} = -\frac{\partial E_{\mathrm{tot}}}{\partial \Phi}.$$

Here, Φ is the total flux piercing the ring structure. This expression was derived neglecting interaction effects like the charging energy. However, a charging energy independent of magnetic field can be added to E_{tot} without affecting the final result for the persistent current.

We wish to emphasize that in this constant-interaction picture which is — as we have shown above — appropriate for our ring, the persistent current can be decomposed into contributions of individual single-particle levels according to

$$I = -\frac{\partial E_{\mathrm{tot}}}{\partial \Phi} = -\sum_i \frac{\partial E_i}{\partial \Phi} = \sum_i I_i,$$

i.e., each state i contributes an amount to the persistent current that is given by the derivative of its energy with respect to magnetic flux.

For the special case of the ideal one-dimensional ring model, we obtain for $m = 0$ the contributions

$$I_\ell = \frac{\hbar e}{2\pi m^\star r_0^2} \cdot \ell = 2.5\,\mathrm{nA} \times \ell.$$

Given that positive and negative angular momentum states are degenerate and therefore successively populated, the contributions of all the low-lying states will cancel and only the highest states will contribute to the total persistent current.

We can now determine the contribution to the persistent current of particular states observed in the Coulomb-blockade regime of our ring. For a particular strongly oscillating state like "State II" depicted in Fig. 12.5, we have a slope of about 500 μeV per flux quantum corresponding to $I \approx 20\,\mathrm{nA}$. States that are flattened by a symmetry-breaking potential or by disorder contribute significantly less to the total persistent currents.

If we assume that currents of all the lower-lying states sum up to zero, this current is also an estimate of the total persistent current in the ring and the determined value is consistent with previous magnetization measurements [310–312].

12.12 Summary

The detailed analysis of quantum rings demonstrates that even in many-electron Coulomb-blockaded systems a detailed understanding of the energy spectrum and interaction effects can be obtained. With advanced fabrication techniques at hand, this opens the path to the understanding of more complex and multiply connected structures on a quantum-mechanical level. Electron–electron interactions beyond the constant-interaction model have been shown to play a minor role in our quantum ring leading to the frequent observation of spin pairs and very small spin splitting. Once ring structures with only one radial mode occupied are available, such quantum rings could be used to investigate spin effects [331] or even Luttinger liquid behavior in a circular 1D system with periodic boundary conditions.

13

Spin filling in quantum dots

13.1 Introduction to spins in quantum dots

In this chapter we investigate the electronic transport through a Coulomb-blockaded quantum dot in which interactions are rather strong, in contrast to the ring structure discussed before. In this dot, which is based on the two-dimensional electron gas in a parabolic quantum well, spin states have been identified in the weak-, intermediate- and strong-coupling regime of the dot to the leads. We observe linear changes in conductance peak spacings with in-plane magnetic field in all regimes, which can be interpreted in terms of a Zeeman splitting of single-particle levels. This allows one to follow the sequence of ground-state spin of the quantum dot as individual electrons are added. A perpendicular magnetic field applied to the dot in the same state allows the investigation of spin-pair candidates under conditions where orbital effects dominate the evolution of conductance peaks. Strong correlations in the position and in the amplitude of neighboring peaks allow the final identification of spin-pairs. The method of combining parallel and perpendicular magnetic fields for identifying spin states and spin-pairs works well for all coupling strengths of dot states to the leads, indicating that the spin degree of freedom is remarkably stable.

The energy spectrum of quantum dots in semiconductor nanostructures can be investigated by Coulomb-blockade experiments [91, 92]. For a circularly symmetric few-electron quantum dot, the measured energy spectrum directly reveals the shell structure of a zero-dimensional system [80]. In quantum dots containing 50 or more electrons, the situation is more involved. The energy spectra have been analyzed primarily on a statistical basis [268, 317, 332, 333]. The reason why we cannot understand such spectra on the basis of simple single-particle models is twofold: on the one hand, the real geometry of such dots is never exactly the intended one and symmetries of the designed dot are lost. Therefore, the single-particle energy levels (or their energy separation) show deviations from the ideal spectrum, so-called fluctuations that can be well described by statistical approaches like the random-matrix theory [93]. In addition, the microscopic structure of the wave functions gives rise to interaction effects (the charging energy is the most prominent one, but also exchange and correlations can contribute) that not only change energy levels significantly but also fluctuate in magnitude from dot state to dot state. Only in very exceptional cases, like the case of the quantum ring shown in the last chapter [315], can the energy spectrum be understood in detail.

In addition to the energy spectrum, the spin of the ground state of tunable quantum dots is of fundamental interest [269, 328, 329, 334, 335] and importance. In a picture of non-interacting electrons, the spin enters the description of the electron dynamics only via the Zeeman term, which becomes important in a magnetic field. In such a model at zero field the orbital motion is completely decoupled from the spin dynamics and each single-particle level is successively occupied by electrons with opposite spin according to Pauli's principle.

The situation becomes more involved if spin-orbit interaction effects are taken into account [336, 337]. They arise from relativistic corrections to the non-relativistic Schrödinger equation. An electron moving with velocity $\boldsymbol{v}$ in an electric field $\boldsymbol{\mathcal{E}}$ experiences an effective magnetic field $\boldsymbol{B}_{\text{eff}} = \boldsymbol{v} \times \boldsymbol{\mathcal{E}}/c^2$ in its rest frame. This field couples to the magnetic moment, i.e., the spin, of the electron and thereby leads to a coupling between the orbital motion and the spin degree of freedom. However, it is generally believed that spin-orbit interactions are very weak in n-type GaAs/AlGaAs heterostructures, and we will therefore neglect this effect in the following.

Coulomb interactions between electrons in principle do not involve the spin. However, they can have an indirect influence on the spin state of a quantum dot via the symmetry requirements for the fermionic many-body wave function. While the Hartree interaction term has no direct influence on the ground-state spin, exchange effects favor the parallel alignment of spins and therefore tend to maximize the total spin of the ground state [cf. eq. (11.3)]. In Ref. [328] it is found that the important parameter is the ratio λ of the interaction strength u to the single-particle level spacing Δ. It has been predicted that for $\lambda < 0.1$ the ground states of a quantum dot mainly have either total spin $S = 0$ or $S = 1/2$. For larger values of λ, higher total spins such as $S = 1$ and $S = 3/2$ have a significant probability. In another theoretical publication it was recently predicted that off-diagonal interaction fluctuations suppress the ground-state magnetization in finite-size systems [335].

In many-electron systems with more than 50 electrons, exchange effects quantified by the energy scale ξ can become comparable to the single-particle level spacing Δ [329]. At low interaction strength the exchange energy scale ξ can be related to the usual electron gas interaction parameter $r_{\text{s}} = 1/\sqrt{\pi n_{\text{s}} a_{\text{B}}^2}$ via [cf. eq.(12.5)] [329]

$$\xi = \frac{\Delta}{\sqrt{2\pi}} r_{\text{s}} \ln \frac{1}{r_{\text{s}}} \quad \text{for } r_{\text{s}} \ll 1,$$

and spin-pairs, i.e., the successive population with spin-up and spin-down electrons in the same orbital state should be rather likely. For stronger interactions, i.e., $r_{\text{s}} > 1$, the exchange contribution becomes comparable to the single-particle level spacing and fluctuations of the exchange contribution from level to level can completely reorder the occupation sequence of levels [278]. In atom physics such considerations lead to the explanation of Hund's rules. In quantum dots with strongly interacting electrons, this is the reason why spin-pairs are expected to occur rather rarely.

Experimentally, the effects of spin can be directly understood in certain few-electron quantum dots [338–341]. In many electron dots based on GaAs/AlGaAs heterostructures, the ratio of electron–electron interaction energy and Fermi energy $r_{\mathrm{s}} = E_{\mathrm{ee}}/E_{\mathrm{F}}$ is typically of the order of 1 leading to very rare occurrence of spin-pairs. For specially designed quantum dots r_{s} can be significantly reduced and spin-pairs are observed [315, 317, 333].

In the following we present a systematic study of quantum dots with a rather large r_{s} value. The host electron gas is a so-called parabolic quantum well (PQW) [342] with a suitably designed back-gate electrode [343]. The position of the electron gas in growth direction can be tuned by front- and back-gate voltages [344]. Quantum dots have been realized on such PQWs using top-gate electrodes, which are nano-fabricated by electron beam lithography [345]. Such systems have been investigated in the regime where the second subband for the confinement in growth (z) direction is occupied and switching behavior of the Coulomb-blockade peaks has been observed, which can possibly be attributed to the occupation of the second subband [345]. Here we focus on the regime where only one subband is occupied and switching events are absent. The quantum dots are as stable as those fabricated on regular two-dimensional electron gases (2DEGs) in heterostructures. With the back gate we drive the system to rather low densities $n_{\mathrm{s}} < 1.5 \times 10^{15}\ \mathrm{m}^{-2}$ such that $r_{\mathrm{s}} \propto 1/\sqrt{n_{\mathrm{s}}}$ can be as large as 1.5, similar as in Ref. [346] but smaller than in experiments on Si [347].

Tuning the quantum dot into the Coulomb-blockade regime, we observe about 40 conductance peaks as a function of plunger-gate voltage. In this gate voltage interval we go from weak via intermediate to strong-coupling of the quantum dot to source and drain as indicated by comparison of the Coulomb-blockade peak width with the single-particle level spacing.

For magnetic fields $B_{\|}$ applied in the plane of the 2DEG we observe a pronounced shift of the Coulomb-blockade peaks similar to previous experiments [348, 349]. The main contribution to this shift is due to the relative diamagnetic shift of energy levels in the dot, which has a $B_{\|}^2$-dependence. It is the same for all conductance peaks and can be eliminated by analyzing peak separations rather than the absolute peak positions. For most conductance peaks in all coupling regimes the Coulomb peak separation is either constant or it changes linearly in $B_{\|}$, which can be explained on the basis of Zeeman splitting of the dot levels.

When the sample is rotated *in situ*, i.e., without warming it up, from the parallel to the perpendicular magnetic field direction, the dot spectrum at $B = 0$ is basically left unchanged. In this perpendicular case, where the peak movement with magnetic field is dominated by orbital effects, about every 10th pair of neighboring peaks shows correlated behavior of amplitude and position. This unique combination of in-plane and perpendicular fields applied successively to a quantum dot in the same state allows a comparison of the dot's energy spectrum and spin splitting over wide ranges of dot-lead coupling.

The results for intermediate coupling can be satisfactorily discussed in terms of a single-particle picture that essentially assumes the absence of any coupling between the spin and the orbital degrees of freedom. Deviations from such a description are observed for weak and strong coupling.

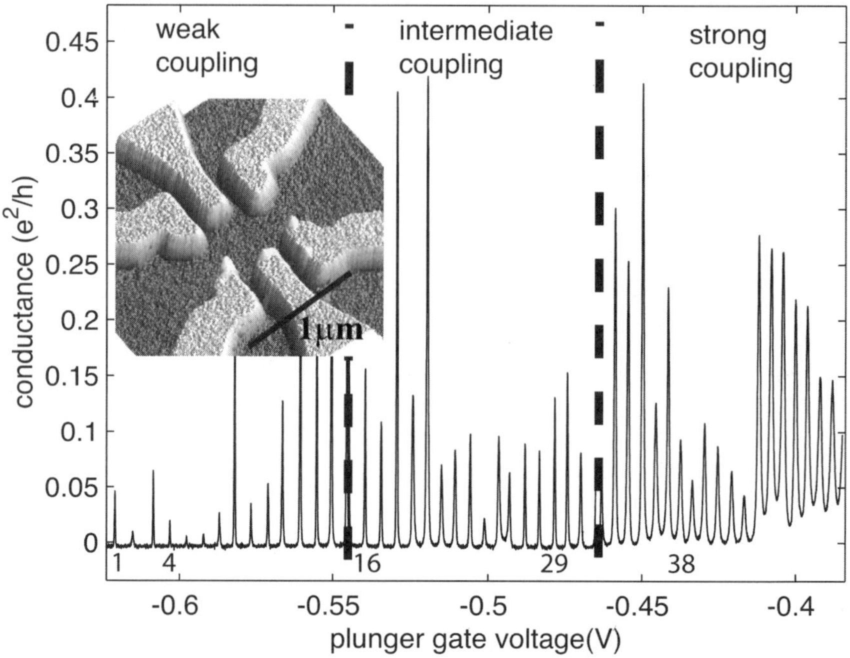

Fig. 13.1. Coulomb-blockade conductance peaks versus plunger-gate voltage taken at $B_{\text{total}} = 0$. The numbers in the bottom of the figure are used to identify peaks. This convention is used throughout the book. The vertical dashed lines divide the data into three regimes, namely, weak, intermediate and strong coupling. The inset in the upper left shows an SFM image of the surface gates defining the quantum dot.

13.2 Samples and structures

The quantum dot samples are based on MBE-grown parabolic $Al_xGa_{1-x}As$ quantum wells (PQWs) with x varying parabolically between 0 and 0.1 [342, 344]. The 760 Å wide wells are sandwiched between 200 Å thick undoped $Al_{0.3}Ga_{0.7}As$ spacer layers and are remotely doped with Si on both sides. A three monolayer thick $Al_{0.05}Ga_{0.95}As$ layer in the center of the well leads to a potential spike that was used to monitor the position of the wave functions with respect to the parabolic confinement in other experiments [344]. This spike is not relevant for the present study. A back-gate electrode consists of a 150 Å thick n^+-doped layer located 1.35 μm below the well. Using the back-gate electrode the density of the two-dimensional electron gas (2DEG) in the PQW can be varied from $n_s = 1 \times 10^{15}\,\text{m}^{-2}$, where only one subband is occupied in the well, up to $5 \times 10^{15}\,\text{m}^{-2}$, where three subbands are occupied. In this density range the mobility changes from $8\,\text{m}^2/(\text{Vs})$ at the lowest to $14\,\text{m}^2/(\text{Vs})$ at the highest densities. The occupation of the second subband starts at $n_s = 2.4 \times 10^{15}\,\text{m}^{-2}$.

The inset in Fig. 13.1 shows the TiAu top-gate electrodes fabricated using electron-beam lithography and a lift-off process. These electrodes define a lateral quantum dot with geometric lateral dimensions of 600 nm $\times$ 600 nm connected to source and drain contacts via the two quantum point contacts (QPCs) QPC1 and QPC2. Two plunger gates allow tuning the number of electrons in the quantum dot by varying the plunger-gate voltage U_{PG}. DC-conductance measurements were carried out with an applied source-drain voltage $U_{\mathrm{SD}} = 8\,\mu\mathrm{V}$ at an electron temperature of less than 140 mK in a dilution refrigerator.

The sample has been studied for a wide range of back-gate voltages. Here we focus on a rather large negative value, $U_{\mathrm{BG}} = -4.5\,\mathrm{V}$, where the density of the 2DEG is $n_{\mathrm{s}} = 1.5 \times 10^{15}\,\mathrm{m}^{-2}$.

13.3 Experiments

With the QPCs in the tunneling regime the Coulomb-blockade effect could be observed as depicted in Fig. 13.1. From an analysis of the Coulomb-blockade diamonds measured in the U_{PG}-U_{SD} plane (see chapter 11, also Ref. [323]) for the weak-coupling regime (see Fig. 13.2), we determine a charging energy of about 920 μeV. Using the self-capacitance of a circular disk, $C_{\Sigma} = 8\varepsilon\varepsilon_0 r$, we find a dot radius of $r = 190\,\mathrm{nm}$. This agrees well with the geometric size if a reasonable depletion length of 100 nm is taken into account. From this dot size the average single-particle level spacing can be estimated to be $\Delta = 2\pi\hbar^2/(m^\star \pi r^2) \approx 60\,\mu\mathrm{eV}$ assuming spin-degeneracy of the levels. This value is in rough agreement with the evaluation of transport through excited states (see Fig. 13.2, compare to Fig. 11.4), which gives a value of about 100 μeV.

In order to determine the 2D electron density, n_{d}, in the dot, which is typically smaller than the density in the unbound 2DEG at the same back-gate voltage, we analyzed magneto-Coulomb oscillations [350] as a function of U_{PG}. All top-gate voltages have to be readjusted when U_{BG} is significantly changed in order to stay in the Coulomb-blockade regime. For example, a more positive U_{BG} will pull the electron distribution in the well toward the back gate. As a consequence the front gate voltages have to be decreased in order to establish the necessary conditions for the observation of Coulomb blockade again (for details see Ref. [345]). From such measurements we estimate the 2D density in the dot to be around $n_{\mathrm{d}} = 0.5 \times 10^{15}\,\mathrm{m}^{-2}$ for the smallest plunger-gate voltages, which leads to about 50 electrons populating the dot. These numbers are confirmed by the filling factor $\nu = 2$-line at about $B = 1.05\,\mathrm{T}$ (see Fig. 7) [339, 340].

The experimental trace in Fig. 13.1 covers a range of about 40 Coulomb-blockade maxima, i.e., the dot population changes by about 40 electrons. In this range the coupling of the dot to its leads quantified by the width Γ of the conductance peaks changes significantly due to capacitive cross talk between plunger gate and the QPCs. We have identified three regimes, named weak, intermediate, and strong coupling. These regimes are marked in Fig. 13.1 and are separated by vertical dashed lines.

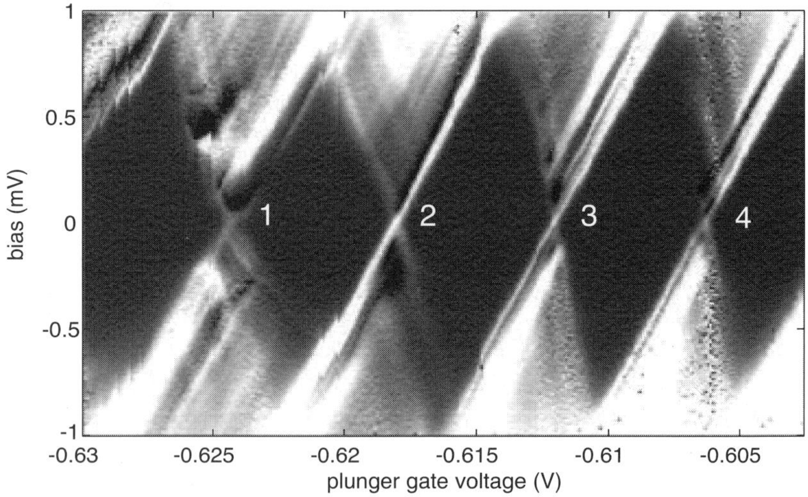

Fig. 13.2. Coulomb diamonds for $B_{\text{total}} = 0\,\text{T}$ in a grayscale plot. The addition energy (between peak 2 and 3) extracted from these measurements is $920\,\mu\text{eV}$, the single-particle level spacing is about $100\,\mu\text{eV}$, and the lever arm $\alpha = C_{\text{g}}/C_{\Sigma} = 0.155$, relating plunger-gate voltage to energy.

The exact position of the boundary between weak, intermediate, and strong coupling can be chosen somewhat arbitrary. Qualitatively, in the weak-coupling regime conductance peaks are thermally broadened, in the intermediate-coupling range we have $\Gamma \approx kT$, and in the strong-coupling regime $\Gamma > kT$. The numbering of the Coulomb-blockade maxima is kept consistent with Fig. 13.1 throughout this chapter.

A careful analysis of Coulomb charging energy, single-particle level spacing and dot size has been performed for all regimes and is summarized in Table 13.1. The capacitance $C_{\Sigma} = e^2/\Delta E_{\text{c}}$ cannot easily be translated into the geometric dimensions of the dot. The model of the self-capacitance of an isolated two-dimensional disk tends to overestimate the dot radius. As mentioned before, we estimate $\Delta \approx 60\,\mu\text{eV}$ assuming a circular dot shape for weak coupling. In this regime the conductance peaks are thermally broadened and their width Γ is independent of gate voltage. The fact that $\Gamma < \Delta$ and $\Gamma < kT$ indicates that we are close to single-level transport. However, when the dot is opened, the width of the conductance peaks is no longer determined by thermal broadening but increases with increasing gate voltage due to increased dot-lead coupling. At the same time, the dot size increases and the single-particle level spacing decreases. In the strong-coupling limit we have $\Delta \sim \Gamma$ and observe a finite conductance between conductance peaks.

The sample is mounted on a revolving stage. For in-plane fields we measure the Hall effect of the underlying 2DEG in order to make sure that the angle is accurate to within $0.01°$. This means that less than one tenth of a flux quantum threads the area of the dot for in-plane fields as high as 13 T. The perpendicular field direction

parameter	weak coupling	intermediate	strong coupling
Γ $(3.5kT)$	43 µeV (142 mK)	52 µeV	83 µeV
$\langle g \rangle$	$0.04e^2/h$	$0.14e^2/h$	$0.18e^2/h$
$\langle \Delta E_\mathrm{c} \rangle$	920 µeV	625 µeV	400 µeV
$\langle \Delta U_\mathrm{PG} \rangle$	5.8 mV	4.7 mV	4.2 mV
$\langle C_\Sigma \rangle$	174 aF	256 aF	400 aF
$\langle \alpha \rangle$	0.159	0.133	0.095
$\langle N \rangle$	50	70	90

Table 13.1. Parameters of the dot in the different coupling regimes. The Coulomb energy $\langle \Delta E_\mathrm{c} \rangle$ is extracted from Coulomb diamonds and the mean peak spacing $\langle \Delta U_\mathrm{PG} \rangle$ from plunger-gate sweeps at $B = 0\,\mathrm{T}$. The quantity $\langle g \rangle$ is the average conductance peak height, $\langle \alpha \rangle = C_\mathrm{pg}/C_\Sigma$ is the electrostatic lever arm of the plunger gate. The typical number of electrons $\langle N \rangle$ is determined from the sheet electron density in the dot (see text) and the dot size for the weak-coupling limit. For the other two regimes, conductance peaks were counted. The Coulomb peaks have been fitted based on thermal smearing. In this sense the Γ values for intermediate and especially large coupling reflect the increased coupling of the dot to source and drain.

can only be determined with about $0.3°$ accuracy, which is enough because orbital effects on the Coulomb-blockade peak position are about two orders of magnitude stronger than spin effects.

Figure 13.3 shows two plunger-gate sweeps before and after the sample has been rotated by $90°$. The rotation creates friction and therefore warms the mixing chamber to about 500 mK. The basic Coulomb-blockade behavior is recovered after rotation, however, and we are confident that we look at the same dot, i.e., the same energy spectrum.

For parallel magnetic fields applied in the plane of the 2DEG, the electron gas in source and drain as well as the dot's energy levels will undergo a diamagnetic shift. We will show below (see section 13.7) that the diamagnetic shift is the same for all dot states. As a consequence this effect leads to a general shift of all Coulomb-blockade resonances along the gate voltage axis. In order to reveal spin effects we analyze differences of peak positions in gate voltage (energy) because they should not depend on orbital effects. The Zeeman splitting for all magnetic fields investigated is much smaller than the Fermi energy in source and drain. We therefore expect that Zeeman effects in source and drain can be neglected and that both spin directions are available for tunneling through the dot at all magnetic fields. If differences of certain neighboring conductance peak positions display a linear magnetic field dependence, we interpret such shifts as arising entirely from the Zeeman effect in the quantum dot.

We can express the Zeeman-splitting arguments in terms of mathematical expressions. In general, an arbitrary ground state of the dot can be characterized by its electron number N and its total spin S_N. Each total spin S_N allows z-components $s = -S_N, -S_N+1, \ldots, S_N-1, S_N$. The Zeeman shift of an individual conductance peak $\Delta E_\mathrm{Z}(B)$ where the dot goes from N to $N+1$ electrons will be the difference

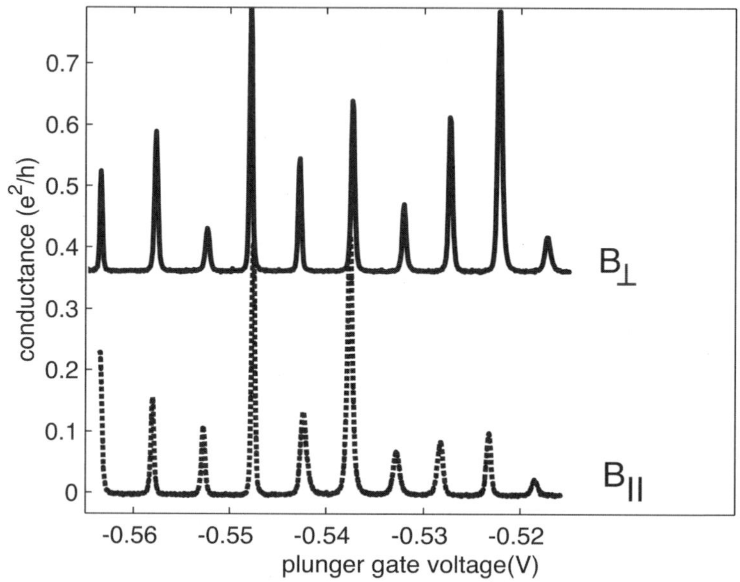

Fig. 13.3. Coulomb peak resonances measured for the intermediate-coupling regime at $B_{\mathrm{total}} = 0$ before and after rotating the sample *in situ*. Rotating leads to slight temporary heating due to mechanical friction, but the sample never warms up above 500 mK. The terms $B_\perp$ and $B_\parallel$ refer to the direction of magnetic field once applied. The two curves have been laterally offset by 17 mV and vertically offset for clarity. The peak amplitudes are slightly different, while the peak positions suggest the quantum states keep their specific character upon rotation of the sample.

of the Zeeman shifts of the two states (N, S_N) and $(N+1, S_{N+1})$, i.e.,

$$\Delta E_{\mathrm{Z}}(B) = (s_{N+1} - s_N) g_{\mathrm{L}} \mu_{\mathrm{B}} B,$$

where μ_{B} is Bohr's magneton. We have assumed here that the g_{L}-factor does not vary from level to level. Because $|S_{N+1} - S_N| = 1/2 + n$, where n is an integer number, the possible slopes of conductance peaks in a magnetic field are determined by the prefactor $s_{N+1} - s_N = \pm 1/2, \pm 3/2, \ldots$. Differences in conductance peak positions $\varDelta(B)$ are then differences of Zeeman shifts $\Delta E_{\mathrm{Z}}(B)$:

$$\varDelta(B) = (s_{N+1} - 2s_N + s_{N-1}) g_{\mathrm{L}} \mu_{\mathrm{B}} B. \tag{13.1}$$

The slope of $\varDelta(B)$ in a magnetic field can only be an integer multiple of $g_{\mathrm{L}} \mu_{\mathrm{B}} B$.

13.4 Weak-coupling regime

We first focus on data taken in the weak-coupling regime. Figure 13.4 shows the conductance through the quantum dot as a function of plunger-gate voltage for a series of parallel magnetic fields $B_{\|}$. The movement of the peak positions with magnetic field is clearly visible. The graph also demonstrates that the sample is stable over the duration of the experiment, i.e., there are no serious charge rearrangements over the course of about 24 hours. This is a precondition for measurements of such small effects as the Zeeman splitting.

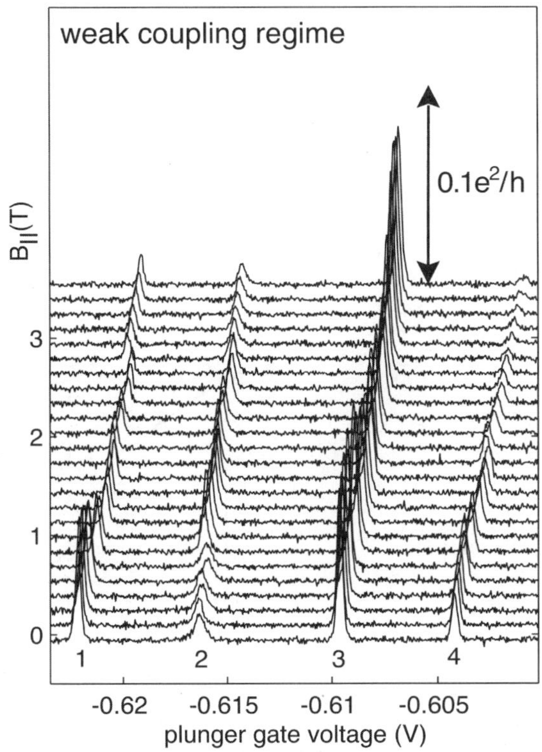

Fig. 13.4. Conductance through the quantum dot as a function of plunger-gate voltage and in-plane magnetic field $B_{\|}$ in the weak-coupling regime. The magnetic field $B_{\|}$ is increased from 0 to 3.6 T in steps of 50 mT and the plunger gate is swept in steps of 40 µV. Only every third measured curve is shown.

Figure 13.5a presents spacings of neighboring Coulomb peaks. As mentioned before, the lever arm relating plunger-gate voltage to energy in the dot is extracted from the measured Coulomb diamonds [323] in Fig. 13.2. The change of lever arm with plunger-gate voltage is explicitly taken into account. The curves are relatively flat up to magnetic fields of about 0.8 T. With the bulk g_L-factor of GaAs $g_L = -0.44$ [351], the Zeeman splitting is about $\Delta E_Z/B_{\|} = 25\,\mu\text{eV/T}$. The electron temperature as determined from the Coulomb-blockade peak width is about 140 mK, resulting in $3.5kT = 42\,\mu\text{eV}$. As long as the Zeeman splitting is small compared to thermal smearing, one does not expect a Zeeman shift. Only when the Zeeman splitting exceeds thermal smearing, a single spin level dominates the conductance peak, which then shifts in accordance with this level. This explains why the Zeeman splitting can only be observed clearly at magnetic fields above 1 T, where the Zeeman splitting is larger than kT. All the peaks have been fitted to a thermally broadened line shape in order to obtain the peak positions very precisely. The effects of g_L-factor tuning as observed in Ref. [352] are not relevant for the present sample design.

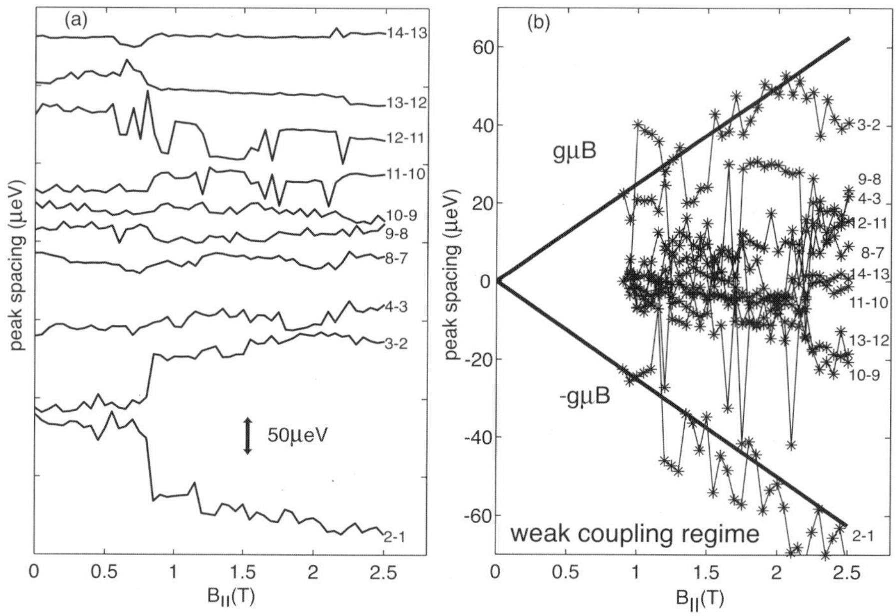

Fig. 13.5. (a) Evolution of peak spacing of the weak-coupling regime with in-plane magnetic field. The peak spacing is extracted from the measured peak motion (see Fig. 13.4), converted into energy using the lever arm α from the non-linear conductance measurements of Fig. 13.2, and vertically offset for clarity. (b) Peak spacings offset to align spacings at $B_{||} = 0\,\text{T}$ and converted into an energy using the corresponding lever arm α extracted from the Coulomb diamonds. The slope vs B corresponds to the change in ground-state spin as each electron is added and hence indicates the change of the spin from one state to the other. The straight lines shows the slope as expected for the bulk GaAs g_L factor $\mid g_L \mid = 0.44$.

We often observe an abrupt change in peak position at $B \approx 0.8\,\text{T}$ (see Fig. 13.5). Background charge rearrangements are a very unlikely cause for these effects because only some peaks are affected and not the entire spectrum. A possible reason could be exchange effects that could suddenly set in once the Zeeman gap exceeds kT (bootstrap effect). However, the magnitude of the jump as well as the extrapolation of the high-field behavior of peak separation down to $B = 0$ do not support this hypothesis.

It is important to note that in this weak-coupling regime only peak spacing 2-1 and 3-2 follow roughly the expected slope for a Zeeman energy-shift, $-g_{\mathrm{L}}\mu_{\mathrm{B}}$ and $+g_{\mathrm{L}}\mu_{\mathrm{B}}$, respectively. All the other peak spacings show a more or less flat behavior or slopes corresponding to a value less than expected for Zeeman splitting. This behavior agrees with the notion that, in a closed few-electron dot, subsequent levels are preferentially filled with parallel spins in analogy to Hund's rules for atoms. Only very rarely do neighboring peaks correspond to opposite spins (see peaks 2-3 and 1-2).

In addition, we observe strong peak spacing fluctuations as a function of parallel magnetic field (see Fig. 13.5). We have no detailed understanding of these fluctuations but speculate that the ground state of the dot may be changed due to correlation effects as a function of parallel magnetic field causing the observed behavior.

13.5 Intermediate-coupling regime

Figure 13.6a shows the positions of Coulomb-blockade peak spacings versus parallel magnetic field in the intermediate-coupling regime. The curves are vertically offset for clarity.

At higher fields all curves show a linear magnetic field dispersion. This is more clearly seen in Fig. 13.6b, where the curves are offset to a common origin at $B = 0\,\mathrm{T}$. The straight lines are calculated with the bulk g_{L}-factor of GaAs, $g_{\mathrm{L}} = -0.44$. The curves fall into three classes, namely, with a negative slope, a positive slope, or a flat behavior. Very similar behavior was observed for GaAs [348, 349] and Si [353] quantum dots.

Flat behavior is expected if successive electrons with the same spin occupy successive orbital states. It is possible to define a population sequence of spin states (see right-hand column in Fig. 13.6a) similar to what is done in Ref. [349]. In this case the sequence is $\uparrow\downarrow\uparrow\downarrow\downarrow\downarrow\uparrow\downarrow\uparrow\downarrow$ for levels 21-30. This sequence is consistent with the experimental observation as presented in Fig. 13.6b. Neighboring levels which are populated with opposite spins are possible candidates for spin-pairs, i.e., states with the same orbital wave functions. Using eq. (13.1) we can work out possible sequences of ground-state spins S_N of the quantum dot. Although such sequences are not unique, we present the ones that are most likely because they involve only low S_N-values in Table 13.2.

In Fig. 13.7a we present Coulomb peak maxima versus perpendicular magnetic field $B_{\perp}$. Spin effects due to Zeeman splitting are expected to be of minor importance for this magnetic field orientation. The movement of the energy levels is rather governed by orbital effects and level crossings. The correspondence of peaks after sample rotation is shown in Fig. 13.3. The corresponding peak amplitudes are shown in Fig. 13.7b.

We can identify pairs of peaks, namely, peaks 28 and 29 as well as 27 and 26, whose position and amplitude dependence is strongly correlated in the magnetic field

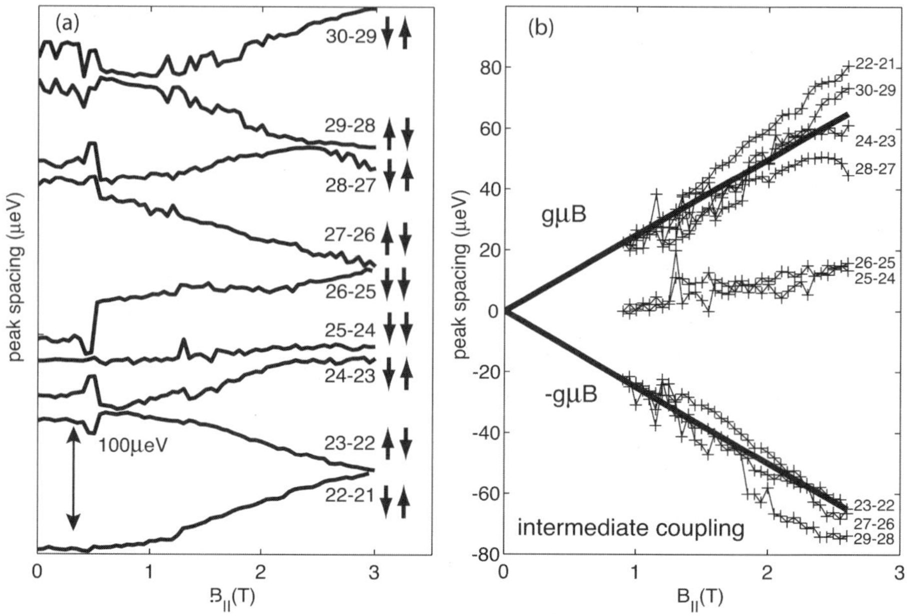

Fig. 13.6. (a) Evolution of peak spacing of the intermediate-coupling regime with in-plane magnetic field. The peak spacing is extracted from the measured peak motion (not shown here), converted into energy using the lever arm α, and shifted together in arb. units. The peak spacing is in most cases flat up to a magnetic field of about 0.8 T (see text for details). (b) Peak spacings offset to align spacings at $B_{\|} = 0\,\mathrm{T}$.

ΔN	$S_{N+\Delta N}$	$s_{N+\Delta N}$	$S_{N+\Delta N}$	$s_{N+\Delta N}$
0	1/2	-1/2	0	0
1	1	-1	1/2	-1/2
2	1/2	-1/2	0	0
3	1	-1	1/2	-1/2
4	1/2	-1/2	0	0
5	0	0	1/2	1/2
6	1/2	1/2	1	1
7	0	0	1/2	1/2
8	1/2	1/2	1	1
9	0	0	1/2	1/2
10	1/2	1/2	1	1

Table 13.2. Two possible sequences of ground-state spins in the intermediate-coupling regime for the range of nine conductance peaks shown in Fig. 13.6a. They were chosen such that total spin quantum numbers higher than 1 are avoided.

range from 0.25 to 1.25 T. We have confirmed this by calculating the cross correlation of amplitude and position between these peaks. For the parallel field data these peak pairs show a linearly decreasing peak separation (see Fig. 13.6a) in tune with the interpretation that the same orbital level is successively populated by spin-up and spin-down electrons (cf. Table 13.2). The combined measurement of the same dot in parallel and perpendicular fields has enabled us to identify spin-pairs in a strongly interacting dot, e.g., neighboring conductance peaks that are governed by transport through the same orbital state with alternating spin.

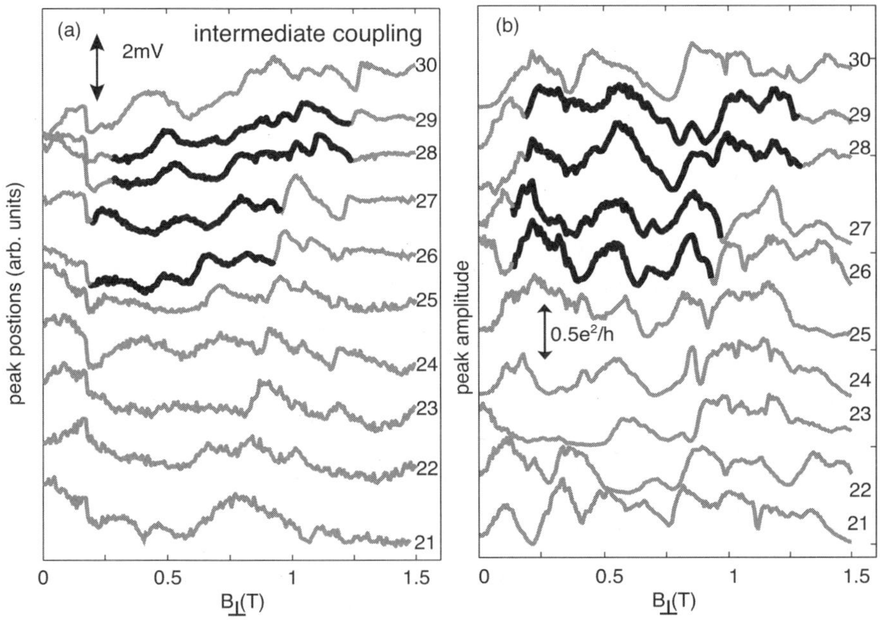

Fig. 13.7. (a) Parametric variation of the peak position in a magnetic field perpendicular to the 2DEG for 10 consecutive Coulomb-blockade peaks. Between the lines a constant of 4.1 mV is subtracted compensating for the contribution of the charging energy to all peak positions. Also indicated is the $\nu = 2$-line. (b) Parametric evolution of the peak conductance, vertically offset by $0.5e^2/h$ with respect to each other. A pair correlation in peak position and peak amplitude is clearly visible for peak 26 and 27 as well as for 28 and 29 (black lines), suggesting the respective electrons occupy the same orbital state forming a singlet.

For magnetic fields where the filling factor in the dot $\nu < 2$ the Coulomb-blockade maxima are known to shift smoothly as a function of perpendicular field (see e.g. Ref. [354]). In this way we identify the position of $\nu = 2$ in Fig. 13.7 and find good agreement with the previously mentioned carrier density in the dot.

At magnetic fields just below the $\nu = 2$-feature there is an odd–even behavior, i.e., the peak position shows an upward cusp for peaks 22, 24, 26 and a flat behavior for peaks 21, 23, 25, and 27. Similar features have been reported before [354] and could be related to ground-state spin rearrangements in the dot.

The Coulomb peaks in the intermediate-coupling regime follow the behavior expected from a single-particle picture strictly appropriate only in the absence of any mechanisms that couple the spin to any orbital degree of freedom. Although we expect that in the intermediate-coupling regime more than one dot level contributes to an individual conductance peak due to level mixing, the transmission of spin-up and spin-down electrons as a function of energy are essentially independent in this case. We can then naively expect independent Zeeman shifts of these transmission functions in opposite directions leading exactly to the observed behavior.

13.6 Strong coupling

For more positive plunger-gate voltages ($U_{\rm PG} > -0.47\,$V in Fig. 13.1) the conductance of the dot increases and the dot becomes strongly coupled to the leads. We first present the Coulomb peak spacing versus parallel magnetic field in Fig. 13.8. Again the traces roughly fall into three categories, namely, linear up or down movement in magnetic field and flat curves almost independent of magnetic field. On the right-hand side in Fig. 13.8b the theoretical expectation based on the bulk $g_{\rm L}$-factor of GaAs is plotted in the same graph. There are clear deviations from these lines, namely, peak spacing 41-40 actually displays a larger slope than expected. Especially for back-gate sweeps (not shown) we find many peak spacing slopes strongly exceeding the expected Zeeman splitting.

Peak position and amplitude are plotted as a function of perpendicular field in Fig. 13.9. The behavior is pretty erratic and no clear spin-pair (with the possible exception of peaks 37 and 38) can be detected. The calculated cross correlation in position is high for all neighboring peaks.

This behavior is in accordance with the expected mesoscopic conductance fluctuations in the strong-coupling regime [94], implying that strong level mixing occurs. Even in this regime the peak spacings in parallel field still collapse reasonably well onto the three branches expected from the single-level transport picture. The occurrence of larger slopes remains an issue to be addressed in the future.

13.7 Diamagnetic shift

For large parallel magnetic fields the energy levels in the dot as well as those in the leads are shifted up following the diamagnetic shift [355]. All previous data in this chapter were shown as a function of plunger-gate voltage. In order to present a comprehensive set of data we show the behavior as a function of back-gate voltage in Fig. 13.10.

One expects that the energy levels follow a parabolic field dependence [355]. The parallel magnetic field is applied along the direction of current flow through the SET.

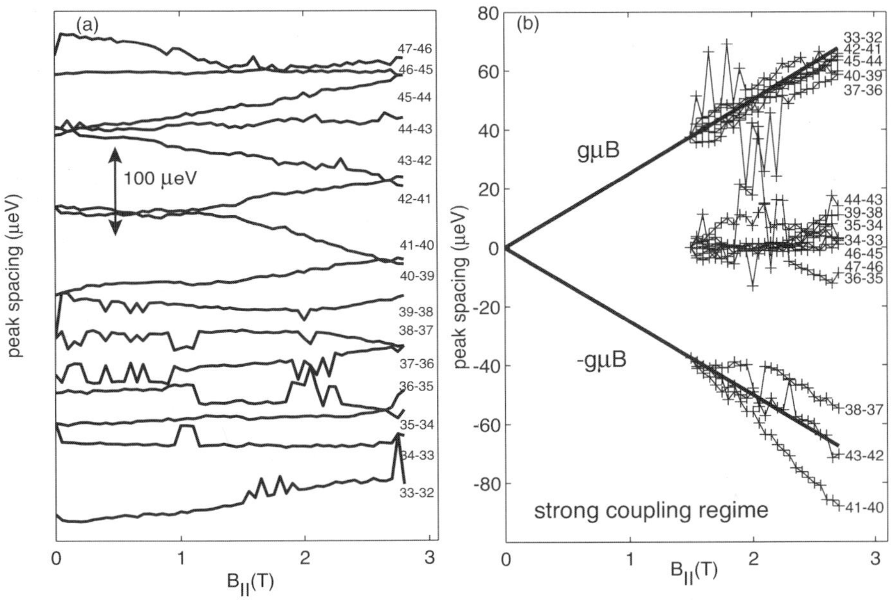

Fig. 13.8. (a) Evolution of peak spacing (strong-coupling regime) with in-plane magnetic field. (b) Peak spacings offset to align spacings at $B_{\|} = 0\,\mathrm{T}$ and converted into an energy using the corresponding lever arm α extracted from the Coulomb diamonds.

For 2DEGs in parabolic quantum wells with similar parameters as the one investigated in this study, such experiments have been done and analyzed in detail [356]. Here, the situation is more involved because the energy levels in the quantum dot are confined in both directions perpendicular to the magnetic field. The energy levels in source and drain also undergo a diamagnetic shift and the net shift of conductance peaks will result from differences in the shift in the dot with respect to source and drain. Because all effects are expected to be parabolic in magnetic field, the overall behavior as observed in Fig. 13.10 is in tune with this picture.

The data indicate that the diamagnetic shift in the quantum dot is stronger than in the surrounding 2DEG, i.e., the Coulomb peak positions move up in gate voltage with increasing in-plane magnetic field. For a perfect parabolic potential one would expect that the lower the Fermi energy is, the narrower is the wave function and therefore the smaller is the diamagnetic shift. In our case a very negative back-gate voltage is applied, which pushes the wave function in z-direction toward the hard wall, which delimits the parabolic potential. In this case the behavior is reversed because a higher Fermi energy leads to a steeper potential via the Hartree interaction and therefore to a narrower wave function. We have simulated the parabolic potential self-consistently and indeed find the two trends in the two regimes as described above. The effectively positive diamagnetic shift as observed in the data of Fig. 13.10 can thus be explained by the wave function probing the hard edges of the parabola in z-direction. Indeed we

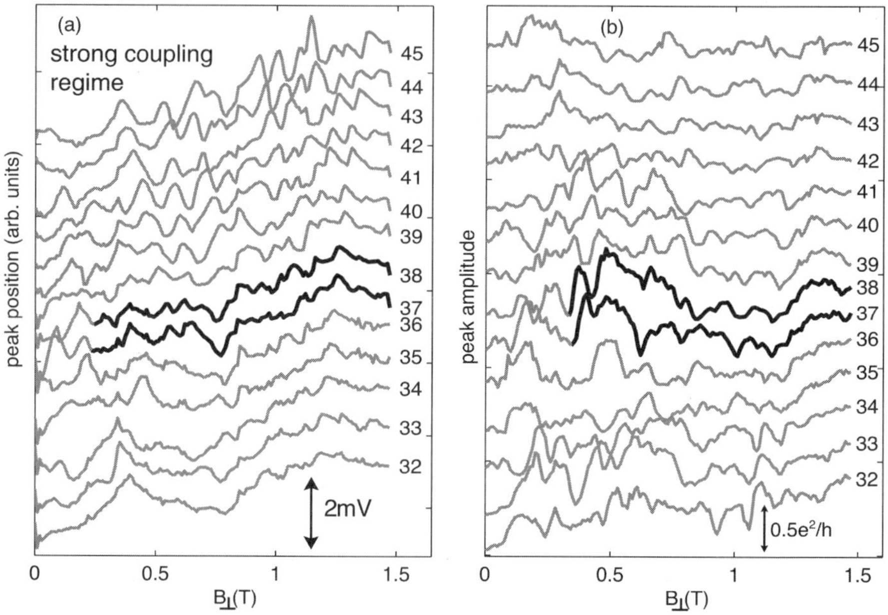

Fig. 13.9. (a) Parametric variation of the peak position (strong-coupling regime) in a magnetic field perpendicular to the 2DEG for 14 consecutive Coulomb-blockade peaks. Between the lines a constant of 3.5 mV is subtracted. (b) Parametric conductance amplitude of the same peaks, offset by $0.4e^2/h$ each. A pair correlation in peak position and peak amplitude is visible for peak 14 and 13 (black lines), suggesting they occupy the same orbital state.

find, that for more positive back-gate voltages and therefore larger carrier densities in the parabolic quantum well, where the electrons reside more in the center of the parabola, the general shift of the Coulomb peaks with magnetic field is reversed.

For a more quantitative analysis we use the following model. For a magnetic field along the x-direction, the confinement in y and z is modified. The potential in these two directions is approximately parabolic and we can write

$$V(y,z) = \frac{1}{2}m^\star \omega_y^2 y^2 + \frac{1}{2}m^\star \omega_z^2 z^2 .$$

The potential in z-direction is given by the as-grown parabola including its hard wall boundaries and modified by electron–electron interactions. The bare potential in y-direction is produced by the voltages applied to the gate electrodes. Obviously, the sample is in the limit $\omega_y \ll \omega_z$. The Schrödinger equation with the above potential and a magnetic field applied along the x-direction can be solved analytically. The resulting energy spectrum is

$$E_{n,\ell} = \hbar\omega_1 \left(n + \frac{1}{2}\right) + \hbar\omega_2 \left(\ell + \frac{1}{2}\right),$$

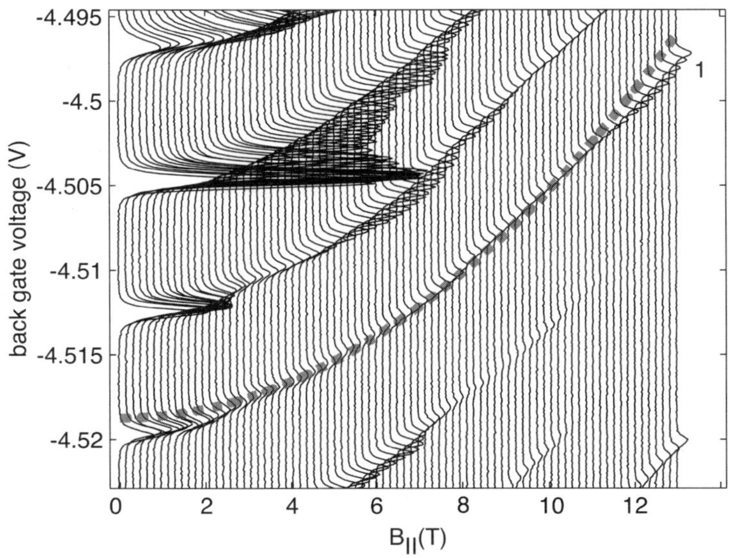

Fig. 13.10. Coulomb-blockade resonances as a function of parallel magnetic fields in the high-field range. The dashed line is a parabolic fit as described in the text.

where ω_1 and ω_2 are functions of ω_y, ω_z, and $\omega_c = eB/m^\star$. For small magnetic fields $B < 5\,\mathrm{T}$, where $\omega_c \ll \omega_z$ we find

$$\omega_1 \approx \sqrt{\omega_z^2 + \omega_c^2} + \mathcal{O}\left(\omega_y^2/\omega_z^2\right),$$

$$\omega_2 \approx \omega_y + \mathcal{O}\left(\omega_y^2/\omega_z^2\right).$$

The dominant contributions to the diamagnetic shift originate from the strong confinement ω_z in z-direction, while the orbital effects governed by the weak confinement ω_y are of the order $(\omega_y/\omega_z)^2$. Because $(\omega_y/\omega_z)^2 = (l_z/l_y)^4 \approx 10^{-4}$, with $l_{y,z}$ being the corresponding lengths $l_i^2 = \hbar/(m^\star \omega_i)$, these effects can be neglected. This result has two consequences:

(1) The dominant contribution for the diamagnetic shift comes from squeezing the wave function in the strong confinement (z-) direction. All energy levels are expected to shift parallel in magnetic field because all electrons in this regime occupy the ground state ($n = 0$) of the ω_z potential. This means that differences of Coulomb peak positions can safely be interpreted as a result of spin effects.

(2) The dependence of the Coulomb peaks as a function of parallel field is governed by the difference of the diamagnetic shifts in source and drain with respect to the diamagnetic shift of the energy levels in the dot. These diamagnetic shifts are different because the 2D carrier density in the quantum dot, i.e., the Fermi energy in the dot, is reduced with respect to the leads. From the direction of the diamag-

netic shift in our dot it follows that ω_z^{2DEG} is smaller than ω_z^{dot}. By fitting parabolas to the observed Coulomb peak dispersion in Fig. 13.10, we find a difference $\Delta z = \sqrt{\langle z^2_{\mathrm{2DEG}} \rangle - \langle z^2_{\mathrm{dot}} \rangle} = 3\,\mathrm{nm}$, a value, which is reasonable if compared to simulations.

13.8 Discussion of the results

The assumption that orbital effects play a minor role for the linear change in Coulomb peak separation as a function of parallel field is on relatively safe grounds (see previous section). In a single-particle picture, where the exchange interaction is neglected, one would expect that orbital states are successively populated by spin-up and spin-down electrons. For our quantum dot we estimate interactions to be important, since the corresponding 2D density is low. In this case exchange interactions are expected to have a significant influence in maximizing the ground-state spin [269, 334, 357] and it has been predicted [329] that only very few spin-pairs occur.

Our experimental observations in all coupling regimes agree qualitatively with these predictions. Not only the observation of states, which move linearly in energy as a function of parallel field but also their sequencing according to the population of spin-up and spin-down states not necessarily in sequential order, supports this view. Another strong experimental argument follows from the data as a function of perpendicular field where spin-pairs can be assigned in agreement with the observations in parallel magnetic fields. The combination of data taken as a function of both parallel and perpendicular magnetic fields for a quantum dot in the same state, a unique feature of our experiments, gives strong evidence for the predictions of random matrix theory [329] describing the population of spin states in quantum dots for various strengths of interactions. Our parallel field data suggest that the behavior of the Coulomb peaks in all coupling regimes are governed by Zeeman splitting with a g_{L}-factor similar to that of an extended two-dimensional electron gas.

It is worth mentioning that we do very rarely observe two successive ascending or descending slopes in a peak position differences plot like Fig. 13.6. A possible theoretical explanation of such an effect is that the ground-state spin of the $N+1$-electron dot and that of the N-electron dot differ by more than $1/2$ indicating that the arrival of the electron in the dot rearranges the spin orientation of other electrons. Such effects are closely related to the spin-blockade effect proposed in Ref. [358] and experimentally investigated in Ref. [353].

In the weak-coupling regime the spin assignment in Fig. 13.5 is in agreement with the above arguments. However, the strong fluctuations in the peak positions, which are most likely not related to background charge rearrangements, make this analysis difficult. It was also observed in Ref. [349] that the weak-coupling regime, which is naively expected to give the best results for the analysis of spin states, because the peaks are narrowest, does not prove itself very valuable for such investigations. The data indicate that many successive electrons occupy states with the same spin direction analog to Hund's rules for atoms.

In the intermediate- and strong-coupling regime, on the other hand, subsequent states of the dot are less frequently occupied with parallel spins as compared to the weak-coupling regime. In the intermediate-coupling regime the occurrence of spin-pairs is therefore more likely, in agreement with the measured data. In the strong-coupling regime the conductance peaks become wider and single-level transport cannot be achieved. This leads to a general enhancement of correlations in position and amplitude between neighboring peaks. We were only occasionally able to find neighboring peaks with correlations significantly stronger than average.

If transport through several levels contributes to the position and amplitude of a given Coulomb-blockade peak and we assign a spin quantum number to each participating orbital level, one would expect averaging of the corresponding Zeeman shifts. This would in average reduce the slope of peak separations as a function of in-plane field in contradiction to the experiment. Generally speaking we find that the spin behavior is more robust than expected from the involved energy scales. A similar statement was recently made by Glazman and coworkers in the context of the Kondo effect in strongly coupled dots in which charge quantization no longer occurs [359]. In our case, the data indicate that it is the over-all interaction of the tunneling electron with the electrons residing in the dot rather than the exact functional form of orbital states, which is relevant for the spin quantization in the dot. Within this interpretation our observations imply that (1) tunneling processes conserve spin, (2) spin-orbit coupling is sufficiently weak, and (3) exchange interactions have a sufficiently small dependence on parallel magnetic field.

13.9 Conclusions

In this chapter we have investigated the behavior of Coulomb-blockade peaks as a function of magnetic fields applied in the plane of and perpendicular to quantum dots in semiconductor heterostructures for a range of coupling regimes between the dot and its leads. In the weak-coupling regime the positions of the Coulomb-blockade resonances show strong fluctuations that inhibit clear assignments of spin to a given state. In the intermediate-coupling regime, the experimental observations are in good agreement with predictions based on a simple single-level transport scenario. This holds for magnetic fields applied parallel and perpendicular to the plane of the electron gas. In the strong-coupling regime the interpretation of the data still follows the single-level transport picture in parallel magnetic field. However, some states show a stronger parallel magnetic field dependence as expected for weakly interacting $s = 1/2$ particles. On the other hand, the perpendicular magnetic field data seem to be consistent with the theory of mesoscopic fluctuations in quantum dots.

Part IV

Local Spectroscopy of Semiconductor Nanostructures

14

Instrumentation: Scanning force microscopes for cryogenic temperatures and magnetic fields

14.1 Introduction: low-temperature scanning force microscopes

Since its invention by Binnig, Quate, and Gerber [360], the scanning force microscope (SFM) has become a standard tool for the investigation of conducting and insulating surfaces on the atomic scale. Low-temperature scanning force microscopes (SFMs) are desirable for many applications in physical research. A particular field of application is the physics of mesoscopic semiconductor structures. In conventional magnetotransport experiments on such systems, the measured conductance does not give direct insight into microscopic properties within the system. On the other hand, detailed microscopic theories, e.g., based on scattering wave functions or on S-matrices, exist for the calculation of the conductance. These theories have to make assumptions about local potential landscapes in the structures in order to lead to quantitative results. It can be expected that scanning probe techniques give more detailed information about the interior of a mesoscopic system.

Although many cryo-SFMs have been built for more than a decade by researchers (see Ref. [361] for a historical overview), the number of successful experiments is comparably small. The first cryo-SFMs have become commercially available only recently, reaching base temperatures around 5 K under UHV conditions. Lower temperatures have been reached with home-built microscopes in non-UHV setups [362–366].

14.2 Design criteria for a low-temperature scanning force microscope for the investigation of semiconductor nanostructures

In our lab cryo-SFMs are built for the investigation of buried two-dimensional electron gases and laterally defined nanostructures on the microscopic scale. The required lateral resolution for these structures is given by the Fermi wavelength of the electrons, which is 30 – 50 nm in typical structures. At the same time, usually only one or two nanostructures with a typical size of 1 μm are located on a large 5 mm

$\times$ 5 mm chip. Finding such a structure at low temperatures without optical access to the microscope is a major challenge. Large scan ranges of several micrometers are therefore important. At the same time, a positioning system is needed for coarse-positioning the structure at low temperatures.

Low temperatures are needed due to the small energy scales relevant for such structures. Elastic impurity or boundary scattering dominate the transport properties of electrons at temperatures below about 4.2 K. At these temperatures quantum phenomena in electron transport such as the Shubnikov–de Haas effect or the quantum Hall effect can be observed. For the investigation of phase-coherence effects such as weak localization or conductance fluctuations, temperatures below 1 K are typically required. For charging effects in quantum dots, i.e., the Coulomb-blockade effect, to be important, even lower temperatures are needed. However, the cooling power of cryostats decreases with base temperature that can be achieved. While the cooling power of a ^{4}He cryostat is typically many milliwatts at 2 K, a ^{3}He cryostat has a cooling power of the order of 0.1 mW and the cooling power of dilution refrigerators can be even smaller. Operating an SFM, however, requires a large number of cables leading to a significant heat load, which diminishes the time for scanning at base temperature. A ^{3}He-system with a large ^{3}He volume allowing scanning at base temperature for about three days, as used in our setup, appears to be a suitable system.

Low-dimensional semiconductor structures show a rich variety of phenomena such as the quantum Hall effect only at elevated magnetic fields. Therefore, a typical transport setup with a cryo-SFM has to include a superconducting magnet supplying magnetic fields of the order of 10 T. At the same time all the SFM components have to be made of non-magnetic materials.

A feedback mechanism is needed for keeping the local sensor at a well-controlled distance from the plane of the electrons or the sample surface. This can be achieved with a conventional cantilever and standard feedback electronics. However, typical semiconductor nanostructures are sensitive to light (persistent photoeffect) and therefore a non-optical cantilever deflection detection method is needed. Such a method has the additional advantage that it leads to less involved setups. Piezoresistive cantilevers have been proposed [367] and applied for low-temperature scanning force microscopy [362, 365, 366, 368–374]. However, these cantilevers are expensive and not easily available. Recently, piezoelectric quartz tuning forks have been employed in a scanning near-field optical microscope designed for operation at low temperatures [375]. In this setup, an optical fiber is glued along one prong of the tuning fork which is mechanically excited to oscillate. The tuning fork is used as a friction-force sensor. The advantages of these piezoelectric sensors are the easy availability, the low cost, and the high quality factors. They have been successfully employed for atomic force microscopy [376], scanning near-field optical microscopy [375, 377–380], magnetic force microscopy [381], and acoustic near-field microscopy [382] at room temperature. In Refs. [375, 377, 379–381] the SFM operation was just used to keep the tip–sample distance fixed while an additional nanosensor measures the physical quantity of interest.

The whole microscope has to be small enough in order to fit into typical cryostats. While the sample space needed for a typical transport experiment at cryogenic temperatures is of the order of 1 cm^3, a SFM is much bigger. The size constraint is mainly dominated by the magnet bore, which in turn is a main factor entering the price of the magnet. Reasonable compromises are found for a magnet bore of 2–3 inch leading to a geometrical constraint for the outer diameter of the microscope of about 40 mm. The length of the microscope is typically limited by the corresponding length of the sample space, which turns out to be of the order of 120 mm in a reasonable system. Most of this length is needed for the long scan piezo, which guarantees the desired large scan range at low temperatures.

A modular design for such a microscope is preferred. Several components with different functionality and maintenance requirements are combined in an SFM. Easy access and exchange possibility has to be provided for all the components. Mechanical stability of the whole assembly is crucial for stable operation. Only the highest quality of all components will produce satisfying performance of the SFM at cryogenic temperatures.

In the following we describe the implementation of a SFM operating at temperatures between 350 K down to below 300 mK and in magnetic fields up to 9 T. The microscope is operated under vacuum; however, samples and tips can not be prepared and exchanged under UHV conditions, in contrast to other setups [361, 383]. Piezoelectric quartz tuning forks were employed for non-optical tip–sample distance control in the dynamic operation mode. Fast response was achieved by employing a phase-locked loop for driving the mechanical oscillator. We describe the scanning force microscope, our implementation of the tuning fork sensors, and analyze the electronics used for the operation of the microscope.

14.3 A scanning force microscope operated in a ^{3}He system

Cryostat setup

Figure 14.1 shows a schematic drawing of the cryostat setup in which the microscope is implemented. The cryostat is lowered into a hole in the concrete lab foundation filled with a few centimeters of sand. We share this foundation with the neighboring lab but the whole concrete block is mechanically decoupled from the rest of the building thereby providing a mechanically very quiet environment. Effective insulation of the cryostat from remaining vibrations is achieved by suspending the whole ^{3}He insert from a special low-vibration platform resting on four sand-filled steel supports into the ^{4}He Dewar, which also hosts the superconducting magnet. The only mechanical coupling between the Dewar and the ^{3}He insert is made via a soft rubber sleeve needed in order to make the cryostat ^{4}He-tight. The ^{3}He insert can be fine positioned parallel to the lab floor and adjusted into an exactly vertical position. Pumping lines and the He-recovery lines are plastic pipes laid through a box of sand in order to minimize mechanical coupling between pumps and the system. For the same reason, pumps are not located on the same concrete platform that hosts the cryostat setup. Microphony of the cryostat is reduced using a rubber mat, which is tightly tied to the cryostat body.

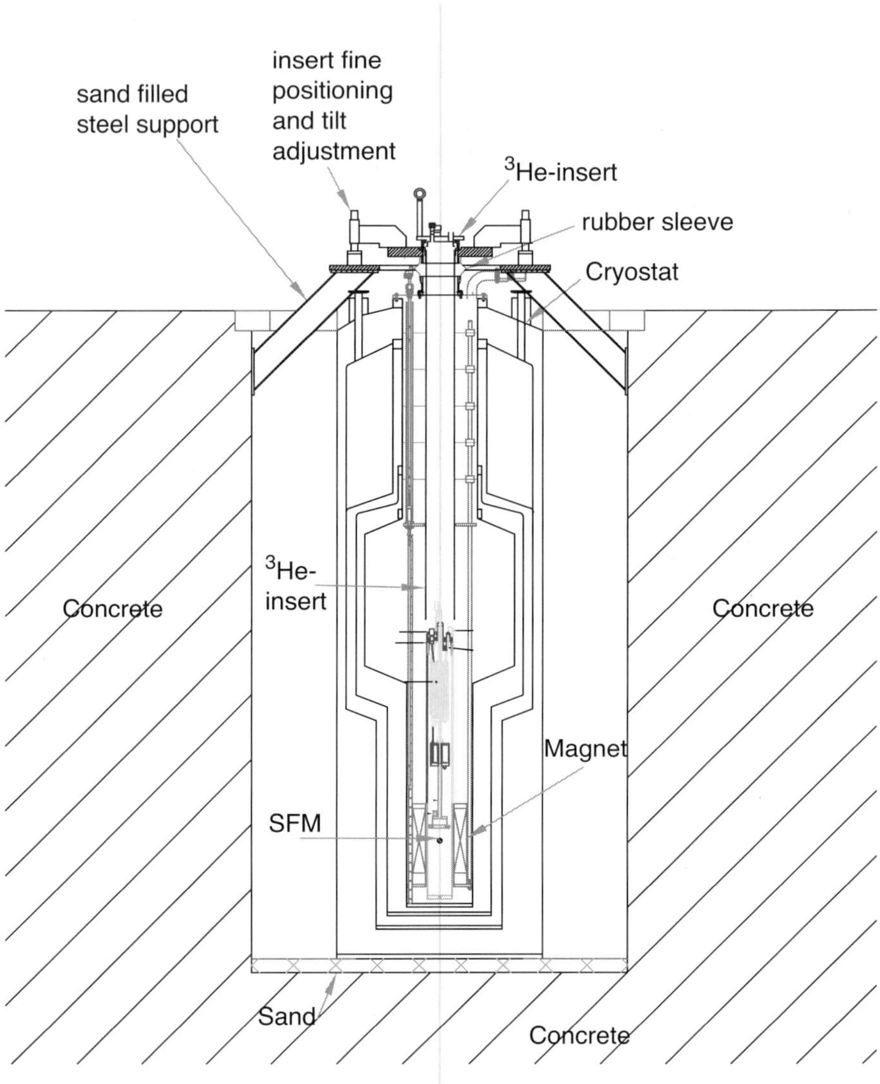

Fig. 14.1. Schematic setup of the whole experimental setup including the cryostat, the ^{3}He-insert, the superconducting magnet and the microscope.

Figure 14.2 shows a schematic drawing of the completely home-built microscope comprising the two main units, the z-module and the x-y table. This assembly is firmly attached to the ^{3}He-pot of the commercial vacuum loading ^{3}He-insert. A vacuum beaker that can be pumped down to about 10^{-5} mbar at room temperature surrounds the refrigerator and the microscope (cf. Fig. 14.1). The microscope can be operated in vacuum at temperatures down to 280 mK. About 50 cables, 6 of which are coaxial lines and 12 are suitable for high voltages up to 1000 V, connect this insert to the electronic setup outside the cryostat. Details about the microscope will be discussed in the following sections.

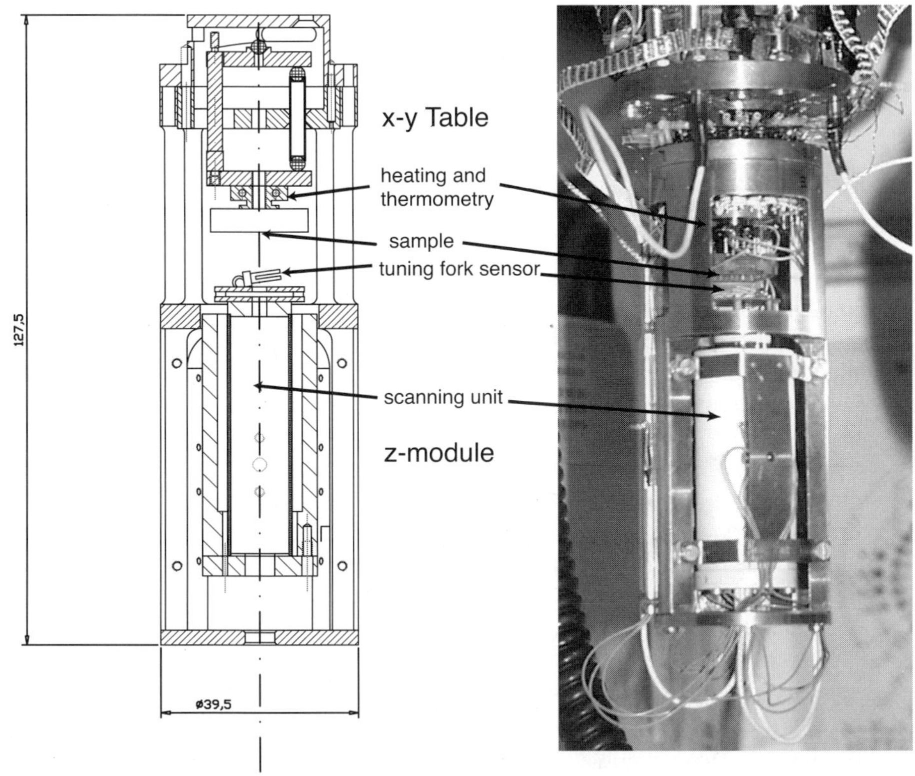

Fig. 14.2. Schematic drawing of the microscope (left) and photography (right).

The z-module

The z-module of the microscope (cf. Fig. 14.2) allows coarse tip–sample approach using a slip-stick drive moving the scan-piezo up or down. It hosts the scan-piezo tube at the end of which the tip–sample interaction sensor is mounted.

The outer frame of the z-module is machined out of non-magnetic CuBe. The slip-stick z-motor has been constructed following the design of Refs. [361, 383]. It uses six shear-piezo stacks, which had to be fabricated by ourselves initially, but are now available commercially.[1] Each stack consists of six shear-piezo plates and a sapphire end-plate. A macor prism is clamped between the six stacks using a CuBe sheet as a spring. AlO plates are glued onto the macor prism in the places where the sapphire end-plates of the stacks touch.

The macor prism has a central hole hosting the scan-piezo tube. The latter is made of a lead-zirconate-titanate ceramic[2] with non-magnetic copper electrodes. The tube scanner has a length of 2 inches and an outer diameter of 0.5 inch. It gives a lateral scan range of 52.2 μm in x-y direction and a z-range of 5 μm at a temperature of 290 K. At base temperature the lateral range is 8.8 μm and the z-range is 0.85 μm.

A sensor mount is attached to the end of the scan-piezo with tapped through holes for screwing the sensor plate onto it. The screws also provide electrical contact of the sensor plate to six coaxial cables.

The x-y table

The x-y-table (cf. Fig. 14.2) is shown in detail in Fig. 14.3. It consists of three main parts:

1. An x-y motor allows coarse positioning of the sample in the plane relative to the tip of the tuning fork sensor. It has a lateral travel range of about 5 mm corresponding to the typical size of a semiconductor chip used in the experiment. The motor is built on a Shapal[3] platform. This ceramic material was used because of its comparably large thermal conductivity of 100 W/(mK) at room temperature, its thermal expansion coefficient of $5.2 \times 10^{-6}\,\mathrm{K}^{-1}$ at room temperature, which is very close to that of piezoceramic materials, and the ability to machine it. Three thin-walled piezo-tubes[4] with sapphire balls at their ends are fed through holes in the Shapal platform and firmly attached with insulating epoxy glue.[5] These tubes stick out above and below the platform. They can bend in x- and in y-direction similar to a conventional scan-piezo due to their four outer electrode segments. The moving x-y platform made of CuBe is clamped with a CuBe spring from top and bottom against the sapphire balls. Suitable sawtooth voltages applied to the electrodes of the piezo-tubes lead to slip-stick motion of the moving platform. At room temperature, about 30 V are sufficient to move the stage, at 100 V a single step amounts to about 1 μm. The motion in x- and y-direction were found to be orthogonal with high accuracy.
2. On top of this moving CuBe platform there is a Cu-block hosting two thermometers and a heating wire. This arrangement is used for measuring and controlling the sample temperature in the whole range between 350 K and 280 mK. Thin flexible Cu wires are used for electrically connecting these components to pins on the fixed Shapal platform. The heater allows us to evaporate the water film from the sample surface before cooldown. It further gives us the possibility to

[1] PI ceramic GmbH, Lindenstrasse, 07589 Lederhose, Germany
[2] EBL-4 by Staveley Sensors Inc, 91 Prestige Park Circle, East Hartford, CT 06108, USA
[3] Supplied by Goodfellow, Ermine Business Park, Huntingdon, PE29 6WR, England
[4] EBL-4 by Staveley Sensors Inc, 91 Prestige Park Circle, East Hartford, CT 06108, USA
[5] EPO-TEK H77 by Polyscience AG, Riedstrasse 13, 6330 Cham, Switzerland

keep the sample warmer than its surroundings during the cooling process in order to avoid freezing contaminations on the sample surface.

3. On top of the thermometry block we have mounted a commercial 32-pin chip socket for ceramic chip carriers. Twelve of the pins are connected to pins on the Shapal platform. The samples are prepared in chip carriers that can then simply be plugged into the socket. The sample mount unit can be used for standard magnetotransport measurements independent of the SFM operation.

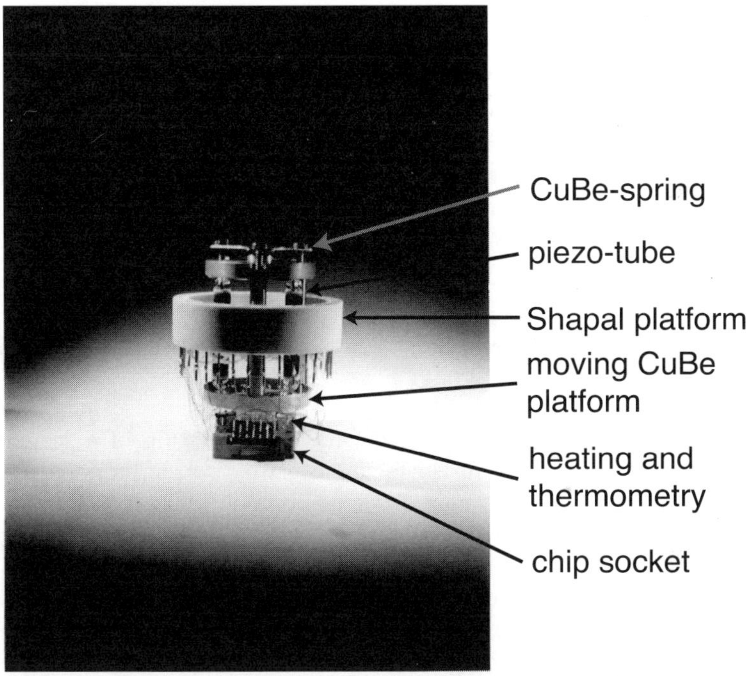

Fig. 14.3. Photography of the x-y-table.

14.4 Scanning Sensors

In this section we discuss sensors used in cryogenic scanning force microscopes for detecting surface topography and/or coupling to buried two-dimensional electrons. We will show which types of sensors have been applied and describe in detail the implementation of piezoelectric tuning fork sensors that we employ in a scanning force microscope operational at 300 mK. It is shown that these sensors can be well modeled by a single harmonic oscillator with very high quality factors at low temperature despite the symmetry-breaking wire attached to one prong. The nested feedback comprising the sensor, a phase-locked loop, and a conventional z-feedback is analyzed in terms of linear control theory and the employed current-voltage converter is identified as the dominant source of noise in the system. It is shown that the nested feedback has an over-all low-pass response. The optimum feedback parameters for the phase-locked loop and the z-feedback can be determined from the knowledge of the tuning fork resonance alone regardless of the tip shape and details of the tip–sample interaction. The bandwidth of the feedback turns out to be independent of the quality factor of the tuning fork, but it is influenced by its stiffness. The advantages of this system compared to pure phase control are discussed.

14.4.1 Introduction

In scanning force microscopy various imaging modes exist that can be classified to be either contact modes or dynamic modes [384]. The latter are characterized by the fact that the tip oscillates above the surface and the parameters of the oscillation are influenced by the tip–sample interaction. True atomic resolution has been achieved in this imaging mode [385–387]. Usually the cantilever oscillation is measured by optical means.

14.4.2 Types of sensors: an overview

There are a number of candidates for the use as sensors in a scanning probe experiment on nanostructures. The most obvious one is the tip of a scanning tunneling microscope [388]. Piezoresistive cantilevers have been proposed [367] and applied for low temperature scanning force microscopy [362, 365, 366, 368–374]. However, these cantilevers are expensive and not easily available. Recently, piezoelectric quartz tuning forks have been employed in a scanning near-field optical microscope designed for operation at low temperatures [375]. They have been successfully employed for atomic force microscopy [376], scanning near-field optical microscopy [375, 377–380], magnetic force microscopy [381], and acoustic near-field microscopy [382]. A very involved and specialized sensor is a scanning single-electron transistor [389]. In the following we will briefly introduce three of the four sensors mentioned above. The piezoelectric tuning fork sensors will be the topic of a more detailed discussion in the next subsection.

STM Tips

Probably the simplest sensor allowing capacitive coupling to a buried electronic nanostructure is a metallic tip as conventionally used for scanning tunneling microscopy (STM). Such tips are commercially available or they can be electrochemically etched from tungsten or platinum/iridium wires. Used in conventional tunneling mode topographic images can be obtained. However, since in typical semiconductor nanostructures the electronic system is buried underneath an insulating barrier of considerable thickness (typically more than 35 nm), relatively high voltages have to be applied in this imaging mode and irreversible local charging of the structure may result at low temperatures.

STM tips have been successfully employed for the so-called "subsurface charge accumulation imaging" [388] which can be — most simplified — be described as a local capacitance measurement. In order to reach a sufficient sensitivity to very small local capacitance changes, stray capacitances have to be minimized by mounting a field-effect transistor (FET) at low temperature in close proximity to the tip. With this FET a low-temperature capacitance bridge can be realized [390], which can measure changes in capacitance as small as about 10 aF. A low-temperature current preamplifier has been employed for the measurement of shot noise in a scanning tunneling microscope [391].

Piezoresistive Cantilevers

Piezoresistive cantilevers have been developed at Stanford University [367] in order to avoid more elaborate methods of cantilever deflection measurement, such as laser beam deflection or optical interferometry. This simplifies the design of a cryogenic scanning force microscope considerably. Such cantilevers are now available from Park Scientific Instruments. Piezoresistive cantilevers are fabricated from silicon wafers. The top layer of the cantilever is doped by ion implantation and shows a piezoresistive effect that is used to measure the cantilever deflection. These cantilevers have typical spring constants between 1 N/m and 20 N/m. The resonance frequencies are between 25 and 240 kHz depending also on the cantilever preparation procedure. The total resistance of the piezoresistive layer is of the order of 2 kΩ and voltages of the order of several volts are used for measuring the resistance. This leads to a considerable joule heating of the sensor with a power of the order of a few milliwatts. The relative resistance change $\Delta R/R$ is proportional to the cantilever deflection Δz, i.e.,

$$\frac{\Delta R}{R} = \alpha_{\mathrm{pr}} \Delta z,$$

where the proportionality constant α_{pr} depends on the geometry of the lever. Typical values are in the range of $1 - 4 \times 10^{-6}\,\mathrm{nm}^{-1}$. A Wheatstone bridge is used to measure the resistance change with a resolution at least in the range of 10^{-6}. On-chip fabrication of the bridge is possible and can help to compensate the temperature dependence of the resistance. Vertical resolution of 0.1 Årms has been demonstrated at a bandwidth of 1 kHz [367]. Force sensitivities of 8.6 fN/$\sqrt{\mathrm{Hz}}$ have been reported [392].

Piezoresistive cantilevers have found many applications in the field of scanning probe techniques. They were, for example, used for magnetic force microscopy [393], infrared imaging [394], biosensing [395], parallel imaging, and lithography [396, 397] and are still subject of research [392]. An interesting self-sensing cantilever concept with an integrated field-effect transistor has been developed in Ref. [398].

For capacitive coupling to a buried electronic system, coated [366] and uncoated [362, 365] cantilevers have been used. A metal coating, e.g., made of iridium, makes sure that the tip surface is an equipotential.

Scanning Single-Electron Transistors

Perhaps the most sophisticated sensors employed for scanning on shallow electronic semiconductor structures are the scanning single-electron transistors. Single-electron transistors (SETs) are tunneling devices with a typical size of 100 nm. The electrical current through SETs is governed by the Coulomb-blockade effect. The physics behind these devices has been reviewed in Ref. [87]. Electrons tunnel from a source electrode onto a small metallic island and subsequently from the island into the drain electrode. The magnitude of the current shows a strong modulation as a function of the electrostatic potential of the island relative to the source and drain electrodes. The period of the oscillation is given by the condition that exactly one elementary charge is additionally induced on the island. SETs can therefore be used as very sensitive electrometers for measuring local electric fields originating from localized charges, or local variations in the electrochemical potential of an electronic system acting as the gate for the island [207, 399].

Scanning SETs have been developed at the Bell Labs in the late 1990s [389, 400–403]. In this approach the SET is fabricated at the end of a glass fiber that has a shallow conical taper terminating at the tip in a flat nearly circular area of about 100 nm diameter [389]. The preparation of such fibers is the same as that for scanning near-field optical microscopy and has been described in Ref. [404]. The three metallic electrodes, i.e., source, drain and island, are fabricated from $10 - 20$ nm thick aluminum. Source and drain electrodes are evaporated onto the fiber from opposite directions onto the side of the fiber. Natural shadowing keeps these electrodes separate. An *in situ* exposure to oxygen creates the aluminum oxide tunnel barriers of a few nanometer thickness. The island is evaporated in a last step from a direction along the fiber axis. Owing to the conical taper of the fiber the aluminum film forms only at the end face of the tip. At low temperatures these aluminum SETs become superconducting. A magnetic field can be applied to suppress the superconductivity.

In a very simplified model the SET current may be modeled as [389]

$$I_{\mathrm{SET}} = a \sin \frac{2\pi Q_{\mathrm{i}}(x,y)}{e},$$

where $Q_{\mathrm{i}}(x,y)$ is the total charge induced on the island and a is the amplitude of the current oscillation at a given source-drain bias. When the SET sensor is scanned at constant distance to the surface of a semiconductor nanostructure in x- and y-direction, the charge induced on the island can be expressed as

$$Q_{\mathrm{i}}(x,y) = Q_{\mathrm{i}}^{(0)}(x,y) + \sum_n C_{n\mathrm{i}}(x,y)(\phi_n - \phi_{\mathrm{i}}),$$

where the $C_{n\mathrm{i}}(x,y)$ are the capacitance coefficients of the SET island to other metallic electrodes in the system (including, e.g., the electronic nanostructure to be investigated, if it is connected to external leads) and $Q_0(x,y)$ is the induced charge on the island due to the presence of fixed charges in the system, e.g., dopants or other charged impurities. The electrostatic potentials ϕ_n are related to the applied voltages U_n and the chemical potentials of the electrodes μ_n via

$$-eU_n = \mu_n(x, y) - e\phi_n(x, y).$$

Inserting these relations into the equation for the SET current we obtain

$$I_{\mathrm{SET}} = a \sin \frac{2\pi \left[Q_{\mathrm{i}}^{(0)}(x, y) + \sum_n C_{n\mathrm{i}}(x, y)(U_{n\mathrm{i}} + \mu_{n\mathrm{i}}(x, y)/e)\right]}{e}, \qquad (14.1)$$

with $U_{n\mathrm{i}} = U_n - U_{\mathrm{i}}$ being the voltage between electrode n and the island and $\mu_{n\mathrm{i}}(x, y) = \mu_n(x, y) - \mu_{\mathrm{i}}$ being the difference in chemical potentials between electrode n and the island (sometimes referred to as the work function difference between the two materials). From this formula it can be seen that the SET current senses local electric charges [via $Q_{\mathrm{i}}^{(0)}(x, y)$], the local chemical potential $\mu_n(x, y)$, and the local capacitance coefficients $C_{n\mathrm{i}}(x, y)$. It is clear that in a well-defined experiment care must be taken that only one of these quantities depends significantly on position.

Because this sensor is not sensitive to the tip–sample force or force gradient, the height (z-) dependence of the capacitances $C_{n\mathrm{i}}$ or the induced charges $Q_{\mathrm{i}}^{(0)}$ can be used for approaching the sensor to the surface [389].

14.4.3 Piezoelectric tuning fork sensors

Piezoelectric tuning forks are an alternative to the piezoresistive cantilevers. They were initially developed for use as very small and stable oscillators in watches. Owing to their use in industry, they are cheap and easily available. The forks are fabricated from wafers of α-quartz with the optical axis (c-axis) approximately normal to the wafer plane. Most of them have a (lowest) resonance frequency $f_0 = 2^{15}$ Hz and quality factors under vacuum between $Q = 20'000$ and $100'000$. In scanning force microscopy and related techniques tuning fork sensors offer the possibility of non-optical detection of the tip oscillation in the dynamic mode via the piezoelectric effect [375–382, 405–407]. It was demonstrated that atomic resolution is possible with these unconventional and very stiff sensors [408]. The fundamental limits to force detection with quartz tuning forks were discussed by Grober and coworkers in Ref. [409]. Owing to the high oscillator quality of tuning forks, the power loss on resonance can be far less than 0.1 µW, which makes it ideal for the use in low cooling power cryostats.

Fabrication of tuning fork sensors with a metallic tip

For experiments on semiconductor nanostructures, a conductive tip is required for coupling capacitively to the buried electron gas. One of our sensors is depicted in Fig. 14.4. It is the same type of tuning fork sensor previously discussed and calibrated in Refs. [407, 410]. The commercially available tuning fork[6] is first prepared by carefully removing its protecting lid on the lathe. The fork is then soft-soldered onto a small circular piece of printed circuit board (PCB) at an angle of about 10°. A 15 µm

[6] NTF3238 from SaRonix, 141 Jefferson Drive, Menlo Park, CA 94025, USA.

Fig. 14.4. Tuning fork sensor used in our cryo-SFM. The casing of a commercial tuning fork (TF) is soldered on a small circular piece of printed circuit board (PCB). The PtIr-wire is glued to the end of one prong of the fork. One end of the wire is connected electrically to the copper post (Cu-post), the other end serves as the tip. The PCB can in addition host small electronic components, like the resistor that can be seen in the foreground.

PtIr wire is glued with conducting epoxy[7] to the end face of one tuning fork prong parallel to the direction of its vibrational bending motion and to the copper post, which serves as the electrical contact to the tip. The free end of the wire is then etched electrochemically resulting in a sharp tip that serves as a probe for the sample surface. Tip radii of about 30 nm have been confirmed using scanning electron microscopy and transmission electron microscopy techniques. The length of the tip measured from the edge of the tuning fork should not be longer than about 300 μm because the thin wire becomes unstable otherwise due to bending vibrations occurring during scans.

The tip prepared on the tuning fork can be used as a tunneling tip (STM mode) or as the tip in dynamic SFM operation mode without any modifications on the scanning head. This gives simultaneous access to two complementary imaging modes at low temperatures without requiring a tip exchange.

Tuning fork sensors: a realistic model

After preparation, the symmetric geometry of the tuning fork is modified, and we have to care about the effects on the oscillation properties. Tuning fork sensors can be classified according to the strength of the mixing between antisymmetric and symmetric mechanical oscillation modes. In the extreme case, one of the tuning fork prongs is firmly attached to a support [406] and very strong mode mixing occurs. In this section we present evidence that our sensors are in the weak mode-mixing limit and that a simple harmonic oscillator model is therefore appropriate.

Figure 14.5(a) shows the current through the tuning fork near resonance measured at a temperature of 4.2 K with an excitation voltage of $U_0 = 10\,\mu$Vrms applied to the tuning fork contacts. The top graph is the magnitude of the current. The experimental details of the admittance measurements will be discussed in section 14.5. In the following, we start with a mechanical model of the tuning fork sensor aiming at

[7] EPO-TEK H20E by Polyscience AG, Riedstrasse 13, CH-6330 Cham, Switzerland.

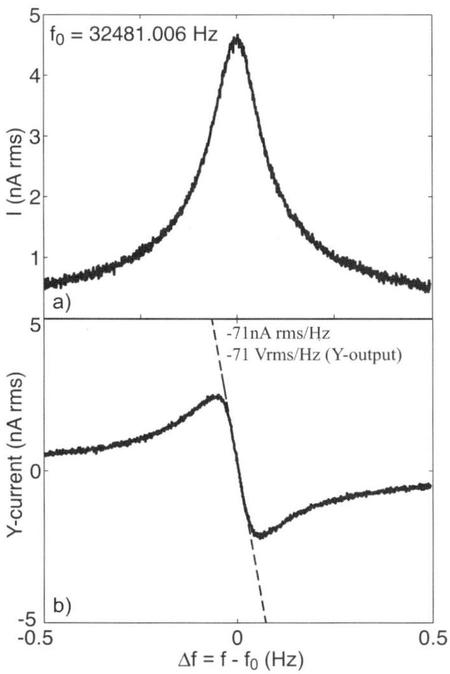

Fig. 14.5. a) Current magnitude near the tuning fork resonance. The frequency at which the current is maximum is the resonance frequency f_0. The width of the resonance is related to the quality factor Q of the tuning fork. b) Y-component of the current (phase signal) near resonance. The linear relation between frequency shift $\Delta f = f - f_0$ and this signal is indicated by the dashed line. The slope of this line determines the sensitivity of the phase signal to shifts in resonance frequency.

the detailed understanding of such resonance curves. In particular, we will focus on the symmetry-breaking effect of the tip preparation.

We describe the mechanical oscillator with a coupled harmonic oscillator model driven by the voltage U_0 via the piezoelectric coupling constant α_p:

$$m_1\ddot{x}_1 + \gamma_1\dot{x}_1 + k_1x_1 + k_\mathrm{c}x_2 + \gamma_\mathrm{c}\dot{x}_2 = -\alpha_\mathrm{p}U_0, \tag{14.2}$$

$$m_2\ddot{x}_2 + \gamma_2\dot{x}_2 + k_2x_2 + k_\mathrm{c}x_1 + \gamma_\mathrm{c}\dot{x}_1 = +\alpha_\mathrm{p}U_0 + F_\mathrm{ts}(x_2). \tag{14.3}$$

The deflection of the two prongs is described by the coordinates x_i. The m_i are the effective masses of the two prongs, the k_i and γ_i are their spring and damping constants, respectively. The tip is attached to prong 2 and interacts with the sample surface via the interaction force $F_\mathrm{ts}(x_2)$. Coupling between the prongs is described by coupling constants k_c and γ_c.

The eigenfrequencies of this oscillator [determined with $F_\mathrm{ts}(x_2) = 0$] are given by

$$\Omega_{1,2}^2 = \omega_0^2 \pm \omega_\mathrm{c}^2\sqrt{1+\kappa^2},$$

where $\omega_0^2 = (k_1/m_1 + k_2/m_2)/2$, $\omega_\mathrm{c}^2 = k_\mathrm{c}/\sqrt{m_1m_2}$, and $\kappa = (k_1/m_1 - k_2/m_2)/(2k_\mathrm{c}/\sqrt{m_1m_2})$. In the case of $\kappa = 0$, in particular when the two prongs are identical, the two eigenmodes are the antisymmetric (the x_i oscillate with zero phase difference) and the symmetric (the x_i oscillate with a phase difference of π) mode. The metallic tuning fork contacts are arranged such that only the symmetric mode is excited by a driving voltage. For $\kappa \ll 1$ the two modes do not mix significantly and the symmetries of the modes are not strongly affected, while for $\kappa \gg 1$ each prong has its own resonance frequency where the other is hardly excited.

In the following we show that our tuning fork sensors are in the regime of negligible mode mixing, because the introduced asymmetry is small, i.e., $\kappa \ll 1$. We can rewrite the expression for κ introducing $m_1 = m$, $m_2 = m + \Delta m$, $k_1 = k$, $k_2 = k + \Delta k$. Expanding up to first order in $\Delta m/m$, $\Delta k/k$, and k_c/k we obtain

$$\kappa = \frac{\Delta m/m - \Delta k/k}{2k_c/k}.$$

It can be seen that the additional mass on one prong can be compensated by increasing the spring constant, a fact exploited in Ref. [411].

We estimate the additional spring constant Δk of a thin wire attached to one prong. It is given by [412] $\Delta k = 3\pi Y r^4/(4L^3)$, where L denotes the length of the wire, Y is Young's modulus of the material, and r is its radius. Typical wires have $r = 7.5\,\mu\text{m}$, $L = 0.5\,\text{mm}$, and $Y = 200 - 400\,\text{GPa}$ leading to $\Delta k = 12 - 24\,\text{N/m}$. With $k = 14'000\,\text{N/m}$, this results in $\Delta k/k \approx 10^{-3}$.

An estimate of the additional mass added to one arm of the tuning fork takes essentially the glue into account. The idealized drop of glue has the shape of a half-sphere with radius r and volume $V = (2\pi/3)r^3$. Using the mass density of the H20E-epoxy $\rho = 2600\,\text{kg/m}^3$ and a radius of $100\,\mu\text{m}$ we obtain a mass of $5.44\,\mu\text{g}$. This weight has to be related to the mass of a single tuning fork arm, which is 1.1 mg. This leads to the ratio $\Delta m/m = 5 \times 10^{-3}$.

The ratio k_c/k can be estimated from finite element simulations of tuning fork sensors to be of the order of several percent.

From these estimates it seems to be reasonable that the employed sensors are in the limit of small mode mixing with κ being typically much smaller than 0.1. An additional observation strongly supports this point of view: only very rarely is a second resonance observed in the admittance. Because the measured current detects only the symmetric component of the oscillation, strong mode mixing would make the second resonance visible.

Motivated by these estimates we approximate in the equations of motions (14.2) and (14.3) $m_1 = m_2 = m$, $k_1 = k_2 = k$, and $\gamma_1 = \gamma_2 = \gamma$. We further neglect the coupling between the center of mass and the relative motion of the prongs, which can be estimated to be small, and obtain the single harmonic oscillator approximation for our tuning fork sensors,

$$\ddot{x} + \frac{\omega_0}{Q}\dot{x} + \omega_0^2 x = \frac{\alpha_p}{m}U_0 + \frac{1}{m}F_{ts}(x), \tag{14.4}$$

where we have introduced $x \equiv x_2$, $(k - k_c)/m = \omega_0^2$, and $(\gamma - \gamma_c)/m = \omega_0/Q$ with Q being the quality factor of the oscillator.

The mechanical oscillation of tuning fork sensors is measured via the piezoelectric effect of the quartz crystal. The induced piezoelectric charge on one tuning fork electrode is given by $q_p = 2\alpha_p x$ and the corresponding piezoelectric current is $I_\mathrm{p} = 2\alpha_\mathrm{p}\dot{x}$. In addition, a current of magnitude $I_\mathrm{c} = C_0\dot{U}_0$ flows through the capacitance C_0 between the tuning fork contacts. The total current as a function of frequency $f = \omega/2\pi$ is therefore given by $I(\omega) = \mathrm{i}\omega C_0 U_0 + 2\mathrm{i}\alpha_\mathrm{p}\omega x(\omega)$. If we neglect the tip–sample interaction force, we can determine $x(\omega)$ from the harmonic oscillator model [eq. (14.4)] and thus $I(\omega)$ is found to be

$$I(\omega) = \mathrm{i}\omega U_0 \left(C_0 + \frac{2\alpha_\mathrm{p}^2/m}{\omega_0^2 - \omega^2 + \mathrm{i}\omega\omega_0/Q} \right). \tag{14.5}$$

Resonance curves like the one shown in Fig. 14.5a can be excellently fitted with this equation.

Near resonance ($\omega \approx \omega_0$) the current is dominated by the piezoelectric contribution and can be expanded to be

$$I(\omega) = \frac{2\alpha_\mathrm{p}^2 Q U_0}{m\omega_0} \left[1 - \mathrm{i}\frac{2Q}{\omega_0}(\omega - \omega_0) \right] + \mathcal{O}\left[(\omega - \omega_0)^2 \right]. \tag{14.6}$$

The imaginary component of the current, which is shifted in the time domain by 90° with respect to the driving voltage, is depicted in Fig. 14.5b. It is proportional to the deviation of the resonance frequency from the excitation frequency. This fact is utilized for the SFM feedback.

Summarizing the discussion of modeling the tuning fork sensors, we have found two reasons why the single harmonic oscillator description [eq. (14.4)] is appropriate for our sensors: first, the added mass (the glue) is small compared to the mass of a single tuning fork arm. Second, the relative mechanical coupling of the two prongs is strong compared to the influence of the added mass and the added spring constant. In this respect these sensors are completely equivalent to other sensors [406] with one prong firmly attached to the support, which act essentially as an extremely stiff piezoelectric cantilever. In other respects there are important differences: we find quality factors Q of up to 250'000 under the UHV conditions occurring in our evacuated sample space at a temperature of 300 mK. These values are one to two orders of magnitude larger than those reported for the other type of tuning fork sensor [406]. The implications of this for the electronic setup to be used will be discussed in section 14.5. An expression has been found for the total current through the tuning fork [eq. (14.5)]. Near the mechanical resonance the current reflects this resonant behavior and the imaginary component of the current is a linear indicator for the deviation of the driving frequency from resonance [eq. (14.6)].

Characterization of tuning fork sensors and amplitude calibration

Equation 14.5 contains a number of parameters characterizing the tuning fork admittance. These parameters are the effective mass m and the effective spring constant $k - k_\mathrm{c}$ entering the resonance frequency $\omega_0 = 2\pi f_0 = \sqrt{(k - k_\mathrm{c})/m}$, the effective damping constant $\gamma - \gamma_\mathrm{c}$ entering the quality factor of the oscillator, and the piezoelectric coupling constant α_p. In this section we show, how these quantities can be measured [410].

In our system the tuning fork is driven by an AC voltage from a Yokogawa function generator FG300. We measure the tuning fork current by using an I-U converter with a current-to-voltage conversion ratio of $K = 10^6$ V/A at 32 kHz. A guard driver neutralizes the huge capacitance $C_K = 1.8$ nF of the long coax cable connecting the tuning fork in the cryostat to the outside and thereby increases the bandwidth of the I-U converter and reduces the output noise.

Figure 14.6b shows a typical resonance in the admittance of a tuning fork measured at room temperature at a pressure of 6×10^{-7} mbar. The admittance exhibits an asymmetric resonance at 32 768 Hz and a sharp minimum about 30 Hz above this

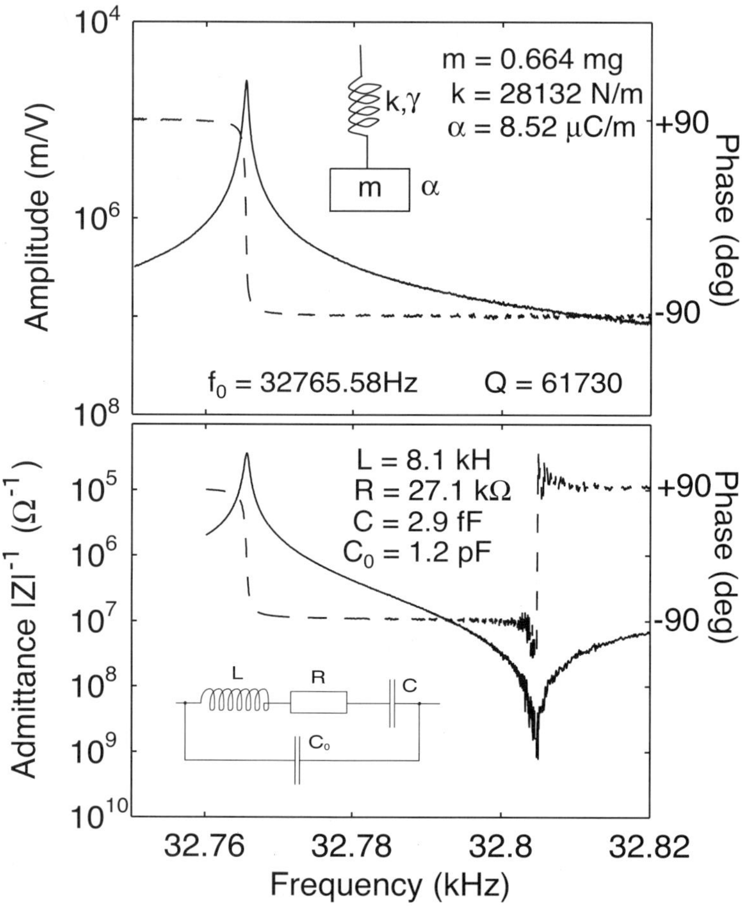

Fig. 14.6. (a) Mechanical resonance measured at room temperature at a pressure of 6×10^{-7} mbar with an optical interferometer. Inset: Image of the tuning fork. (b) Electrical tuning fork resonance measured simultaneously. Inset: Equivalent circuit for the piezoelectric quartz tuning fork resonator. The solid and dashed lines are the respective amplitude and phase.

resonance. The behavior of the admittance is described by eq. (14.5). Equivalently it can be modeled with the equivalent circuit shown in the inset of Fig. 14.6b [410].

Using the mechanical model [eq. (14.4) with $F_{\text{ts}}(x) = 0$] one can relate L, R, and C to the effective mass of one tuning fork prong m, the damping constant $\gamma - \gamma_{\text{c}}$, the spring constant $k - k_{\text{c}}$, and the driving force $\alpha_{\text{p}} U_0$ via $L = m/(2\alpha_{\text{p}}^2)$, $C = 2\alpha_{\text{p}}^2/(k - k_{\text{c}})$, and $R = m(\gamma - \gamma_{\text{c}})/(2\alpha_{\text{p}}^2)$. The capacitance C_0 is mainly determined by the geometrical arrangement of the contacts on the crystal, the dielectric properties of the quartz and the cable capacitances. The fit to the measured admittance in Fig. 14.6 (which could not be distinguished in the plot from the measured curve) leads to $C_0 = 1.2129\,\text{pF}$, $C = 2.9\,\text{fF}$, $L = 8.1 \times 10^3\,\text{H}$, $R = 27.1\,\text{k}\Omega$, $f_0 = 32765.58\,\text{Hz}$, and $Q = 61730$.

In addition to the electrical resonance we measured the mechanical resonance amplitude x of one of the tuning fork arms [see Fig. 14.6a] utilizing an interferometer setup as it is usually used for optical cantilever deflection detection in a scanning force microscope[361]. From a combination of both measurements [Figs. 14.6a and b] and using the relation $I_{\text{p}} = 4\pi f \alpha_{\text{p}} x$ [413], we determined the effective mass $m = 0.332\,\text{mg}$, the quality factor $Q = 61734$, the spring constant $k - k_{\text{c}} = 14066.4\,\text{N/m}$, and the piezoelectric coupling constant $\alpha_{\text{p}} = 4.26\,\mu\text{C/m}$. The effective mass calculated from the density of quartz and the dimensions of a tuning fork arm turns out to be 0.36 mg, in good agreement with our measured value. A linear relation between the driving voltage and the oscillation amplitude was found in the interferometer measurement down to amplitudes of 1 nm as well as in large amplitude measurements performed under an optical microscope up to amplitudes of about 100 mm.

With the experimentally determined parameters of the tuning fork at hand, one can now set up a well-controlled scanning force operation with the tuning fork sensors. The knowledge of the piezoelectric constant α_{p} allows one to translate measured currents on resonance directly into oscillation amplitudes of the tip. We have verified with low-temperature force-distance measurements with varying oscillation amplitude that α_{p} has at 300 mK within a few percent the same value that we determined at room temperature.

Force calibration

The idea behind the dynamic mode SFM operation of the tuning fork is the same as for normal cantilevers. The elastic interaction of the tip with the sample surface will shift the resonance frequency via the presence of force gradients. Inelastic tip–sample interactions will alter the Q value of the oscillator. In this section we will show how sensitivity of the sensor to force gradients can be calibrated experimentally [414].

In order to translate measured frequency shifts into a quantity describing the tip–sample interaction, we use the force gradient approximation as the basic assumption, i.e.,

$$\Delta f = \eta \left(\frac{\partial F_z}{\partial z} \right), \tag{14.7}$$

where $\partial F_z/\partial z$ is the gradient of the tip–sample interaction force and η is the proportionality constant, which is to be determined by the calibration procedure. This approximation is valid for small tip-oscillation amplitudes. It is the limiting case of more general treatments discussed in the literature [415]. The calibration constant η is independent of the details of the tip shape and depends only on the mechanical properties of the tuning fork. In the following we will compare three different methods of calibration and report the determined values of the calibration constant η.

Harmonic oscillator model

If we apply the harmonic oscillator model [eq. (14.4)] for the resonance of our tuning fork, we can calibrate the sensor simply by determining the effective spring constant $k - k_c$ from the electrical resonance curve. We obtain the relation

$$\frac{\Delta f}{f_0} = \frac{1}{4(k - k_c)} \left(\frac{\partial F_z}{\partial z} \right), \tag{14.8}$$

which differs by a factor of 1/2 from the result obtained for conventional cantilevers. This factor reflects the fact that only one prong of the fork senses the interaction but both prongs are oscillating. This method gives $\eta = f_0/(4k)$. For the sensor used for the studies described here we measured the resonance frequency $f_0 = 32596.793\,\text{Hz}$ and the stiffness $k - k_c = 14.06\,\text{nN/pm}$ and therefore obtained a value of $\eta = 0.6\,\text{Hz/(N/m)}$. Because the resonance frequency of our sensors can be determined with an accuracy better than 1 mHz, the main source of error in this method may be found in the determination of $k - k_c$. It is calculated from $k - k_c = 4\pi^2 f_0^2 m$, thus any error in the assumption of an unaltered mass m at low temperatures will cause an error in η.

Calibration with the electrostatic force

The two other calibration methods make use of the attractive electrostatic force acting between the tip and the sample if a constant tip–sample voltage U_{ts} is applied. It is given by the expression $F(z) = 1/2(\partial C/\partial z)(U_{ts} + \Delta(\mu_{ch}/e))^2$, where $C(z)$ is the tip–sample capacitance and $\Delta\mu_{ch}$ is the difference of the chemical potentials of the two materials, which is usually taken to be equal to the work function difference between the two materials. Using eq. (14.7) we obtain for the frequency shift

$$\Delta f = \frac{\eta}{2} \left(\frac{\partial^2 C}{\partial z^2} \right) (U_{ts} + \Delta(\mu_{ch}/e))^2 \,. \tag{14.9}$$

The measurement of $\Delta f(U_{ts})$ shown in Fig. 14.7 was performed at a large tip–sample separation (of the order of 100 nm) and with a nominal tip-oscillation amplitude of 20 nm. For this tip–sample separation the electrostatic force is by far the dominant interaction. The data can be fitted extremely well by a parabola and therefore allows us to determine η. From the measurement in Fig. 14.7 the calibration constant η could be determined using eq. (14.9) if the second derivative of the capacitance were known at the same distance. For the determination of this quantity we have used two distinct techniques, which will be described in the following.

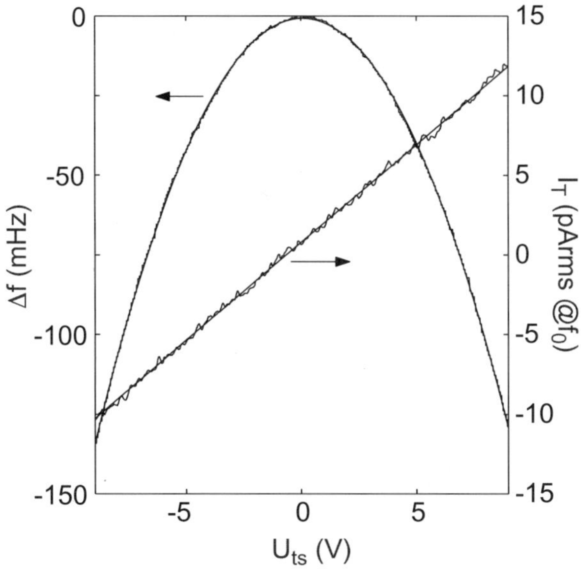

Fig. 14.7. $\Delta f(U_{\mathrm{ts}})$ measured at 4.5 K (left axis) and Kelvin current (right axis).

1. We have measured the capacitance depicted in Fig. 14.8 by applying an AC tip–sample voltage of 100 mV(rms) at a frequency of 95 kHz. The measurement was fitted with a smooth analytical curve in order to allow a better determination of the second derivative. Inserting the result in eq. (14.9), we obtain $\eta \approx 1$ Hz/(N/m). The accuracy of this method is mainly determined by the limited accuracy of the determination of the very small tip–sample capacitance, which results in an uncertainty of the η value of more than a factor of 2 for the measurement shown.
2. The second approach for determining the second derivative of the capacitance employs the Kelvin method in which the gradient of the tip–sample capacitance is directly measured by oscillating the tip over the sample surface at a constant bias voltage and recording the current induced on the tip by the motion. This current is given by

$$I_{\mathrm{T}} = 2\pi \mathrm{i} f_0 a_{\mathrm{T}} \left(\frac{\partial C}{\partial z} \right) (U_{\mathrm{ts}} + \Delta\mu_{\mathrm{ch}}/e).$$

   Fig. 14.7 shows an example of an curve measured at the resonance frequency and a nominal tip-oscillation amplitude for a large tip–sample distance (of the order of 100 nm). A linear fit to the curve allows the determination of $\partial C/\partial z$. The second derivative can be obtained by repeating this measurement as a function of z. Inserting the result into eq. (14.9) leads us to the calibration $\eta = 0.8$ Hz/(N/m). For this calibration method the tip-oscillation amplitude a_{T} has to be known and any error in this number will lead to an error in η.

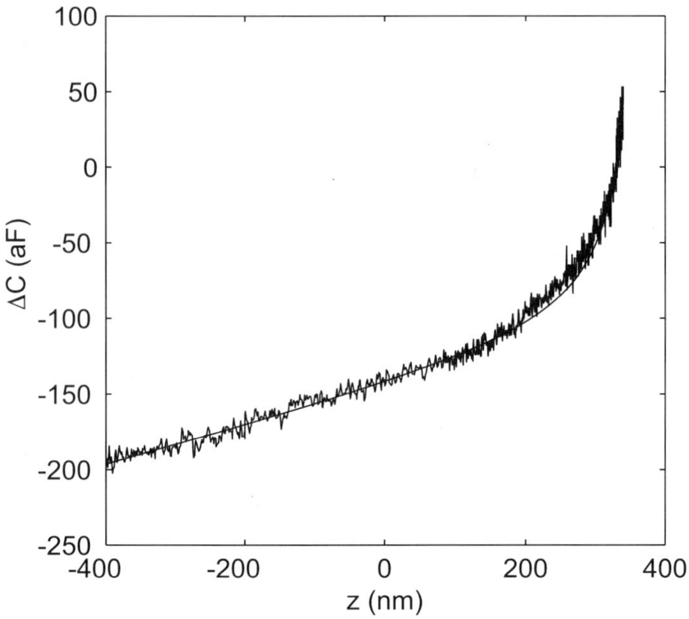

Fig. 14.8. Change of tip–sample capacitance, $\Delta C(z)$, as a function of the tip–sample separation z measured at 1.7 K with a HOPG sample.

14.5 Electronics for a high-Q tuning fork sensor

In this section we describe our nested SFM feedback system comprising the tuning fork sensor, a phase-locked loop and a commercial digital z-feedback.[8] This setup is rather versatile and can, in principle, be used in any kind of non-contact scanning force microscopy setup.

14.5.1 Tuning fork admittance and frequency demodulation

It was shown in Ref. [384] that the bandwidth of a sensor that can be modeled with a harmonic oscillator is inversely proportional to Q. This limits the bandwidth for the detection of frequency shifts via the phase signal to $f_0/(2Q)$, where f_0 is the resonance frequency of the sensor. Because our tuning fork sensors have $f_0 \approx 32\,\mathrm{kHz}$ and Q values up to 200'000, we end up with a bandwidth below 1 Hz. It was shown by Albrecht and coworkers in Ref. [384] that the force gradient can be measured with

[8] TOPS3-system by Oxford Instruments

larger bandwidth at the expense of increased noise as compared to pure phase detection (also called slope detection) if the cantilever is driven by a self-exciting positive feedback loop with automatic gain control. A different technique for increasing the bandwidth was introduced in scanning force microscopy by Dürig and coworkers in Ref. [416]. They used a phase-locked loop for this purpose. Phase-locked loops are one among many ways of demodulating frequency-modulated signals and are an established method for frequency measurement [417–421]. Employed in scanning force microscopy they allow the simultaneous measurement of conservative force gradients leading to frequency shifts and dissipative effects causing amplitude damping. Digital versions of phase-locked loops have been developed by Edwards [376] and later by Loppacher and coworkers [422, 423]. The electronic setup used in our system for controlling the tuning fork sensors is very similar to the one developed by Edwards in Ref. [376].

However, a topic that has — to our knowledge — been insufficiently discussed in the literature is the question, why it is difficult to scan with high-Q cantilevers and how the interplay between a phase-locked loop and the z-feedback influences the performance of a microscope during scanning. Dürig and coworkers did present a theoretical description of complete SFM-feedback systems using linear control theory [424], but analyzed their phase-locked loop system only in terms of its noise limits. Frequently, the paper of Albrecht and coworkers is interpreted in the sense that with high-Q cantilevers fast scanning with pure phase control (slope detection) is not possible owing to the bandwidth limitation given by the time constant $\tau = 2Q/\omega_0$. However, there is a second effect of high Q shown in their paper, namely, the transient beat term, leading to restrictions of the dynamic range of slope detection. Although there is experimental evidence that a frequency tracking feedback does indeed allow faster scanning with high quality (see, e.g., [376]), it is not apparent whether the bandwidth limitation or the beat term are responsible for the limitations of slope detection. In particular, it has not been discussed in the literature how the response of high-Q sensors performs within a complete feedback system including a PLL.

In order to clarify these issues, we performed detailed experimental performance tests with our system and analyzed the results within linear control theory. Special attention is paid to the propagation of noise in the nested phase-locked loop and z-control feedbacks. We show how we determine the optimum settings for feedback parameters. They are found to be independent of the tip shape and the details of the tip sample interaction. The advantages of using a feedback with phase-locked loop for the very-high-Q tuning fork sensors will be discussed and a comparison is made with the phase control mode (slope detection).

Admittance measurement

The inner dark gray region in Figure 14.9 shows the setup for the admittance measurement of the tuning fork as part of the whole nested feedback scheme. The tuning fork (TF) is driven by the voltage controlled oscillator (VCO) with typical excitation amplitudes between $10\,\mu\text{V}$ and $1\,\text{mV}$ depending on the Q value of the tuning fork. The tuning fork current is converted into a voltage with the I-U converter (IUC).

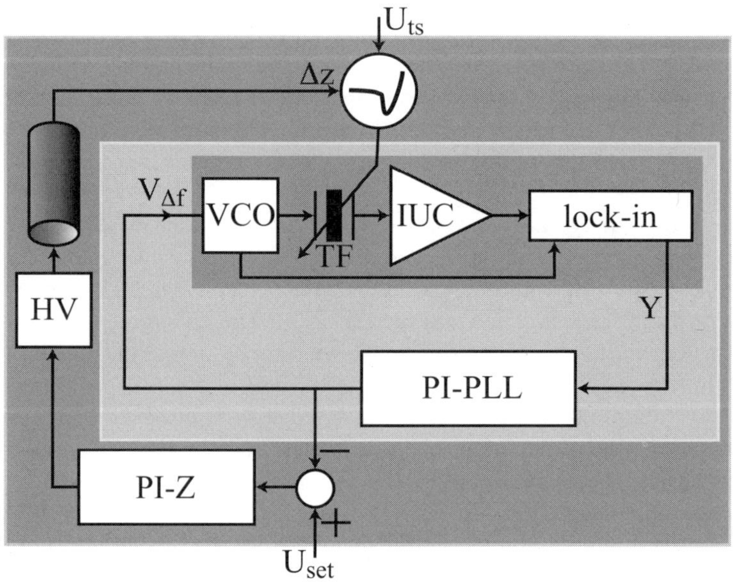

Fig. 14.9. Schematic drawing of the SFM electronic setup comprising the tip–sample interaction sensor (inner dark gray area), the phase-locked loop (light gray area), and the z-feedback (outer dark gray area). The tuning fork (TF) is driven by a voltage controlled oscillator (VCO). The current through the fork is converted into a voltage with a special I-U converter (for details see text) labeled IUC. This signal is then demodulated with the lock-in amplifier that delivers the amplitude (not shown for simplicity) and phase signal (Y). The phase-locked loop is made complete by feeding the phase back into the VCO via a P-I controller (PI-PLL). The output $U_{\Delta f}$ of this P-I controller is the input to the z-feedback loop. The latter determines the deviation of $U_{\Delta f}$ from a user-defined setpoint U_{set} and feeds the resulting error signal into another P-I controller (PI-Z). The z-position of the tip is then set via the high-voltage amplifier (HV) and the z-piezo. The tip–sample separation Δz acts on the tuning fork resonance frequency via the tip–sample interaction (circle with stylized force-distance curve).

The AC output voltage of the IUC is demodulated by the lock-in amplifier with sensitivity S_{LI}, which determines the in-phase (X) and 90° (Y) current components corresponding to the real and imaginary components in eq. (14.6). The steady-state output voltages up to first order in the frequency shift are given by

$$X_{\mathrm{res}} = \frac{10\,\mathrm{V}}{S_{\mathrm{LI}}} K \frac{\alpha_{\mathrm{p}}^2 Q U_0}{\pi m f_0},$$
$$Y_{\mathrm{res}} = -X_{\mathrm{res}} \frac{2Q}{f_0} \Delta f,$$

where we have introduced $f_0 = \omega_0/(2\pi)$ and the frequency shift $\Delta f = f - f_0$. The Y component of the signal is a linear indicator of the deviation of the driving frequency from the resonance frequency which can be utilized for the SFM feedback.

Noise in the admittance measurement

It turns out that the output noise of the I-U converter, which was found in this setup to be $n_{\mathrm{IUC}} = 800\,\mathrm{nV}/\sqrt{\mathrm{Hz}}$ at 32 kHz, dominates the noise in the whole SFM feedback. The spectral noise density on the output signals of the lock-in, $n_{\mathrm{Y}}(\omega)$, is given by the amplified spectral noise density of the I-U converter, i.e., $n_{\mathrm{Y}}(\omega) = 10\,\mathrm{V}/S_{\mathrm{LI}} n_{\mathrm{IUC}}(\omega)$. The integrated output noise $\delta Y_{\mathrm{res}} = \int \mathrm{d}f\, n_{\mathrm{Y}}^2(2\pi f)$ depends on the lock-in time constant τ and on the order n of its low-pass filter via

$$\delta Y_{\mathrm{res}} = c_n \times \frac{10\,\mathrm{V}}{S_{\mathrm{LI}}} \times \sqrt{\frac{n_{\mathrm{IUC}}^2}{2\pi\tau}}. \tag{14.10}$$

The constants c_n are of order unity and can be calculated and measured ($c_1 = 1.25$, $c_2 = 0.89$, $c_3 = 0.77$, $c_4 = 0.70$). In our setup we use a lock-in time constant of $\tau = 100\,\mu\mathrm{s}$ corresponding to a bandwidth of 1.6 kHz and a fourth-order low-pass filter with slope 24 dB/octave. With $S_{\mathrm{LI}} = 10\,\mathrm{mV}$ this leads to $n_{\mathrm{Y}} = 800\,\mu\mathrm{V}/\sqrt{\mathrm{Hz}}$ and to typical values of $\delta Y_{\mathrm{res}} = 22\,\mathrm{mV}$ corresponding to a frequency noise of 0.3 mHz. Relative frequency shifts $\Delta f/f_0$ of the order of 10^{-8} can therefore be detected.

Demodulation of the modulated tuning fork signal

For the SFM operation it is crucial to know the bandwidth of the tip–sample interaction sensor, i.e., the time it takes to respond to a given step in Δz or U_{ts}. The step response of our sensor can be determined similar to Albrecht [384] (see Appendix D). If U_{ts} (or Δz) is modulated with frequency ω and the resulting frequency shift $\Delta f(\omega) = \alpha U_{\mathrm{ts}} \ll f_{\mathrm{G}} = f_0/(2Q)$, one finds

$$Y_{\mathrm{res}}(\omega) = k(\omega)\Delta f(\omega) = -X_{\mathrm{res}}\frac{2Q}{f_0}\Delta f(\omega)\frac{1}{1 + \mathrm{i}\omega/(2\pi f_{\mathrm{G}})}, \tag{14.11}$$

i.e., a first-order low-pass behavior with bandwidth f_{G}. In the step response, relaxing the condition of small frequency step Δf leads away from pure low-pass characteristics to an oscillatory component in the time domain with frequency Δf (see Appendix D). This fact may be of crucial importance for the stability of a feedback system. For example, too large scan speeds in phase-control mode may lead to an unstable feedback. This effect is more serious, the larger the Q value of the sensor is. We suggest that this is the reason why fast scanning is not generally possible with high-Q cantilevers unless surfaces are smooth enough that the frequency shifts encountered are sufficiently small.

A measurement of the low-pass behavior is shown in Fig. 14.10. It was performed at a temperature of 4.2 K with the tip at a distance of 80 nm from the surface of a GaAs/AlGaAs heterostructure in which a two-dimensional electron gas (2DEG) resides 34 nm below the surface. The tuning fork was driven on resonance with an AC voltage of $U_0 = 10\,\mu\mathrm{V}$ applied to the tuning fork contacts. The 2DEG was employed as the metallic counter electrode of the tip. A DC tip–sample voltage of $U_{\mathrm{ts}} = -9\,\mathrm{V}$ was applied to the 2DEG while keeping the tip grounded. Following the idea of eq. (14.19) a low-frequency AC voltage $\Delta U_{\mathrm{ts}} = 300\,\mathrm{mV}$ was added in

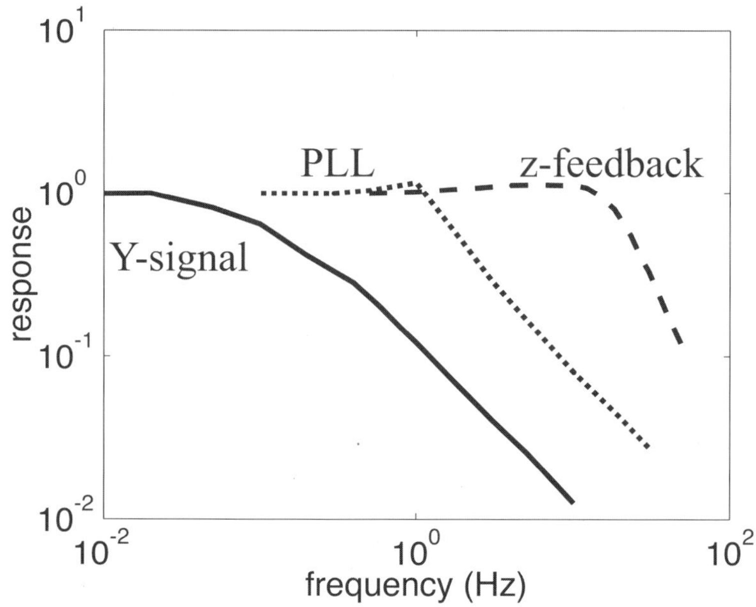

Fig. 14.10. Various response functions of our feedback system. The curve labeled "Y-signal" is the response function of the bare tuning fork sensor demodulated by the lock-in amplifier. The curve labeled "PLL" is the response function of the phase-locked loop. The curve labeled "z-feedback" is the response function of the z-feedback including the phase-locked loop. Details are explained in the text.

order to modulate the resonance frequency of the tuning fork. Fitting the low-pass behavior (not shown) results in the characteristic frequency of the tuning fork $f_{\mathrm{G}} = 0.13\,\mathrm{Hz}$ corresponding to $Q = 116'000$. In other experiments at low temperatures we achieved Q values of up to 250'000.

In order to summarize the results of this section we want to emphasize the following points: the Y output of the lock-in amplifier is used as the frequency detector in our setup. For small changes in resonance frequency $\Delta f \ll f_{\mathrm{G}} = f_0/(2Q)$ the response shows low-pass behavior. Larger changes in resonance frequency lead to an oscillatory step response of the phase signal. High-Q sensors have a stronger tendency to an oscillatory response due to their smaller f_{G} as compared to lower Q sensors.

14.5.2 Frequency detection with a phase-locked loop

Owing to the high Q values of our sensors, the response to changes in resonance frequency, in particular during the approach of the tip to the surface, is too slow to allow reasonable approach speeds. This is one reason why it is advantageous to employ a phase-locked loop in our setup. As we will show below, the phase-locked loop (PLL) increases the bandwidth of the frequency-shift detection compared to that of pure phase (slope) detection (f_{G}) shown before at the expense of increased noise.

Linear response

The PLL setup is schematically shown in Fig. 14.9 within the light gray box. The output of the lock-in amplifier, i.e., the phase signal Y, is fed into a P-I controller (PI). Its output $U_{\Delta f}$ is used as the input of the VCO. In the following we describe the PLL by the linear equations

$$Y(\omega) = k(\omega)(f_{\mathrm{res}}(\omega) - f_{\mathrm{d}}(\omega)) + n_{\mathrm{Y}}(\omega), \tag{14.12}$$

$$U_{\Delta f}(\omega) = R(\omega)Y(\omega), \tag{14.13}$$

$$f_{\mathrm{d}}(\omega) - f_0 = \eta U_{\Delta f}(\omega), \tag{14.14}$$

where f_{res} is the resonance frequency of the tuning fork given by external parameters such as Δz and U_{ts}, f_{d} is the frequency output of the VCO driving the tuning fork, and f_0 is the center frequency of the VCO around which f_{d} can be modulated. The function $k(\omega)$ is the response [eq. (14.11)] of the interaction sensor shown in Fig. 14.9 as the inner dark gray region including the lock-in demodulator with its frequency response defined by the lock-in time constant and filter characteristics. It transfers the frequency shift $f_{\mathrm{res}}(\omega) - f_{\mathrm{d}}(\omega)$ into the lock-in output signal $Y(\omega)$ [eq. (14.12)]. The additional term $n_{\mathrm{Y}}(\omega)$ is for a later discussion of the PLL noise and will be assumed to be zero here. The function $R(\omega)$ is the response function of the P-I controller transferring the lock-in output $Y(\omega)$ into the input voltage of the VCO, $U_{\Delta f}(\omega)$ [eq. (14.13)]. The constant η is the response function of the VCO, which transfers the input voltage of the VCO into a frequency shift [eq. (14.14)]: it is assumed to be frequency independent. The total response of the PLL depends crucially on the characteristic frequency of the P-I controller, f_{PI}, its gain P, on the characteristic frequency of the tuning fork, f_{G}, and on the bandwidth and filter characteristics of the lock-in amplifiers. Parameters are optimized, if

$$f_{\mathrm{PI}} = f_{\mathrm{G}}. \tag{14.15}$$

If $f_{\mathrm{PI}} < f_{\mathrm{G}}$, the response of the PLL is unnecessarily retarded, while for $f_{\mathrm{PI}} > f_{\mathrm{G}}$ the response function develops an overshoot and tends to become unstable. If we regard $U_{\Delta f}$ as the output and f_{res} as the input of the PLL, we find the solution of the linear response equations

$$U_{\Delta f}(\omega) = PLL(\omega)\,(f_{\mathrm{res}}(\omega) - f_0) \tag{14.16}$$

with the PLL-response function

$$PLL(\omega) = \frac{1}{\eta}\frac{L(\omega)}{1 + L(\omega)}$$

and the open loop response defined as $L(\omega) := R(\omega)k(\omega)\eta$. The system is unstable, if the complex poles of $PLL(\omega)$, given by the solutions of $1 + L(\omega) = 0$, are located in the positive imaginary half plane. Stable feedback is only achieved if the gain P of the P-I controller is kept below a critical value P_c. For example, in our setup this value depends on the filter characteristics of the lock-in amplifier. Smaller time constants (larger bandwidths) or a lower order of the output filter lead to a larger P_c.

At very low frequencies $R(\omega)$ is huge and $L(\omega) \gg 1$. In this limit $PLL(\omega)$ is constant with the value $1/\eta$. At frequencies much larger than $f_{\rm PI} = f_{\rm G}$, the function $R(\omega)$ is constant but $k(\omega)$ becomes very small such that $L(\omega) \ll 1$ and the response function decays like $k(\omega)$. For frequencies below the bandwidth of the lock-in amplifier (typically 1.6 kHz in our experiments) $k(\omega)$ behaves like a first-order low-pass as discussed before. For higher frequencies the low pass at the lock-in output leads to a much stronger decay (typically fifth order).

We now discuss the behavior of $PLL(\omega)$ for the case $f_{\rm G} = f_{\rm PI}$. In this case it reduces to

$$PLL(\omega) = \frac{1}{\eta}\frac{1}{1+{\rm i}f/(Pa\eta f_{\rm G})} = \frac{1}{\eta}\frac{1}{1+{\rm i}f/f_{\rm PLL}},$$

where we have introduced $a = X_{\rm res}2Q/f_0$, i.e., to a low-pass behavior with the new characteristic frequency $f_{\rm PLL} = Pa\eta f_{\rm G}$. The product $a\eta$ acts like an additional P-factor and the effective P-factor is $P_{\rm eff} = Pa\eta$. The bandwidth of the PLL is a factor of $P_{\rm eff}$ larger than the bandwidth of the interaction sensor itself. This is shown in Fig. 14.10 as the curve labeled "PLL," which was measured with $\eta = 0.1$ Hz/V, $a = 71$ V/Hz, and $P = 2$. Higher bandwidths can be achieved with higher $P_{\rm eff}$; however, the stability condition $P_{\rm eff} < P_c$ and noise considerations constitute the upper experimental limits.

Noise considerations for the PLL

As discussed before, the dominating noise from the I-U converter, $n_{\rm IUC}(\omega)$, can be represented as an equivalent noise source $n_{\rm Y}(\omega)$ at the Y output of the lock-in amplifier. We now want to discuss how this noise transfers to the output of the phase-locked loop, i.e., to a noise component $n_{\rm PLL}(\omega)$ on the signal $U_{\Delta f}(\omega)$ (see Fig. 14.9). The linear response analysis starts from eqs. (14.12–14.14) taking the additional noise term $n_{\rm Y}(\omega)$ in eq. (14.12) into account. The system can again be solved for $U_{\Delta f}(\omega)$. The solution is the sum of one term describing the desired controlling function [see eq. (14.16)] and an additional noise term. The noise term resulting for $f_{\rm G} = f_{\rm PI}$ is

$$n_{\rm PLL}(\omega) = P\frac{f - {\rm i}f_{\rm G}}{f - {\rm i}f_{\rm PLL}} n_{\rm Y}(\omega). \tag{14.17}$$

At frequencies $f \gg f_{\rm PLL}, f_{\rm G}$, the noise gain is simply given by P, while for $f \ll f_{\rm PLL}, f_{\rm G}$ it is given by $Pf_{\rm G}/f_{\rm PLL} = 1/(a\eta) < 1$. The low-frequency noise is reduced by the PLL. However, because $f_{\rm G}$ is small compared to the lock-in amplifier bandwidth of about 530 Hz, the low-frequency noise suppression is not relevant in practice for the integrated noise spectrum and to a good approximation the integrated output noise of the PLL $\delta U_{\Delta f}$ is enhanced by the factor P over the noise on the Y output $\delta Y_{\rm res}$ [cf. eq. (14.10)]:

$$\delta U_{\Delta f} = P\delta Y_{\rm res}. \tag{14.18}$$

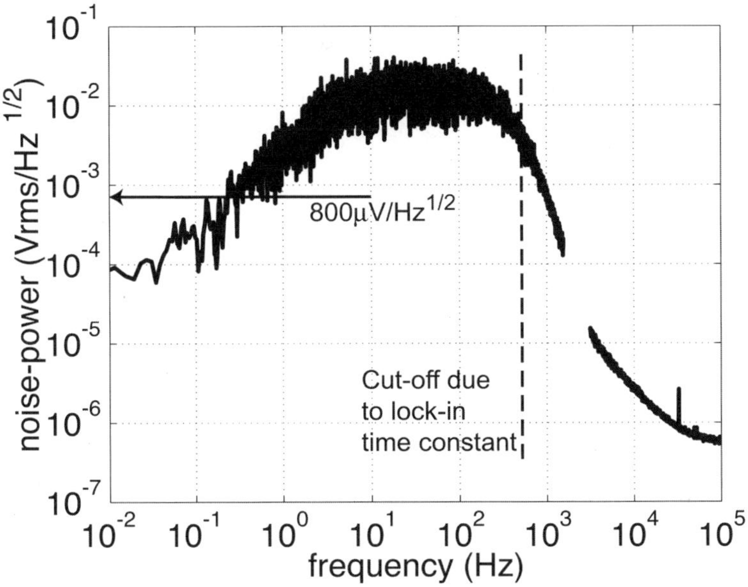

Fig. 14.11. Output noise of the PLL in the full range of frequencies. At high frequencies the response is cut off due to the low-pass filter of the lock-in amplifier. At low frequencies the noise figure is suppressed according to eq. (14.17), while it is enhanced at intermediate frequencies by the proportionality constant P of the P-I controller.

A measurement of the PLL noise is shown in Fig. 14.11. Here, $\eta = 0.1$ Hz/V, the tuning fork excitation was 10 µVrms. The lock-in time constant $\tau = 300$ µs at 24 dB/oct is responsible for the strong cut-off of the noise spectrum above 500 Hz. The I-U converter output noise n_{IUC} was amplified by a factor 1000 by the lock-in and therefore has a value of 800 µV/$\sqrt{\mathrm{Hz}}$ at the lock-in output. In the frequency range between 5 and 500 Hz the PLL noise is enhanced by the factor $P = 12$. At the lowest frequencies the noise is reduced in agreement with the above discussion. The integrated noise spectrum corresponds to an effective frequency noise of about 20 mHz/$\sqrt{\mathrm{Hz}}$. If desired, parameters can be chosen such that the frequency noise is suppressed well below 1 mHz/$\sqrt{\mathrm{Hz}}$ at the cost of bandwidth. For scanning we typically work with a PLL noise of 10 mHz/$\sqrt{\mathrm{Hz}}$.

Amplitude feedback

As an additional refinement we can use the measured amplitude of the tuning fork oscillation to feed it back into the amplitude modulation input of the Yokogawa oscillator. This additional feedback keeps the amplitude of the tuning fork oscillation at a fixed value by adjusting the amplitude of the driving voltage. We omit the amplitude control in the discussion of the feedback systems in order to avoid unnecessary complications.

14.5.3 Frequency shift and tip–sample interaction

If the tip is brought close to the sample surface, the resonance frequency of the tuning fork sensor depends on the tip sample interaction via the tip–sample separation Δz and the tip–sample voltage U_{ts}. The action of these quantities on the admittance measurement are indicated in Fig. 14.9 by the circle enclosing a stylized force-distance curve. In order to relate them to the frequency shift we apply the Hamilton–Jacobi perturbation result [425]

$$\begin{aligned}\Delta f(\Delta z, U_{\text{ts}}) &= \\ &-\frac{f_0}{2\pi A k}\int_0^{2\pi} \mathrm{d}x\, F_{\text{ts}}\left(\Delta z + A\sin x, U_{\text{ts}}\right)\sin x \\ &\approx \alpha \Delta U_{\text{ts}} + \beta \Delta z,\end{aligned} \tag{14.19}$$

where $\alpha = \partial \Delta f(0, U_0)/\partial U_{\text{ts}}$ and $\beta = \partial \Delta f(0, U_{\text{ts}})/\partial \Delta z$. The approximate expression of the frequency shift has been obtained by an expansion to first order in the tip–sample separation Δz and the tip–sample voltage U_{ts}, which generates electrostatic tip–sample forces.

14.5.4 The z-feedback

Linear response

The PLL feedback is part of the z-feedback loop as shown in Fig. 14.9. The PLL-output voltage $U_{\Delta f}$ is fed into the z-feedback P-I controller (PI), which drives the high-voltage amplifier (HV). This amplifier supplies the voltage for the z-piezo electrodes and thereby determines the tip–sample separation Δz. As a result of the force-distance interaction characteristic, Δz and the tip–sample voltage U_{ts} are translated into a resonance frequency of the interaction sensor [see eq. (14.19)], which is part of the PLL. For controlling the tip at constant frequency shift Δf_{set} above the sample surface, a certain $U_{\text{set}} = \Delta f_{\text{set}}/\eta$ is chosen. The feedback will then keep the tuning fork resonance frequency shift at Δf_{set} by controlling Δz during a scan.

In the following we discuss the z-feedback in a similar fashion to the PLL. Here the relevant linear equations are (cf. Fig. 14.9)

$$\Delta f(\omega) = \alpha U_{\text{ts}}^{\text{AC}}(\omega) + \beta \Delta z(\omega), \tag{14.20}$$
$$\Delta z(\omega) = \mu \Delta U_{\text{HV}}(\omega), \tag{14.21}$$
$$U_{\Delta f}(\omega) = PLL(\omega)\Delta f(\omega) + n_{\text{PLL}}(\omega), \tag{14.22}$$
$$e(\omega) = U_{\text{set}} - U_{\Delta f}(\omega), \tag{14.23}$$
$$\Delta U_{\text{HV}}(\omega) = PI(\omega)e(\omega). \tag{14.24}$$

Equation (14.20) describes the resonance frequency shift of the sensor caused by the tip–sample voltage and the tip–sample separation [cf. eq. (14.19)]. The constant μ is the (temperature-dependent) calibration of the z-piezo tube (e.g., 425 nm/70 V at

4.2 K) converting the applied high-voltage signal $\Delta U_{\rm HV}(\omega)$ into the z-position of the tip, $\Delta z(\omega)$ [eq. (14.21)]. Equation (14.22) describes the response of the phase-locked loop [cf., eq. (14.16)]. The added noise term $n_{\rm PLL}(\omega)$ is again for later discussion of the noise properties of the system and will be set to zero for the moment. The voltage $U_{\rm set}$ is the setpoint of the feedback. The function $PI(\omega)$ is the response of the z-feedback P-I controller including the high-voltage amplifier. It transfers the error signal [eq. (14.23)] into the voltage applied on the z-piezo, $\Delta U_{\rm HV}(\omega)$ [eq. (14.24)]. We are interested in the response of Δz as a result of surface roughness on the sample. We can simulate this roughness by applying a sinusoidal AC tip–sample voltage $U_{\rm ts}^{\rm AC}$. The solution is

$$\Delta z(\omega) = Z(\omega)\frac{\alpha}{\beta}U_{\rm ts}^{\rm AC}(\omega).$$

Here, the quantity $\alpha/\beta U_{\rm ts}^{\rm AC}$ plays the role of an effective surface roughness $\Delta z_{\rm eff}$ and

$$Z(\omega) = -\frac{M(\omega)}{1+M(\omega)},$$

with the open loop response $M(\omega) := \beta\mu PI(\omega)PLL(\omega)$.

In the limit of low frequencies $PLL(\omega)$ is constant, $PI(\omega)$ dominates, and $\beta\mu PI(\omega)PLL(\omega) \gg 1$ such that $Z(\omega)$ is constant and given by $-z_{\rm eff}$. In the limit of large frequencies this response decays like $PLL(\omega)$.

Again, the system is unstable if the complex poles of $Z(\omega)$, given by the solutions of $1 + M(\omega) = 0$, are located in the positive imaginary half plane. Stable feedback is achieved only if the gain of the z-feedback P-I controller, P_z, is kept below an upper limit $P_{\rm c}^{(z)}$. The value of $P_{\rm c}^{(z)}$ depends — among other quantities — on β and therefore on the characteristics of the tip–sample interaction.

If we assume again that the characteristic frequency of the z-feedback P-I controller is equal to the PLL-bandwidth $f_{\rm PLL}$, i.e.,

$$f_{\rm PLL} = f_z, \tag{14.25}$$

the z-feedback response simplifies to the low-pass

$$Z(\omega) = -\frac{1}{1+{\rm i}f/f_{\rm FB}}$$

with the z-feedback bandwidth

$$f_{\rm FB} = \frac{P_z\beta\mu}{\eta}f_{\rm PLL} = P_z P\beta\mu a f_{\rm G}. \tag{14.26}$$

Here the product $\beta\mu a$ acts as an additional P-factor and the effective value is $P_{\rm eff}^{(z)} = P_z P\beta\mu a$ such that $f_{\rm FB} = P_{\rm eff}^{(z)} f_{\rm G}$, i.e., the bandwidth of the z-feedback is again larger than the bandwidth of the PLL. This is demonstrated in Fig. 14.10 where the curve labeled "z-feedback" was measured by varying the frequency of $U_{\rm ts}^{\rm AC}(\omega)$. The figure summarizes the resulting frequency characteristics of the two nested feedback loops, i.e., the PLL and the z-feedback. It is demonstrated that the total system response is a low-pass (below the bandwidth of the lock-in amplifier) if all parameters are properly set. The bandwidth of the z-feedback can be increased by using higher $P_{\text{eff}}^{(z)}$; however, the stability criterion $P_{\text{eff}}^{(z)} < P_{\text{c}}^{(z)}$ and noise considerations constitute the experimental limits.

Noise considerations for the z-feedback

We now consider the contribution of the dominant I-U-converter noise, $n_{\text{IUC}}(\omega)$, which transfers into the output noise of the PLL, $n_{\text{PLL}}(\omega)$ [see eq. (14.17)]. The relevant equations for the transfer into noise on $\Delta z(\omega)$ are given by eqs. (14.20–14.24) with $n_{\text{PLL}}(\omega)$ in eq. (14.22) being non-zero. The response of the z-feedback to the PLL-output noise determined from these equations under the condition $f_{\text{PLL}} = f_z$ is

$$n_z(\omega) = -\mu P_z P \frac{f - \mathrm{i} f_{\text{G}}}{f - \mathrm{i} f_0} n_{\text{Y}}(\omega),$$

where $n_z(\omega)$ is the spectral noise density of the z-position of the tip. It is interesting to note that the bandwidth of the PLL does not enter $n_z(\omega)$. The integrated noise of the z-feedback is approximately given by

$$\delta z = \mu P_z P \delta Y. \tag{14.27}$$

Higher bandwidth of the z-feedback will lead to larger z-noise and for high quality imaging it is important to keep δz as low as possible (typically of the order of 1 Å). On the other hand, reasonable scan speeds require high bandwidths of the order of several hundred Hz. This naturally leads to the question, how the optimum feedback parameters can be found from the above analysis for a given setup.

14.5.5 About feedback parameters

The optimum feedback parameters are given by the following conditions:

1. The characteristic frequency of the PLL-PI-controller is identical to the characteristic frequency f_{G} of the (tuning fork) sensor.
2. The characteristic frequency of the z-feedback P-I controller is identical to the PLL bandwidth.
3. The P_{eff} parameter of the PLL is well below the critical value P_{c} where the PLL feedback becomes unstable.
4. The $P_{\text{eff}}^{(z)}$ parameter of the z-feedback is well below the critical value $P_{\text{c}}^{(z)}$ where the z-feedback becomes unstable.
5. The P_{eff} parameter of the PLL is small enough to give acceptable frequency noise.
6. The $P_{\text{eff}}^{(z)}$ parameter of the z-feedback is small enough to give acceptable z-noise.

We repeat the corresponding equations (14.15), (14.25), (14.18), and (14.27) below:

$$f_{\rm PI} = f_{\rm G} = \frac{f_0}{2Q}, \tag{14.28}$$

$$f_z = f_{\rm PLL} = P\eta X_{\rm res}, \tag{14.29}$$

$$\delta U_{\Delta f} = P\delta Y_{\rm res}, \tag{14.30}$$

$$\delta z = \mu P_z P\delta Y_{\rm res}. \tag{14.31}$$

From these equations the optimum feedback parameters can be found as follows: First, the tuning fork resonance curve is measured and $f_{\rm G}$ is determined from its half width at the $\sqrt{2}$-maximum value. This frequency determines the characteristic frequency to be set on the P-I controller of the PLL via eq. (14.28). The gain P of this P-I controller is set such that the output noise of the PLL [$\delta U_{\Delta f}$, eq. (14.30)] corresponds to a reasonable frequency noise (typically 10 mHz for scanning in our setup). Next, the P-I parameters of the z-feedback are set such that eq. (14.29) is fulfilled. Finally, the gain P_z of the z-feedback is adjusted according to the tolerable z-noise with the help of eq. (14.31). We typically keep the z-noise around 1 Å for normal scanning operation.

We emphasize that, within the linear feedback analysis presented here, the optimum feedback parameters determined in this way do not depend on any details of the tip–sample interaction influenced, e.g., by the tip shape. Knowledge of the resonance curve of the sensor alone is sufficient for the optimum settings of the feedback. This allows one to set appropriate parameters for the feedback *before* the tip has ever come close to the surface and fatal tip crashes due to unstable feedback can be successfully avoided.

For a given z-noise there is a certain freedom concerning the choice of the parameters P and P_z, because according to eqs. (14.30) and (14.31) only their product needs to be kept constant. This freedom allows us to choose the bandwidth of the PLL either relatively high or relatively low in the span between $f_{\rm G}$ and $f_{\rm FB}$. If the PLL bandwidth is chosen to be high, e.g., close to $f_{\rm FB}$, the PLL will be able to track the resonance frequency with the same bandwidth as the z-feedback and $U_{\Delta f}$ will be constant throughout a scan. However, if the PLL bandwidth is chosen to be low, e.g., closer to $f_{\rm G}$, the PLL cannot track the resonance frequency with the same bandwidth as the z-feedback and appreciable error signals arise whenever there is a step on the surface. This is demonstrated in Fig. 14.12. The two images in the top row are topographic images of a detail of a Hall bar structure with SFM-written oxide lines in the bottom right corner. The left (right) image was taken with small (large) PLL bandwidth at fixed bandwidth of the z-feedback. The bottom row shows the corresponding frequency error signal. It can be seen that with a slow PLL, an additional image can be obtained in which the contours of edges appear very sharp. With a fast PLL this contrast disappears completely. This "differential" image of the surface can be very useful because it eliminates sample tilt and amplifies edges of small height nicely.

Effect of Q and tuning fork spring constant on the bandwidth

It becomes clear from the above discussion that the bandwidth of the z-feedback is given by eq. (14.26):

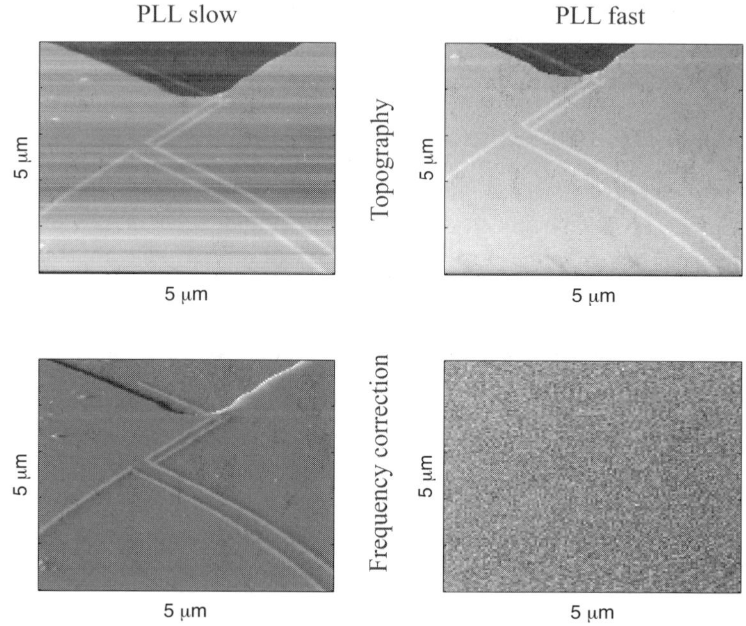

Fig. 14.12. Left column: Topographic image (top) and frequency error signal (bottom) with small PLL bandwidth. In this case the error signal enhances edges of the topography without deteriorating the performance of the z-feedback. Right column: the same with large PLL bandwidth. In this case the error signal is just noise.

$$f_{\mathrm{FB}} = P_z P \beta \mu a f_{\mathrm{G}} = P_z P \beta \mu X_{\mathrm{res}}, \tag{14.32}$$

where we have used the relations $f_{\mathrm{G}} = f_0/(2Q)$ and $a = X_{\mathrm{res}} 2Q/f_0$. Most remarkably, the Q-factor cancels in the result for f_{FB} due to the fact that the slope a of the tuning fork response Y-signal around resonance (see Fig. 14.5) is proportional to Q, while the bandwidth of the tuning fork response f_{G} is proportional to $1/Q$. It is important to state here that this result is not at all in contradiction with the paper by Albrecht and coworkers [384] in which the authors state that for high-Q cantilevers the phase or amplitude response is slowed down. This fact is fully considered in our description. However, these authors did not analyze a complete SFM-feedback system and their result must not be erroneously transferred to this case.

In order to prevent misinterpretation of our eq. (14.32), we recall that the low-pass characteristic of a high-Q cantilever (or tuning fork) entering its derivation is valid *only if the changes of the resonance frequency encountered during a scan (i.e., the error signal in the bottom row of Fig. 14.12) are small compared to* $f_{\mathrm{G}} = f_0/(2Q)$. This fact makes clear that for given PLL- and z-feedback parameters, the scan speed cannot be made arbitrarily high, because, depending on the surface roughness, sooner or later the feedback may become unstable. However, increasing the gain of the PLL can help to allow faster scanning by faster tracking of the varying cantilever resonance.

In contrast to Q, the stiffness $k - k_c$ of the sensor directly enters the quantity β, i.e., the slope of the $\Delta f(z)$ curve at the operating point [cf., eq. (14.8)] and therefore has a direct influence on f_{FB} within the linear response approximation. Softer sensors will lead to larger values of β and thereby increase the bandwidth of the z-feedback.

Similar arguments play a role for the comparison of scanning in the attractive or repulsive part of the $\Delta f(z)$ curve. In the attractive part typical values of β are much smaller than in the repulsive part. Therefore the bandwidth tends to be considerably larger for controlling in the repulsive regime.

It is worth mentioning that the feedback performance characterized by f_{FB} does indeed depend on the force-distance relationship via the quantity β, i.e., it depends, for example, on the tip shape. Nevertheless, the optimum feedback parameters can be found without the knowledge of β, i.e., they are independent of the force-distance relationship. As shown above, these results are a consequence of the linear response analysis and we find them confirmed in the experimental results.

Comparison to phase control (slope detection)

We want to discuss briefly why the use of the PLL is advantageous compared to phase control, where the phase change measured as the Y output of the lock-in in our case is directly used for the z-controller without the intermediate PLL.

From a similar linear response analysis as performed in this chapter for our nested system, we find that the same bandwidths of the z-feedback can be achieved with pure phase control. However, the same restrictions as above apply again for the tolerable frequency shifts. Sensor response becomes oscillatory, if the scan speeds are chosen too high, and with phase control one does not have the ability to reduce the encountered frequency shifts by tracking the varying resonance frequency faster. It seems to us that this effect imposes the limit on scan speeds with high-Q cantilevers operated in phase-control mode.

Enhancement of the bandwidth for scanning operation is therefore not the compelling argument for the use of the additional PLL feedback. It is rather the reduction of the encountered frequency-shift error signals during a scan by fast tracking of the varying resonance frequency with the PLL that makes these systems superior to pure phase control. We can therefore summarize several reasons why we use the PLL:

1. The PLL increases the bandwidth of the tuning fork response when the z-feedback is not yet controlling, e.g., during the approach of the tip. The PLL allows the use of reasonable approach speeds.
2. The PLL increases the dynamic range of the frequency detection. The Y signal of the lock-in depends linearly on the frequency shift only in a limited range of frequencies around resonance. With high-Q sensors this range can become so narrow (100 mHz) that during tip approach or when scanning with large scan speeds on rough surfaces the frequency shift runs out of this range. The PLL avoids this problem by tracking the resonance. The frequency range that can be

used with PLL is mainly determined by the VCO settings. We typically use a range of 2 Hz.

3. The PLL allows to image with an additional adjustable frequency contrast (see Fig. 14.12).
4. When one measures $\Delta f(z)$-curves, the PLL allows the clear distinction between conservative frequency shifts and dissipative effects [414, 424].

14.6 Force-distance studies on HOPG with piezoelectric tuning forks at 1.7 K

Introduction

In this section piezoelectric quartz tuning forks have been employed as the force sensor in a dynamic mode scanning force microscope operating at temperatures down to 1.7 K at He-gas pressures of typically 5 mbar. This microscope was a modified commercial low-temperature scanning tunneling microscope that we used before we built the more advanced ^{3}He-microscope setup described above. Here, an electrochemically etched tungsten tip glued to one of the tuning fork prongs acts as the local force sensor. Its oscillation amplitude can be tuned between a few angstroms and tens of nanometers using the same phase-locked loop setup as described above. The sensor is used for a study of the interaction between the tip and a HOPG substrate [414]. Force gradient and dissipated power can be recorded with the phase-locked loop setup simultaneously. It is found that during the approach of the tip to the sample considerable power starts to be dissipated although the force gradient is still negative, i.e., the tip is still in the attractive regime. This observation concurs with experiments with true atomic resolution that seem to require the same tip–sample separation.

The interaction of the tip with the surface of a sample is of crucial importance for the operation of scanning force microscopes [360]. A renewed interest in this subject has arisen since atomic resolution was achieved in the dynamic mode on the surfaces of insulators measured in UHV that were previously not accessible for local investigations on the atomic scale [385, 426–429]. Recently, atomic resolution was achieved in the dynamic mode on highly oriented pyrolytic graphite (HOPG), a so-called van der Waals surface, at temperatures of 10 K [415]. The low temperatures were necessary in order to reduce the thermal noise and provide stable imaging conditions necessary for atomic resolution.

In the following we show force-distance studies on HOPG substrates at temperatures down to 1.7 K using a tuning fork sensor [414]. We discuss simultaneous measurements of the force gradient and of the power dissipation vs. tip–sample separation.

Microscope Setup

Here a commercial cryo-SXM from Oxford Instruments was used that has been modified for working as a scanning force microscope utilizing a piezoelectric quartz tuning fork sensor for operation in the dynamic mode. The microscope is run in the

He-gas atmosphere of the variable temperature insert of a cryostat at a base temperature of 1.7 K and pressures of typically 5 mbar. The microscope is designed for the local investigation of semiconductor nanostructures in high magnetic fields, which requires a large scan range at low temperatures at the cost of lateral spatial resolution. Hence, atomic resolution cannot be achieved. Further details about the microscope can be found in Ref. [407]. The electronic setup including the phase-locked loop is identical to the one describe above. The PLL delivers the frequency shift of the tuning fork sensor and the dissipated power caused by tip–sample interactions. In our experiments below 4.5 K, nominal tip oscillation amplitudes between 0.56 nm and 20 nm have been used. The quality factors of the oscillators ranged between 10'000 and 120'000 depending on temperature and pressure [410].

Force-distance curves

Figure 14.13(b) shows the frequency shift as a function of the tip–sample separation z at zero bias voltage for different tip-oscillation amplitudes between 0.56 nm and 2.0 nm. All these curves have been measured for both sweep directions and no significant hysteresis was found. At large z, all the curves exhibit the well-known slow decrease of Δf with decreasing z until a minimum is reached. Beyond this minimum, Δf increases steeply with a rate of up to 10 Hz/Å at the smallest tip-oscillation amplitude. At a given z, larger amplitudes lead to smaller frequency shifts in accordance with the theoretical results in Ref. [425, 430]. The minimum in Δf shifts to larger z and becomes more shallow when the amplitude is increased. For the smallest amplitudes below 0.4 nm the curves are almost identical. We therefore conclude that for these amplitudes the force-gradient approximation is still valid even when the tip is very close to the sample.

Figure 14.13(a) shows the corresponding measurements of the dissipated power P as a function of the tip–sample separation z. It is calculated from $P = U_0 I_{\mathrm{TF}}$, where U_0 is the excitation voltage applied to the tuning fork and I_{TF} is the tuning fork current measured in phase with the excitation. We use the fact that the power dissipated in the system, in particular the power dissipated during tip–sample interaction, has to be supplied by the electrical excitation of the tuning fork. At large z the dissipated power is constant. When the tip approaches the surface, the power starts to increase into the picowatt range. For larger tip-oscillation amplitudes the onset of the dissipation is at larger tip–sample separations. Even at the smallest tip-oscillation amplitudes, the onset of the power dissipation occurs at a tip–sample separation at which the frequency shift is negative and the tip experiences an overall attractive force. We believe that at these tip–sample separations the foremost tip atoms start to come into intimate contact with the surface atoms and in addition to the conservative tip–sample interaction forces, there occur dissipative phenomena. We speculate that phonon emission into the tip and the sample, or electrical losses due to the induced Kelvin current might be responsible for this effect. Further investigations are necessary to clarify the origin of the power dissipation in the system. However, it is interesting to note that the power increase occurs at a position in the curves where other authors have reported true atomic resolution [385, 415, 426–429]. At smaller

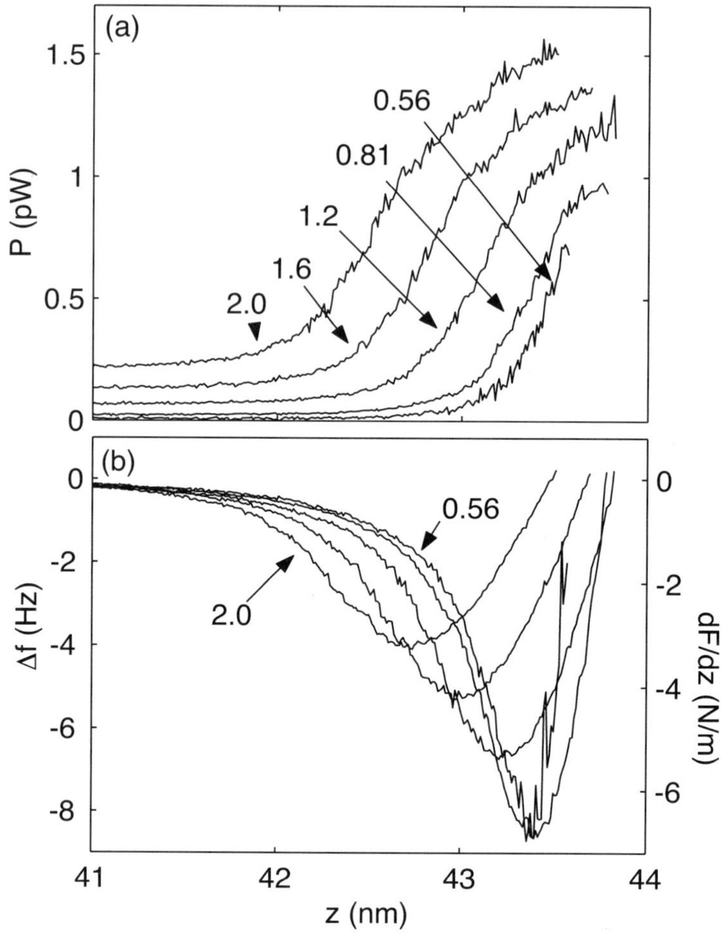

Fig. 14.13. (a) Dissipated power measured with a tungsten tip on a HOPG substrate at 1.7 K for various tip-oscillation amplitudes. Numbers next to arrows indicate amplitudes in nm. (b) Frequency shift vs. distance for different amplitudes.

tip–sample separations, the tip must deform massively but the reproducibility of our curves suggest that this deformation is elastic.

Conclusions

In this section we have described force-distance studies measured at 1.7 K using piezoelectric tuning fork sensors. Very high quality factors up to 120'000 allow the detection of frequency shifts of less than 1 mHz. Small tip-oscillation amplitudes down to 0.5 nm have been realized. Measurements were performed in which the force gradient and the dissipated power were measured simultaneously as a function of the tip–sample separation on a HOPG surface. The data suggest that already at relatively large tip–sample separations when the tip experiences a net attractive force, the foremost tip atoms touch the surface and power is dissipated. At smaller tip–sample separations the tip is significantly elastically deformed and at the same time the power dissipation increases strongly.

15

Local investigation of a two-dimensional electron gas with an SFM at cryogenic temperatures

In this chapter a method for the local characterization of two-dimensional electron gases will be described, which is based on experiments using the scanning probe microscope described previously. While conventional magnetotransport experiments, i.e., the Hall effect and the measurement of Shubnikov–de Haas oscillations in the magnetoresistance, allow the determination of an average electron density in the system, the method described here leads to the measurement of a local electron density at the position of the conducting SFM tip.

15.1 Samples and structures

The layer sequence of the samples used for this study is shown in Fig. 2.1. It is a GaAs/AlGaAs heterostructure with a shallow two-dimensional electron gas (2DEG) confined at the heterointerface, which is 34 nm below the sample surface. The electrons are supplied by a Si δ-doped plane 17 nm remote from the heterointerface.

Figure 2.2 shows the corresponding self-consistent conduction band profile acting as the effective potential for the electrons in the growth direction. The electrons occupy the lowest electric subband in the triangular well formed by the conduction band offset and the smoothly bent GaAs conduction band edge. The Fermi energy is pinned at the surface by a large density of surface states energetically located in the center of the bulk band gap. This leads to the appearance of the built-in potential barrier at the surface with a height of about half the band gap.

Four ohmic gold–germanium contacts have been diffused into the samples at the edges of a $5 \times 5\,\mathrm{mm}^2$ piece of wafer giving electrical access to the two-dimensional electron gas. The remaining parts of the surface were not further treated. They are therefore covered by a few nanometers of native oxide.

15.2 Kelvin-probe measurements

The sample was cooled down to the base temperature of 300 mK in the SFM setup. A variable DC voltage was applied at the 2DEG contacts while the tip was kept

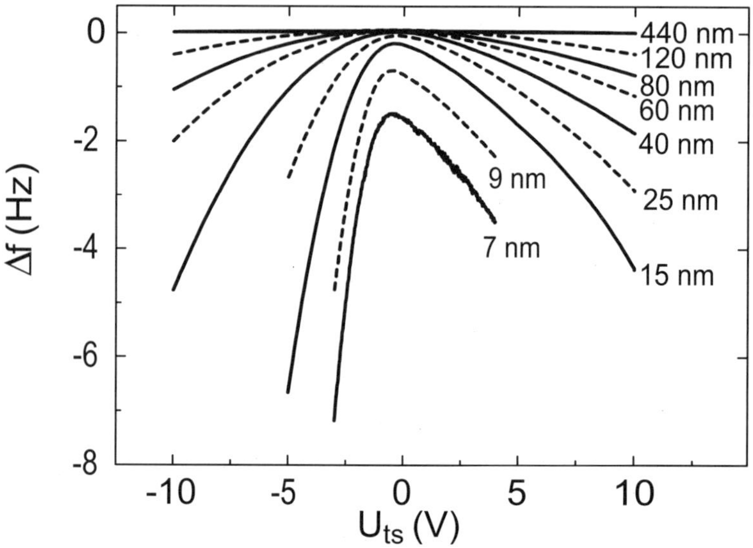

Fig. 15.1. Frequency shift of the tuning fork resonance as a function of the voltage applied between the PtIr tip and the buried two-dimensional electron gas for various tip–sample separations.

grounded. In this way the electrostatic force acting between the metallic tip and the 2DEG can be controlled. As in Kelvin-force microscopy, this electrostatic force can be measured with the SFM tip. In our setup the frequency shift of the tuning fork sensor measures the additional force gradient created by the applied voltage, similar to the measurement shown in Fig. 14.7. In contrast to a metallic sample (also the HOPG-sample used for the experiments in 14.4.3), the 2DEG in the heterostructure has a relatively small density of states and a small Fermi energy. Therefore, relatively small voltages applied between the 2DEG and the tip can change the local electron density significantly, leading locally to a total depletion of the electron gas in the extreme case.

Figure 15.1 shows measured frequency shifts as a function of the applied tip–sample voltage for various tip–sample separations. Compared to the curve measured on a HOPG substrate (Fig. 14.7), a clear asymmetry can be observed leading to a weaker (stronger) frequency shift at positive (negative) voltages. For a given voltage, the frequency shift decreases with increasing tip–sample separation. Comparing to eq. (14.9) we can state that the measured curves cannot be fitted with a parabola, i.e., under the assumption of voltage independent $\partial^2 C/\partial z^2$.

Qualitatively, we expect a locally enhanced electron density under the tip for negative voltages (in this case the tip is positive relative to the 2DEG) and a reduced density for positive voltages eventually leading to complete local depletion. Therefore we expect an ideal capacitor model with a voltage-independent capacitance to work only for negative voltages. For positive voltages depletion of the 2DEG leads

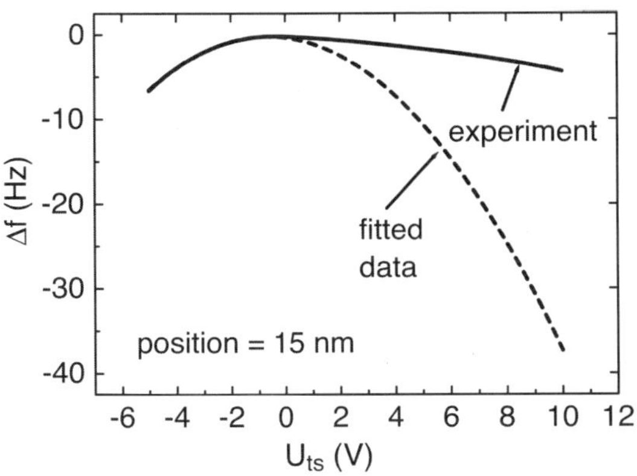

Fig. 15.2. Frequency shift as a function of tip–sample voltage for a tip–sample separation of 15 nm. A parabola was fitted to the left branch of the measured curve.

to a hole in the capacitor plate, which is a change in geometry of the capacitor and must therefore lead to a reduced capacitance. A voltage dependence of $\partial^2 C/\partial z^2$ is the natural consequence.

Following this qualitative picture we assume a voltage-independent capacitance for negative voltages and fit the left branches of the curves with a parabola according to eq. (14.9). This is shown in Fig. 15.2 for a tip–sample separation of 15 nm. The information deduced from these fits is twofold. On the one hand, we obtain the tip–sample voltage at which the parabola has its maximum. This quantity will be discussed later. On the other hand, we get the curvature of the parabola, which is proportional to $\partial^2 C(z)/\partial z^2$. A plot of this quantity as a function of tip–sample separation is shown in Fig. 15.3. It can be seen that the curve can be well characterized by having a z^{-2}-dependence. According to a comprehensive comparison of different models for tip–sample arrangements with different tip geometries performed by Belaidi and coworkers [431], such a behavior is consistent with the model of a metallic sphere above a metallic plane. Given that this slope persists up to tip–sample separations of about 700 nm we have to conclude that the tip diameter is at least of the same order of magnitude in this experiment.

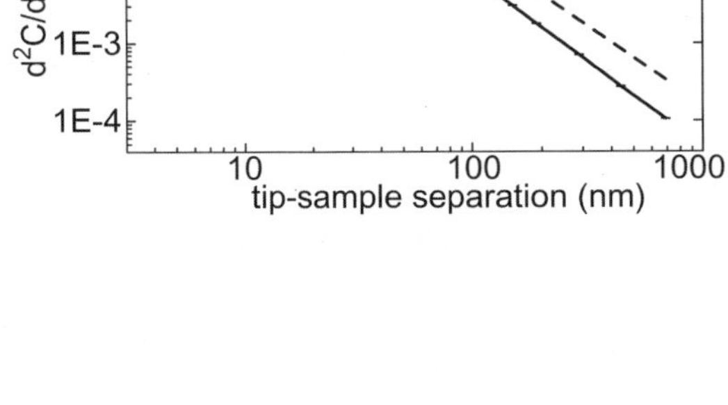

Fig. 15.3. Second derivative of the tip–sample capacitance as a function of tip–sample separation as determined from the curvature of the fitted parabolas (cf. Fig. 15.2). The straight solid line has the slope of a z^{-2}-dependence.

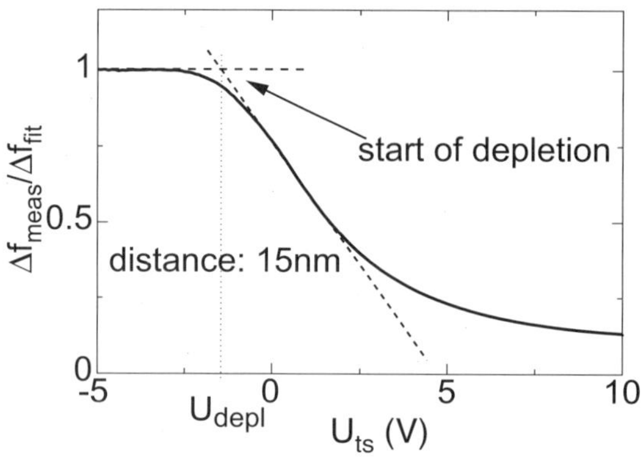

Fig. 15.4. Normalized $\partial^2 C(V)/\partial z^2$. At negative tip–sample voltages this quantity is unity. The reduction at positive voltages is caused by total depletion of the 2DEG under the tip. Such curves allow the definition of a depletion voltage.

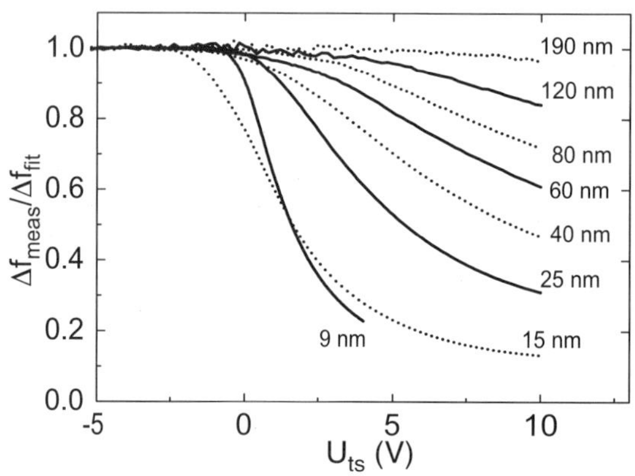

Fig. 15.5. Normalized $\partial^2 C(z)/\partial z^2$ as a function of the tip–sample voltage for various tip–sample separations showing the depletion of the two-dimensional electron gas under the tip.

We now turn to the discussion of the asymmetry of the Kelvin measurement. The explicit parabolic voltage dependence of the frequency shift in eq. (14.9) can be eliminated from the curves by normalizing $\Delta f_{\rm meas}$ by $\Delta f_{\rm fit}$, i.e., by plotting the ratio of the two as a function of voltage. Figure 15.4 shows the result of the normalization procedure for the curve taken at 15 nm tip–sample separation. By virtue of the normalization procedure and in accordance with our qualitative picture, $\partial^2 C/\partial z^2$ is constant for negative voltages. At a voltage of about -5 V, depletion starts under the tip. With increasing voltage the area of the depleted region grows continuously reducing $\partial^2 C/\partial z^2$. We empirically define the onset voltage for depletion, $U_{\rm depl}$, in the way indicated in Fig. 15.4. At this voltage a linear fit made in the turning point of the decaying $\partial^2 C/\partial z^2$-curve crosses unity. The same analysis has been applied to all curves in Fig. 15.1 for different tip–sample separations and the result is shown in Fig. 15.5. From these curves we can extract the dependence of $U_{\rm depl}$ on tip–sample separation (see Fig. 15.6). The series of points can be quite well fitted with a linear function. This finding can be interpreted within a suitable model (see below).

We briefly come back to the second type of information that can be obtained from the parabolic fit shown in Fig. 15.2, namely, the voltage at which the parabola has its maximum. Conventionally this voltage is interpreted as the difference in work functions of the two materials involved. Figure 15.7 shows this quantity as a function of tip–sample separation. Although the error bars are rather large, a slight trend for an increase at smaller tip–sample separation becomes evident, in contradiction of the notion of a constant material-dependent work function difference $\Delta\mu_{\mathrm{ch}}$. The origin of this behavior will be traced back to the presence of fixed localized charges in the system (surface charges, doping plane) below.

15.3 General electrostatic consideration

A general model for the experimental system (see Appendix A) starts from Poisson's equation

$$\nabla\left[\varepsilon(\boldsymbol{r})\varepsilon_0\nabla\phi(\boldsymbol{r})\right] = -\rho_{\mathrm{ion}}(\boldsymbol{r}) \tag{15.1}$$

with the spatially varying dielectric function $\varepsilon(\boldsymbol{r})$, fixed ionic charges $\rho_{\mathrm{ion}}(\boldsymbol{r})$ accounting, e.g., for ionized donors or surface charges, and the electrostatic potential $\phi(\boldsymbol{r})$. The two-dimensional electron gas is, for simplicity, considered to be a perfect

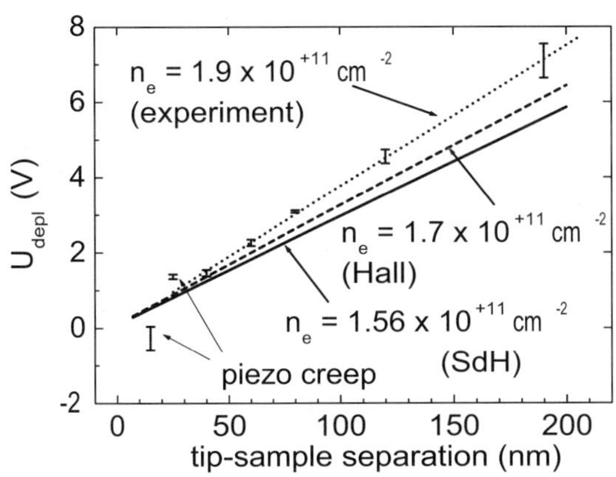

Fig. 15.6. Local electron densities compared to the Hall and Shubnikov–de Haas densities.

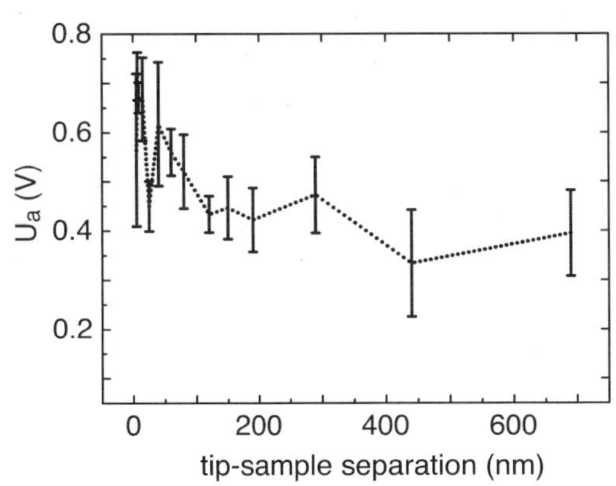

Fig. 15.7. Contact voltage difference between the tip and a GaAs sample as a function of tip–sample separation.

metal plate and the tip is also regarded as a perfect metal of arbitrary shape. In order to be most general, we even allow other metallic electrodes to be present in the system, e.g., metallic gates or more remote metallic surfaces. The boundary conditions for the problem are then given on the metallic surfaces S_i by

$$\phi(\boldsymbol{r})|_{S_i} = \phi_i,$$

where the ϕ_i is the electrostatic potential of electrode i. A general solution of this problem can be formally expressed using the Green's functions of the system (see Appendix A) [284]. The resulting charge density induced on electrode i can be written as

$$\sigma_i(\boldsymbol{r}) = \sigma_i^{(0)}(\boldsymbol{r}) + \sum_j c_{ij}(\boldsymbol{r})\phi_j, \tag{15.2}$$

where the surface charges $\sigma_i^{(0)}$ and σ_i and the coefficients c_{ij} are given by integral expressions involving Green's function of the problem and therefore depend on position $\boldsymbol{r}$ on the electrode (see Appendix A). If we consider our experiment with two electrodes, one being the two-dimensional electron gas and one the tip, and we set the potential of the electron gas to zero without a loss of generality, we obtain for the charge density induced in the two-dimensional electron gas

$$\sigma_{\mathrm{2DEG}}(\boldsymbol{r}) = \sigma_{\mathrm{2DEG}}^{(0)}(\boldsymbol{r}) + c_{\mathrm{2DEG-tip}}(\boldsymbol{r})\phi_{\mathrm{2DEG-tip}}. \tag{15.3}$$

Between the electrostatic, the chemical, and the electrochemical potential there is the general relation $\mu_{\mathrm{elch}} = \mu_{\mathrm{ch}} - e\phi$, such that $\phi = (\mu_{\mathrm{elch}} - \mu_{\mathrm{ch}})/(-e)$. The applied voltage between two electrodes is the difference of their electrochemical potentials: $U_{ij} = \mu_{\mathrm{elch},i} - \mu_{\mathrm{elch},j}$. Applying these relations to eq. (15.3), we obtain

$$\sigma_{\mathrm{2DEG}}(\boldsymbol{r}) = \sigma_{\mathrm{2DEG}}^{(0)}(\boldsymbol{r}) + c_{\mathrm{2DEG-tip}}(\boldsymbol{r})(U_{\mathrm{ts}} + \Delta\mu_{\mathrm{ch}}/e), \tag{15.4}$$

where $\Delta\mu_{\mathrm{ch}}$ is the work function difference between the tip material and the two-dimensional electron gas and U_{ts} is the applied voltage. At this point we can nicely discuss the influence of the tip on the system. If $U_{\mathrm{ts}} + \Delta\mu_{\mathrm{ch}}/e = 0$, the electron density is simply given by $\sigma_{\mathrm{2DEG}}^{(0)}(\boldsymbol{r})$. This quantity depends, however, on the presence and the exact position of the tip via the boundary conditions of the problem. This means that even if the difference in the work functions between the materials is exactly cancelled by the applied voltage, the presence of the tip can change the local electron density in the 2DEG via the dependence of $\sigma_{\mathrm{2DEG}}^{(0)}(\boldsymbol{r})$ on tip position.

Application of a suitable voltage U_{ts} at constant tip position will change the local electron density proportional to the differential capacitance function $c_{\mathrm{2DEG-tip}}(\boldsymbol{r})$. This function depends on the exact shape of the tip, but will for a typical conical tip be similar to a Lorentzian distribution with an extremum below the tip apex [362]. A negative voltage on the tip relative to the (grounded) electron gas will deplete the electrons, and the electrostatic model with a voltage independent $c_{\mathrm{tip-2DEG}}(\boldsymbol{r})$ will be applicable until complete depletion is reached. Beyond this point $c_{\mathrm{tip-2DEG}}$ will become voltage dependent, because the boundary conditions of the problem change:

the 2DEG will behave like a metallic plate with a hole under the tip whose size depends on U_{ts}.

We now return to a further discussion of the general case of many electrodes. The integrated charge on electrode i is [cf., eq. (A.6)]

$$Q_i = Q_i^{(0)} + \sum_j C_{ij}\phi_j, \tag{15.5}$$

where the C_{ij} are the capacitance coefficients between electrodes i and j and the $Q_i^{(0)}$ are the charges induced on electrode i when all electrodes are grounded. This induced charge is due to the presence of the fixed ionic charges in the system. The capacitance coefficients have the property $C_{ij} = C_{ji}$ and $\sum_j C_{ij} = \sum_i C_{ij} = 0$. They will start to depend on voltage as soon as depletion starts in the 2DEG, similar to the $c_{ij}(\boldsymbol{r})$.

The measured frequency shift in the experiment is proportional to the force gradient in z-direction felt by the scanning tip. It can be calculated within our model from the rate of change of the total electrostatic energy of the system, when the tip-electrode ($i = 0$) is moved in z-direction. The resulting force gradient is [cf., eq. (A.7)]

$$\frac{\partial F_z}{\partial z} = \frac{1}{2}\sum_{ij}\phi_j \frac{\partial^2 C_{ji}}{\partial z^2}\phi_i + \sum_j \phi_j \frac{\partial^2 Q_j^{(0)}}{\partial z^2}.$$

For a description of the experiment we again consider a system with only two electrodes, namely, the tip and the two-dimensional electron gas. The capacitance matrix is then given by $C_{11} = C_{22} = C$ and $C_{12} = C_{21} = -C$. Again we set the electrostatic potential of the electron gas to zero, i.e., $\phi_1 = 0$. We then find for the force gradient:

$$\frac{\partial F_z}{\partial z} = \frac{1}{2}C''(z)\left(\phi_0 + \frac{Q_0^{(0)''}(z)}{C''(z)}\right)^2 - \frac{(Q_0^{(0)''})^2}{2C''(z)}, \tag{15.6}$$

where the double primes indicate the second derivative with respect to the z-position of the tip. Including the relation between electrostatic potential, chemical potential, and electrochemical potential to eq. (15.6), we find for the force gradient

$$\frac{\partial F_z}{\partial z} = \frac{1}{2}C''(z)\left(U_{\mathrm{ts}} + \frac{\Delta\mu_{\mathrm{ch}}}{e} + \frac{Q_0^{(0)''}(z)}{C''(z)}\right)^2 - \frac{(Q_0^{(0)''})^2}{2C''(z)},$$

where $\Delta\mu_{\mathrm{ch}}$ is the difference in the chemical potentials between the tip material and the electron gas (also referred to as the work function difference between the two materials) and U_{ts} is the tip–sample voltage. According to this equation, the force gradient (and thus the measured frequency shift) is parabolic in the voltage and the apex of the parabola is at

$$U_{\mathrm{a}} = \frac{\Delta\mu_{\mathrm{ch}}}{-e} - \frac{Q_0^{(0)''}(z)}{C''(z)}. \tag{15.7}$$

This result makes clear that the apparent work function difference given by the apex of the parabola may in fact depend on the z-position of the tip if fixed charges are present in the system, in agreement with the results shown in Fig. 15.7.

In analogy to the remarks made above about the depletion of the electron gas, $C(z)$ and also $C''(z)$ will become functions of U_{ts} as soon as the two-dimensional electron gas begins to deplete under the tip.

15.4 Plate capacitor model

In order to illustrate the above somewhat abstract considerations, we illustrate them with a simple parallel-plate capacitor model sketched in Fig. 15.8. It is expected to lead to reasonable results, if the tip radius is large compared to the tip–sample separation. Two planes of fixed charges are present in this model: the doping plane with charge density σ_{d} is located at a distance d_1 from the two-dimensional electron gas and a surface charge σ_{s} is located directly at the dielectric interface of the sample (ε_2) to vacuum (ε_1). The separation of the tip from the surface is z. In such a system the charge density in the two-dimensional electron gas is given by

$$\sigma_{\mathrm{2DEG}} = -\underbrace{\frac{\varepsilon_1\varepsilon_2\varepsilon_0}{\varepsilon_1(d_1+d_2)+\varepsilon_2 z}}_{C(z)/A}(U_{\mathrm{ts}}+\Delta\mu_{\mathrm{ch}}/e) + \underbrace{\frac{\varepsilon_1\sigma_{\mathrm{d}}d_2+\varepsilon_2(\sigma_{\mathrm{d}}+\sigma_{\mathrm{s}})z}{\varepsilon_1(d_1+d_2)+\varepsilon_2 z}}_{Q^{(0)}_{\mathrm{2DEG}}(z)/A:=-en_{\mathrm{s}}(z)},$$

which has exactly the structure of eq. (15.4). We start the discussion assuming $U_{\mathrm{ts}} + \Delta\mu_{\mathrm{ch}}/e = 0$. For very large tip–sample separation z, the charge density in the two-dimensional electron gas is given by $-en_{\mathrm{s}}^{\infty} = \sigma_{\mathrm{d}} + \sigma_{\mathrm{s}}$. For very small z it changes to $n_{\mathrm{s}}^{0} = \sigma_{\mathrm{d}}d_2/(d_1+d_2)$. The total voltage that one has to apply across this capacitor in order to deplete an electron density $n_{\mathrm{s}}(z)$ is given by

$$U_{\mathrm{depl}} = \frac{\Delta\mu_{\mathrm{ch}}}{-e} + \frac{en_{\mathrm{s}}(z)A}{C(z)} = \frac{\Delta\mu_{\mathrm{ch}}}{-e} - \frac{\varepsilon_1\sigma_{\mathrm{d}}d_2 - \varepsilon_2 en_{\mathrm{s}}^{\infty}z}{\varepsilon_0\varepsilon_1\varepsilon_2}. \tag{15.8}$$

The depletion voltage depends linearly on the tip–sample separation in accordance with the measurement depicted in Fig. 15.6. From the slope $\mathrm{d}U_{\mathrm{depl}}/\mathrm{d}z = en_{\mathrm{s}}^{\infty}/(\varepsilon_0\varepsilon_1)$

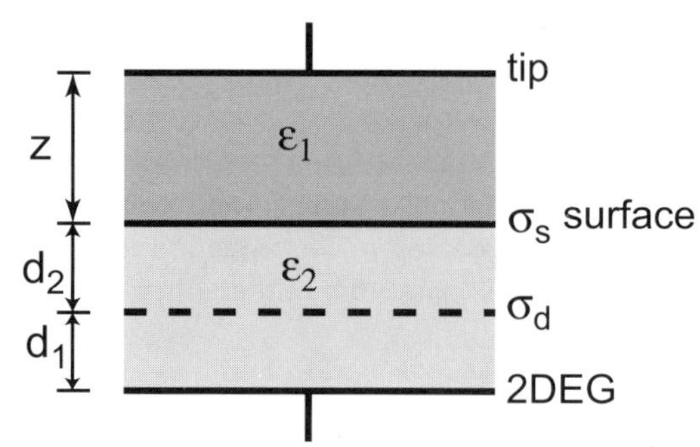

Fig. 15.8. Plate capacitor model for the tip–sample system. Part of the capacitor is filled with the AlGaAs/GaAs dielectric which has a dielectric constant of $\varepsilon_2 = 12.53$ [351], the other part is the vacuum gap characterized by $\varepsilon_1 = 1$.

the local electron density can be determined to be $1.9 \times 10^{11}\,\mathrm{cm}^{-2}$. As a comparison we plot curves into Fig. 15.6, which are predicted by eq. (15.8) with densities n_s determined from Hall and Shubnikov–de Haas measurements. Given the simplicity of the model, the agreement is reasonable, especially if we take into account that the 2DEG density can be inhomogeneous across a macroscopic sample.

If we evaluate the z-dependence of the apparent work function difference given by eq. (15.7) for the plate capacitor model, we find

$$U_\mathrm{a} = \frac{\Delta\mu_\mathrm{ch}}{-e} - \frac{\sigma_\mathrm{s}(d_1 + d_2) + \sigma_\mathrm{d} d_1}{\varepsilon_1 \varepsilon_0}$$

independent of z. Inserting our sample parameters, $\varepsilon_1 = 1$, $d_1 + d_2 = 34\,\mathrm{nm}$, $d_1 = d_2 = 17\,\mathrm{nm}$, $\sigma_\mathrm{d} = 1.6 \times 10^{-3}\,\mathrm{Cm}^{-2}$ (estimated), and $\sigma_\mathrm{s} = -1 \times 10^{-3}\,\mathrm{m}^{-2}$ (estimated), we obtain a value of $-0.76\,\mathrm{V}$ for the second term, indicating that the presence of fixed charges in the sample can have an appreciable influence on the apparent work function difference measured in the experiment. However, given the uncertainties in the estimated values, one should not take the exact value too quantitatively, but rather consider it as an estimate of the order of magnitude of the effect. This z-independent result has to be compared with the experimental results shown in Fig. 15.7, which indicate a small z dependence. Obviously, a more realistic model including the actual tip shape would solve this problem. In fact, the tip shape does influence the geometric capacitance and the dependence of $\partial^2 C/\partial z^2$ on the tip–sample separation. Looking at the order of magnitude of the apparent work function difference in the measurement, we can see that the estimated influence of the fixed charges in the system can be quite significant and is of the same order as typical work function differences between materials.

More involved models for the tip–sample system would take a more realistic shape of the tip into account, as we mentioned above. Further improvement can be gained by taking the screening properties of the two-dimensional electron gas into account self-consistently. This would include density of states contributions to the capacitance which were not taken into account in the presented analysis.

16

Local investigation of edge channels

In this chapter we discuss measurements aiming at the local investigation of edge channels in the quantum Hall regime. Several techniques will be discussed, which have been reported by a small number of groups worldwide. An experiment will be looked at in more detail in which the tunneling resistance between edge channels was measured. By using a local potential perturbation induced by the tip of a scanning force microscope, it was possible to enhance the tunneling coupling and map its strength along the edge of the two-dimensional electron gas.

16.1 Brief introduction to edge channels

The importance of the sample edge for the quantum Hall effect [48] was pointed out soon after its discovery on the basis of a model of non-interacting electrons [432]. Later interactions were considered by using self-consistent descriptions of the sample edge [50, 433, 434]. Figure 16.1 shows the general notion of the self-consistent potential and the electron density at the sample edge in equilibrium. In the bulk of the electron gas (left in the figure) the Landau levels have an energetic spacing of $\hbar\omega_c$, where $\omega_c = eB/m^\star$ is the cyclotron frequency, B is the magnetic field normal to the plane of the electron gas, and $m^\star$ is the effective mass of the electrons. The electron density in the bulk is constant. Near the edge the Landau levels have to follow the increasing electrostatic confinement potential and the electron density has to go to zero. Owing to the density of states in a magnetic field with peaks in energy at the positions of the ideal Landau levels, the density of states at the Fermi energy oscillates strongly as a function of position leading to oscillating screening properties of the electron gas at the edge. Whenever an Landau level crosses the Fermi energy screening is good and the potential flattens out. At the same time the electron density shows a strong gradient. These regions are called compressible strips because they have a finite density of states at the Fermi energy. The regions between compressible strips are called incompressible, because the density of states at the Fermi energy vanishes there. In these strips the potential has a strong gradient while the density is flat. Screening in the incompressible strips is poor. Geller and Vignale have calculated the equilibrium currents in the compressible and incompressible strips self-consistently [434]. They found that in the compressible regions the current proportional to the density gradient and the current direction leads to a diamagnetic magnetic moment. In the incompressible regions the equilibrium current is proportional to the potential gradient and produces a paramagnetic effect. The subtle interplay between these two spatially separated types of current lead naturally to the oscillatory magnetization of a two-dimensional electron gas as a function of magnetic field [435].

Since the theoretical prediction of the existence of self-consistent edge-channels and compressible and incompressible strips in the quantum Hall regime, there has been a series of experimental attempts to measure local properties at sample edges in high magnetic fields. Electron–phonon interaction was used in first experiments [436]. Optical techniques with a spatial resolution of about 1 μm in the best case also supported the notion of edge channels [437–439]. Later experiments tried to detect the edge currents inductively, but evidence for bulk currents was found [440, 441]. Recently, edge strips were imaged using a metallic single-electron transistor fabricated near the edge of a 2-dimensional electron gas (2DEG) [442].

16.2 Scanning probe experiments

It is obvious to employ scanning probe techniques with their unprecedented potential of spatial resolution for such investigations and several experiments have been

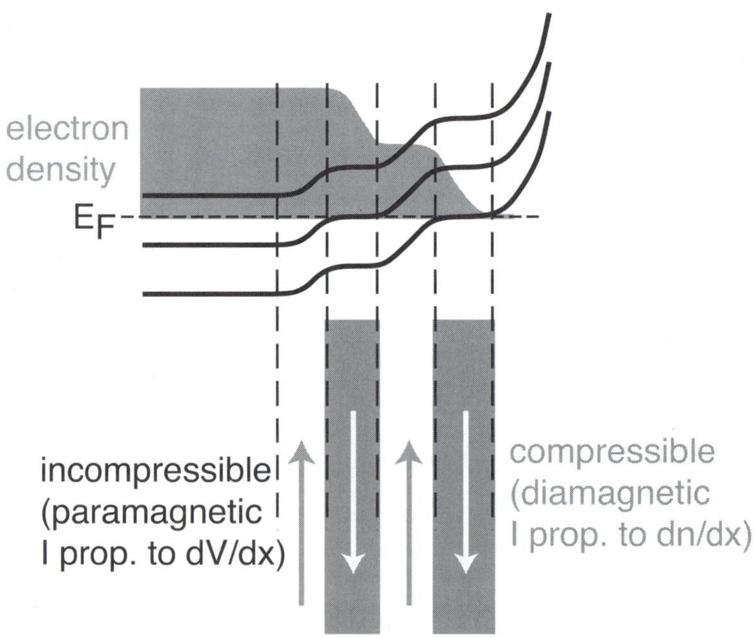

Fig. 16.1. Self-consistent edge channel structure

reported during the past few years using a scanning single-electron transistor [401], scanned potential microscopy [443], Kelvin probe techniques [368], and subsurface charge accumulation measurements [444–446]. In the following we give a brief overview of these experiments.

16.2.1 Scanning SET measurements

The scanning single-electron transistor (SET) sensors described before have been employed in Ref. [401] for an investigation of GaAs/AlGaAs heterostructures with a two-dimensional electron gas 100 nm below the surface. The SET sensor was kept at a distance of 100–200 nm above the sample surface. Spatial resolution is estimated to be a few hundred nanometers.

The scanning SET was operated as part of a feedback loop keeping the SET current [eq. (14.1)] and thereby the induced charge on the SET island constant at the operating point by controlling the voltage between the sensor and the 2DEG. This voltage is the measured output signal. Low-frequency (below 100 Hz) lock-in techniques were used in order to extract the small signals from transport through the 2DEG or the capacitance between the SET and the backgate from the huge unwanted signals of biased surface gates and fluctuating surface dopant charges.

In the sample a backgate 5.4 μm below the 2DEG was used for tuning the electron density, which was in the range of 10^{11} cm^{-2} at liquid-He temperatures. Mobilities of about 4×10^6 cm^2/(Vs) were determined.

In so-called transparency measurements, the local compressibility of the two-dimensional electron gas, i.e., the local density of states at the Fermi level, was measured. This was accomplished by modulating the voltage between the backgate and the SET. The electron gas between these two "metallic" electrodes is kept at constant potential with respect to the sensor and therefore screens the AC field more or less, depending on the (local) filling factor. Compressible strips are found that screen the signal from the backgate well, while incompressible strips do not screen. The inhomogeneous local electron density can be deduced from such measurements by plotting contours of constant filling factors. In some places closed incompressible contours lead to isolated regions in the 2DEG that cannot be charged. Steplike features of the local chemical potential difference between 2DEG and SET could be measured with the sensor at a fixed position, while an incompressible strip was made to pass the region below the tip by changing the backgate. Such steps reflect the step like local self-consistent potential expected in the regions of the strips (see Fig. 16.1). It was also attempted to measure the Hall-voltage distribution between opposite edges of the sample. Around filling factor $\nu = 2$ a change from a linear to a flat voltage distribution was found. However, not the full Hall voltage drop could be measured due to screening effects by the top-gate electrodes defining the edges. It remains to be discussed how far the presence of the metallic SET influences the distribution of potentials and currents, compressible and incompressible strips, e.g., by screening or by an insufficiently compensated work-function difference between 2DEG and SET.

16.2.2 Scanned potential microscopy

In another experiment, the quantum Hall effect has been locally investigated using a gold-coated piezoresistive cantilever coupling capacitively to the two-dimensional electron system [443]. The 2DEG was again located in a GaAs/AlGaAs heterostructure, 77 nm below the surface. Measurements were performed at temperatures between 0.7 and 1 K. An AC voltage was applied to the source contact of the Hall bar sample while the drain contact was grounded. This leads to a local electrostatic potential $U_{\mathrm{AC}}(x,y)$ in the two-dimensional electron gas. The measured quantity in this experiment is the force acting on the cantilever. It is dominated by the electrostatic force contribution given by

$$\begin{aligned} F = & \frac{1}{2}C'(x,y)\left(U_{\mathrm{tip}} - U_{\mathrm{2DEG}} + \Delta\mu_{\mathrm{ch}}(x,y)/e\right)^2 \\ & + Q_{\mathrm{tip}}^{(0)\prime}(x,y)\left(U_{\mathrm{tip}} - U_{\mathrm{2DEG}} + \Delta\mu_{\mathrm{ch}}(x,y)/e\right) \end{aligned} \tag{16.1}$$

in a classical capacitive model. Here, $C'(x,y) = \partial C(x,y,z)/\partial z$ is the z-derivative of the position-dependent capacitance between tip and 2DEG, $Q_{\mathrm{tip}}^{(0)\prime}$ is the z-derivative of the charge induced on the tip at zero bias due to fixed charges in the system (e.g., surface charges and ionized dopants), $\Delta\mu_{\mathrm{ch}}(x,y)$ is the local work function difference between tip and sample, U_{tip} is the voltage on the tip, $U_{\mathrm{2DEG}} = -\mu_{\mathrm{elch}}(x,y)/e$ is the local voltage in the 2DEG under the tip. In equilibrium the latter quantity is position independent. The idea of the measurement is to apply an AC transport current through the Hall bar sample at frequency ω which induces a position-dependent $U_{\mathrm{2DEG}}^{\mathrm{AC,tr}}(x,y)$ reflecting the Hall voltage profile. The resulting AC force

$$\begin{aligned} F_{\mathrm{AC}}^{\mathrm{tr}} = & -\left[C'(x,y)\left(U_{\mathrm{tip}} + \Delta\mu_{\mathrm{ch}}(x,y)/e\right)\right. \\ & \left. + Q_{\mathrm{tip}}^{(0)\prime}(x,y)\right] U_{\mathrm{2DEG}}^{\mathrm{AC,tr}}(x,y) \end{aligned}$$

is used for driving the cantilever at its resonance frequency at 120 kHz. The measured oscillation amplitude is proportional to the driving force (enhanced by the Q-factor of the oscillating cantilever on resonance) and therefore to the local electrochemical potential in the two-dimensional electron gas. At a fixed magnetic field the other quantities (capacitance and induced charge) may also contribute to the position dependence of the measured force, especially in the quantum Hall regime, where localization of electronic charge in the bulk of the Hall bar plays a crucial role and the time constants for establishing an equilibrium electrochemical potential in the bulk can be very long. A suitable reference measurement, in which the AC voltage is applied to all contacts of the Hall bar, implying that no transport current flows and no Hall voltage builds up, can be used for eliminating the undesired prefactor. In this case, the measured AC force, $F_{\mathrm{AC}}^{\mathrm{eq}}$, will be given by the above expression for $F_{\mathrm{AC}}^{\mathrm{tr}}$ with the same prefactor, but a position-dependent equilibrium voltage $U_{\mathrm{2DEG}}^{\mathrm{AC,eq}}$. Normalizing $F_{\mathrm{AC}}^{\mathrm{tr}}$ by $F_{\mathrm{AC}}^{\mathrm{eq}}$ gives

$$\frac{F_{\rm AC}^{\rm tr}}{F_{\rm AC}^{\rm eq}} = \frac{U_{\rm 2DEG}^{\rm tr}(x,y)}{U_{\rm 2DEG}^{\rm eq}(x,y)}.$$

The normalized Hall voltage profile measured in this experiment at a given position along the Hall bar is found to be periodic in the inverse magnetic field $1/B$, in agreement with a general consideration of the effect of B on the voltage $U_{\rm 2DEG}^{\rm AC}(x, y; B)$. The normalized Hall voltage profile at different filling factors ν around $\nu = 6$ shows a transition between a linear behavior above and below the plateau in the Hall resistance and a profile with a flat region in the bulk with all the voltage dropping near the edges of the sample. This implies that sample edges and bulk do not equilibrate at these fields, implying that the sample edges are effectively decoupled from the bulk. High DC transport currents with creating Hall voltages above the Landau level spacing $\hbar\omega_{\rm c}$ lead to equilibration of edge channels. No compressible or incompressible strips were observed in this measurement, but the width of the edge state regions is found to increase with the number of edge channels with a width of about 300 nm per channel. In the flat region of the Hall resistance plateau, the normalized Hall voltage profile was found to develop significant gradients within the bulk of the sample.

Related experiments for the measurement of Hall voltage profiles were performed by another group using a different technique [366, 368, 374, 447]. In contrast to the previous method the force [eq. (16.1)] is not measured in this experiment, but the force gradient

$$\begin{aligned} F' = & \frac{1}{2}C''(x,y)\left(U_{\rm tip} - U_{\rm 2DEG} + \Delta\mu_{\rm ch}(x,y)/e\right)^2 \\ & + Q_{\rm tip}^{(0)''}(x,y)\left(U_{\rm tip} - U_{\rm 2DEG} + \Delta\mu_{\rm ch}(x,y)/e\right) \end{aligned}$$

This is accomplished by driving the iridium-coated piezoresistive cantilever externally on resonance and measuring the resonance frequency shift caused by the electrostatic force gradient with a feedback loop. Apart from this difference, the technique corresponds to the one presented above: a few Hz AC transport current is applied to the source contact of the 2DEG with the drain contact grounded. This induces a local low-frequency variation of the electrochemical potential, which in turn leads to a position-dependent low-frequency modulation of the frequency shift that can be detected using lock-in techniques. The position-dependent prefactor of $U_{\rm 2DEG}^{\rm AC,tr}$ is again eliminated by normalizing to the result of a reference measurements in which the AC voltage is applied to all the contacts of the electron gas. An advantage of not driving the cantilever with the tip-sample voltage lies in the fact that the very low modulation frequency of 3.4 Hz (compared to 100 kHz used above) can be used which minimizes RC effects in the charging of regions in the 2DEG with localized states.

Three different types of Hall potential profiles are observed in these experiments depending on the value of the filling factor ν: linear Hall potential drops are observed up to $\nu = 1.76$ when $\nu = 2$ is approached from lower filling factors. At filling factors around $\nu = 2$ the Hall potential drops rather arbitrarily in the bulk of the Hall bar. At filling factor $\nu = 2.14$ the potential becomes flat in the bulk and all the potential drop occurs near the edges of the sample. Proceeding to lower filling factors the same scenario repeats around filling factor $\nu = 4$. The behavior is consistent with the experimental results discussed before and it is related to the notion of compressible and incompressible regions in Ref. [374]. It is shown that the equilibrium center positions of the innermost incompressible strip does indeed move with ν according to a theoretical prediction [448].

With this technique the Hall potential distribution was also investigated near potential probe contacts [447]. An incompressible strip is found at filling factors slightly above integer values, which is not only present along the mesa edges, but also along the metal contact. It is conluded that the bulk is strongly decoupled from the edge and the ohmic contacts at these filling factors.

16.2.3 Subsurface charge accumulation imaging

The third set of experiments to be discussed here uses the so-called subsurface charge accumulation imaging in the quantum Hall regime at a temperature of 300 mK [388, 444–446, 449, 450]. This technique measures the local accumulation of charge in a two-dimensional electron gas in response to an AC voltage (100 kHz) applied between the conducting tip scanned at constant height above the surface and the 2DEG. The DC perturbation of the tip on the 2DEG is minimized by compensating the work function difference between the tip and the sample with a DC tip voltage of 0.4 V. The tip is connected to a sensitive charge detector, which relies on a field-effect transistor mounted at cryogenic temperatures [390]. Signals in-phase and out-of-phase with the AC excitation are detected with lock-in techniques. Because most of the electric field lines between 2DEG and tip terminate far away from the tip apex, there is a large background signal adding to the local signal of interest. Detailed fine structure on a length scale of about 100 nm can be resolved on the resulting images acquired in the quantum Hall regime, if the background signal is subtracted employing a bridge circuit [390]. Some of the observed structure in the experimental images can be explained using the idea of compressible and incompressible strips. The suppression of the local charging signal can arise due to two distinct reasons: either the local compressibility of the electron gas below the tip apex is low (low density of states at the electrochemical potential) or the conductivity of the 2DEG surrounding the tip position is too low to allow charging under the tip within half a period of the excitation (RC effect).

Possible models needed for an interpretation of the data are discussed in Ref. [450]. The simplest model entirely neglects RC effects and describes the measured capacitive signal as arising from a series connection of a plate capacitor with a geometric capacitance C_{g} and a quantum capacitance contribution $C_{\mathrm{q}} = e^2\mathcal{D}(B)$, which is proportional to the local magnetic-field-dependent density of states $\mathcal{D}(x, y; B)$. The most involved model treats the electron gas on a discrete lattice with given local conductances and capacitances between neighboring sites. This conductance and capacitance pattern has to be optimized until the simulation fits the measured image.

The charge sensor is sensitive enough to detect individual charges entering the region under the tip. At suitable filling factors and tip-sample voltage, a quantum dot can be formed under the tip which contains only a small number of electrons separated from the surrounding electron gas by a circular incompressible strip. Single-electron charging of this dot has been observed [445], and the varying number of electrons in this dot as it is dragged along with the tip has been used for mapping the local potential in the Hall liquid.

16.2.4 Local modification of inter-edge-channel tunneling with a scanning force microscope

In this section we discuss another type of local edge state imaging in the quantum Hall regime, which is also based on low-temperature scanning probe techniques [451, 452]. In this approach the tunneling current is measured between edge channels that are separately contacted. The tunneling current is locally modified using a potential perturbation induced by the conducting tip of a scanning force microscope.

The samples are based on an GaAs/AlGaAs heterostructure with the two-dimensional electron gas residing 34 nm below the sample surface. Figure 16.2 shows a photograph of the structure, which was prepared using photolithographic techniques. The structure is essentially a circular mesa with a diameter of about 500 μm with a central hole of 20 μm in diameter. The sample is connected to the measurement setup via the internal ohmic contacts C1 and C2. A star-shaped gate electrode splits the two-dimensional electron gas into four ungated sectors.

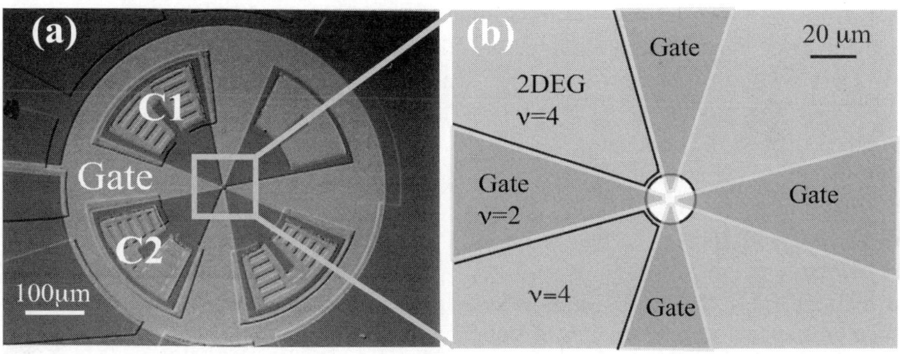

Fig. 16.2. (a) Photography of the circular mesa structure with the internal contacts C1 and C2. A small circular hole is etched into the center of the structure. (b) Schematic blow-up of the central part of the sample. For appropriate choice of U_G and B the indicated filling factors can be achieved leading to the distribution of edge channels as indicated by the black lines.

The sample is mounted in a home-built low-temperature atomic force microscope utilizing a piezoelectric quartz tuning fork force-sensor. The basic principle of the microscope and the characteristics of the tuning fork sensors have been described above and can be found in Refs. [407, 410, 414]. Experiments were performed in the variable temperature insert of a standard ^{4}He cryostat at a temperature of 1.9 K. A voltage of 10 μV was applied between C1 and C2 and the two-terminal conductance was measured with a current-voltage converter.

Figure 16.2(b) shows a schematic magnification of the central part of the structure for illustrating the basic concept of our experiment. Using the appropriate combination of gate voltage, U_G, and magnetic field, B, we are able to set up a situation in which the bulk of the ungated electron gas has a Landau level filling factor $\nu = 4$, i.e., two spin-degenerate edge channels exist, while the gated regions have a filling factor of $\nu = 2$ supporting only one spin-degenerate edge channel. The latter will be able to circulate around the central hole of the structure as shown in the figure. The other edge channel is not allowed under the gate. It will therefore come from C1 (or C2) along the edge of the gate, run in parallel with the $\nu = 2$ edge channel along the edge of the central hole, and then return to contact C1 (or C2) along the other edge of the gate. Since under these conditions the bulk regions of the gated as well as of the ungated 2DEG are insulating, the current from C1 to C2 involves tunneling processes between the two edge channels. Spatially, tunneling has to occur in regions where both edge channels run in parallel, i.e., in the vicinity of the edge of the central hole. Conceptually, this tunneling transport is somewhat related to recent experiments on tunneling through the edge states around an antidot [453], however, the quantization of edge states around the central hole (the "antidot") plays no role in our structure. The central idea of the present experiment is that the tunneling current between the two edge channels can be locally enhanced (or suppressed) by applying a local potential perturbation capacitively induced by the conducting tip.

In order to find the appropriate settings for U_G and B we characterized the structure by measuring the two-terminal conductance between contacts C1 and C2 as a function of these two parameters with the tip withdrawn from the surface of the structure. The result is shown as a gray scale image in Fig. 16.3.

It can be seen that at $U_G < -0.08$ V the regions under the gate are completely pinched off and the conductance vanishes. At $U_G > -0.05$ V the gray scale plot is dominated by the Landau fan given by the 2DEG under the gate. The conductance oscillates as a function of B due to the Shubnikov–de Haas effect. The contribution of the ungated 2DEG can barely be seen in this range of U_G. However, in a very small range of gate voltage around $U_G \approx -0.07$ V vertical cuts through the plot reveal $1/B$-periodic oscillations, which are independent of U_G but can hardly be seen on the gray scale plot, i.e., we observe the Shubnikov–de Haas oscillations of the ungated regions. They lead to the bright horizontal lines in the figure where the filling factors for the ungated regions are indicated. It can be seen from the figure that at $B \approx 5$ T and $U_G \approx 0.02$ V we achieve the desired $\nu = 2$ under the gate, while in the ungated regions we have $\nu = 4$.

While keeping the sample under these conditions, the tip is now scanned across the sample surface near the edge of the inner hole. The tip was kept at a constant voltage $U_t = 0$ V, but owing to the work function difference between the tip material PtIr and the sample there exists an effective electrostatic potential drop between tip and 2DEG, which locally depletes the electron gas. Figure 16.4(a) shows a 5 μm×5 μm topographical image of the edge of the central hole of the sample. In Fig. 16.4(b) the conductance image can be seen measured simultaneously with the topography. The

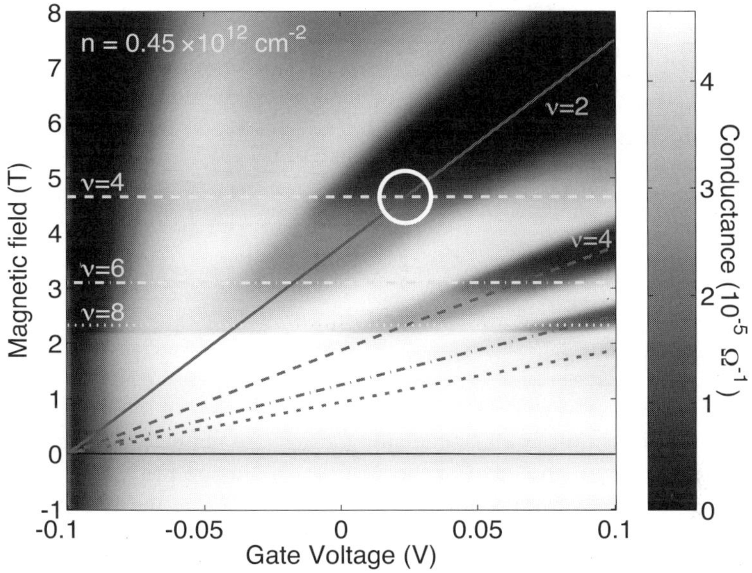

Fig. 16.3. Conductance measured as a function of magnetic field and gate voltage. The bright horizontal lines indicate the filling factors in the ungated regions of the 2DEG. The gray Landau fan indicates filling factors in the gated regions of the 2DEG.

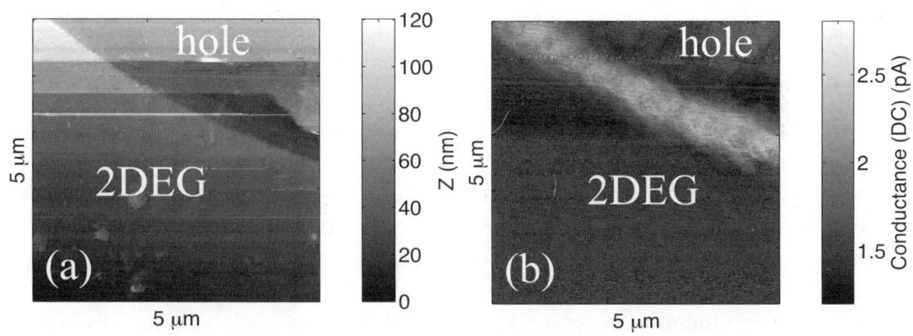

Fig. 16.4. (a) Topographic image of the mesa edge defining the central hole etched into the 2DEG. (b) Conductance as a function of the tip position. The tip induced local potential leads to an enhanced tunneling coupling of the two parallel edge channels and thereby to an increase in the conductance.

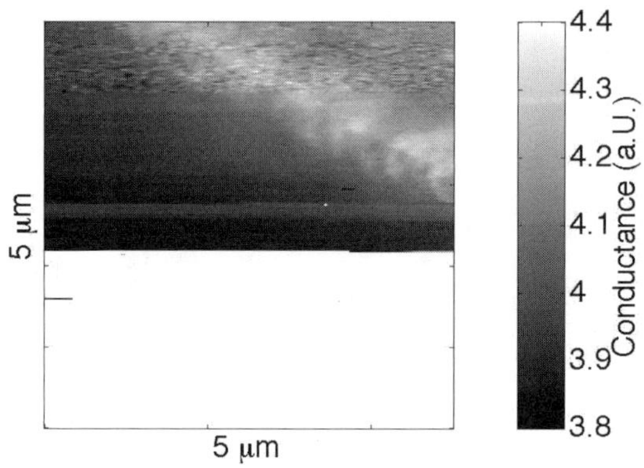

Fig. 16.5. Image of the edge-channel coupling at enhanced spatial resolution.

conductance of the sample is enhanced along a stripe of about 700 nm width, which follows the curvature of the edge of the 2DEG. The image proves in a direct way that we can indeed enhance the tunneling coupling between the two parallel edge channels by applying a local potential perturbation.

It is well established that the conductance or resistance of a sample in the quantum Hall regime is highly non-local [49, 454] and the phase-coherence length of electrons in edge channels can be macroscopically large. Therefore, it is not *a priori* clear that the local perturbation induced by the presence of the tip will change the tunneling coupling of the edge channels *locally*. Yet we argue that the self-consistent nature of the edge channel formation may well lead to a local enhancement of the tunneling coupling. According to the adiabatic edge channel description [30], edge channels will follow equipotential lines along the sample edge. The local perturbing potential will affect the spatial run of the equipotential lines locally (i.e., on the scale of the screening length) and therefore we expect the edge channels to follow this perturbation on a local scale. This scenario is schematically depicted in Fig. 16.6. Edge states form along the mesa edge as well as around the tip-induced antidot-like potential. When the tip is sufficiently remote from the edge [Fig. 16.6(a)], there is no mutual interaction between the two types of edges. When the tip is close enough to the mesa edge, the edge channel structure will change locally as illustrated in Fig. 16.6(b). The tunneling coupling between the edge channels depends on the exponential overlap of the wave functions and therefore on the width of the incompressible strip separating them. A spatial rearrangement of the self-consistent strips will change the inter-edge-channel coupling. We suggest that the *local* change of the tunneling coupling creates the observed conductance contrast in Fig. 16.4.

The local nature of the conductance contrast makes the presented imaging method a promising tool for the local investigation of edge channel coupling and inter-edge-channel scattering (see Ref. [455] for a review, also [456]). As a matter of fact it can already be observed in Fig. 16.4(b) that the bright stripe is not homogeneous but exhibits some internal structure. In order to illustrate this further, we show in Fig. 16.5 the conductance image obtained in the course of the same cool down of the sample but with the tip in better condition. Unfortunately the tip shape can change

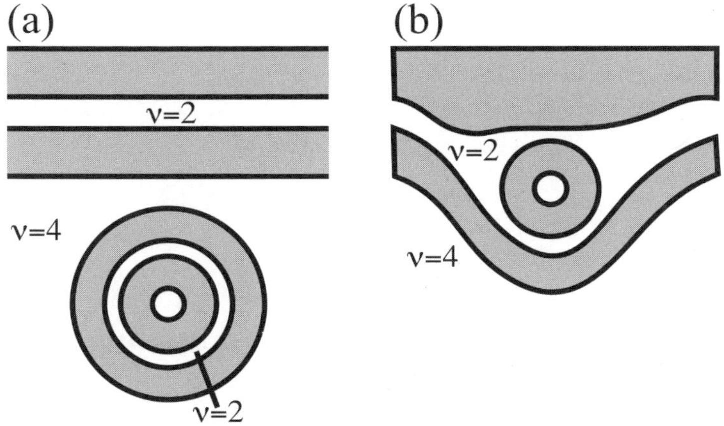

Fig. 16.6. (a) Tip far away from the edge. To edge states encircle the tip-induced potential hill. States at the sample edge are not affected. (b) Tip close to the edge of the sample. The tip induced potential locally changes the run of the edge currents and thereby modifies the tunneling coupling between them.

during a cool down, e.g., due to contaminations picked up from the sample surface. In some fortunate cases, however, the spatial resolution of the images is improved. In Fig. 16.5 it is clearly visible that the extent to which the edge channel coupling can be enhanced depends strongly on the position of the tip even within the strip of enhanced conductance. The image shows fringes leading away from the center of the stripe reminding of the fascinating fine structure observed with other methods in the bulk of 2DEGs in the quantum Hall regime [388]. More investigations will help to further establish the presented imaging technique and to deepen our understanding of its interpretation.

17

Scanning gate measurements on a quantum wire

In this chapter we present measurements on a semiconductor quantum wire in which we induce a local potential perturbation with the metallic tip of a scanning force microscope. Measurement of the sample resistance as a function of tip position results in an electrical map of the wire in real space. We find the fingerprint of potential fluctuations in the wire, which appear as local resistance fluctuations in the images. In a local transconductance measurement we observe small oscillations on the scale of the Fermi wavelength of electrons, which may arise from interference of electron waves. Related scanning gate measurements of other groups will be briefly addressed in the following introduction.

17.1 Introduction to scanning gate measurements on mesoscopic systems

Electronic transport in mesoscopic semiconductor structures at low temperatures is non-local and phase coherent [34]. In a conventional transport experiment, in which the resistance or conductance of a sample is typically measured as a function of gate voltage or magnetic field, the measured conductance is given by the transmission probabilities between the leads connecting the system to the measurement circuit [33]. In general, it is not possible to reconstruct the scattering potential within the mesoscopic system from the knowledge of the transmission matrix. However, additional information about the quantum states within the device can be obtained from measurements that apply a local perturbation to the system [344, 362, 363, 365, 369–373, 457]. While in Ref. [344] the local potential perturbation was fixed and the two-dimensional electron gas in a parabolic quantum well could be displaced, in Refs. [362, 363, 365, 369–373, 457] the local electrostatic potential induced by the tip of a scanning force microscope was utilized. In the following, we briefly review the experimental work published in recent years.

In the first paper reporting this "scanning gate technique" Eriksson and coworkers [362] investigated the ballistic electron flow through a quantum point contact at a temperature of 4.2 K. They found that the resistance change caused by the tip-induced potential decay exponentially with distance from the point contact in agreement with elastic scattering by the disorder potential in the collector electron gas. In addition, they measured the angular distribution of the transmitted electrons showing the collimation of the ballistic electron beam close to the point contact. Calculations were presented in this paper indicating that the spatial resolution of the technique is comparable to the electron gas depth below the surface. Similar experiments on a quantum point contact have been reported in Ref. [369].

Another measurement on a quantum point contact was performed by Crook and coworkers [371] at 4.2 K, where the authors were able to map the transmitted lateral modes with remarkable spatial resolution of about 40 nm although their electron gas was buried 100 nm below the surface. This experiment is the analog of the "wave-function mapping" experiment reported in Ref. [344].

Recently, the electron flow past a quantum point contact was imaged with high spatial resolution achieved at 4.2 K on a GaAs/AlGaAs heterostructure with an only 57 nm thick cap layer above the electron gas [373]. It was shown in the experiment and also modeled theoretically that the classical contribution of the electron flow branches in the two-dimensional collector, a result which had not been anticipated by other experiments. In addition, quantum interference effects were observed in these highly sensitive measurements.

Experiments on two constrictions in series for a "magnetic steering experiment" were reported in Ref. [365, 370]. Classical cyclotron orbits of ballistic electrons in a magnetic field with the radius $R_{\mathrm{c}} = \hbar k_{\mathrm{F}}/(eB)$ could be mapped out with a spatial resolution of about 200 nm.

Scanning gate microscopy (SGM) has been applied to the investigation of electron transport in carbon nanotubes [458–460] at room temperature. The authors were able to "directly image individual scattering sites in semiconducting single-wall nanotubes" [459] and metallic single-wall nanotubes [460] or an inhomogeneous potential landscape [458], again in semiconducting single-walled tubes. They showed that the resistance image of a nanotube shows hot spots in locations where the electric field drop was found to be high with electrostatic force microscopy (EFM) [459]. In some cases these hot spots occurred in positions where the nanotube had a kink allowing one to correlate information from three different sources (topography, EFM, and SGM) to extract a unified picture of the system.

A very recent publication by Woodside and McEuen reports low-temperature (0.6 K) scanning gate studies of single-walled carbon nanotubes in which one or more quantum dots form leading to the Coulomb-blockade effect in the conductance [457]. These authors find Coulomb-blockade oscillations as a function of the position of the scanning tip allowing them to locate the quantum dots along a certain nanotube and to estimate their sizes.

In our experiments we adopt the scanning gate approach for the investigation of the transport properties of a semiconductor quantum wire at cryogenic temperatures.

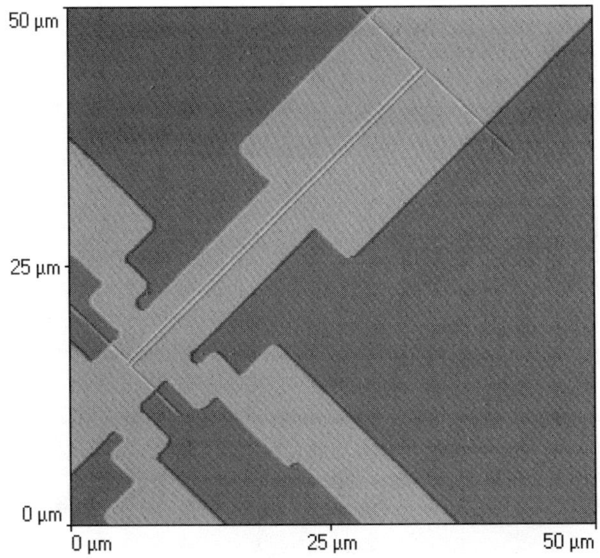

Fig. 17.1. SFM image of the Hall bar with the oxide lines defining the quantum wire.

17.2 Samples and structures

The sample is based on a GaAs/AlGaAs heterostructure in which the heterointerface is buried 34 nm below the surface. A Hall-bar geometry, a part of which can be seen in Fig. 17.1, was patterned using photolithography. At the temperature $T = 1.7\,\mathrm{K}$ the bulk electron density measured in this system is $n_\mathrm{s} = 4.3 \times 10^{11}\,\mathrm{cm}^{-2}$, corresponding to a Fermi wavelength of 38 nm. The mobility at $T = 1.7\,\mathrm{K}$ is $\mu = 10^6\,\mathrm{cm}^2/(\mathrm{Vs})$ corresponding to a mean free path of $l_\mathrm{e}^{2\mathrm{D}} = 10.8\,\mu\mathrm{m}$ and a diffusion constant of $D = 1.52\,\mathrm{m}^2/\mathrm{s}$.

The quantum wire has been patterned using the so-called AFM lithography[1] [316] directly onto the Hall-bar structure. In Fig. 17.1 the white lines on the Hall bar are the SFM-written oxide lines that deplete the 2DEG below them and thereby form a long quantum wire flanked by two separately contacted regions of the 2DEG, which we used as in-plane gates. The wire length is $L = 40\,\mu\mathrm{m}$ and its lithographic width is $W = 400\,\mathrm{nm}$. The lateral depletion length is typically $15 - 20\,\mathrm{nm}$ such that the number of modes in the channel can be estimated to be $N = 20$. This number can be confirmed by Shubnikov–de Haas measurements that count the number of modes which become successively depopulated as the magnetic field is increased.

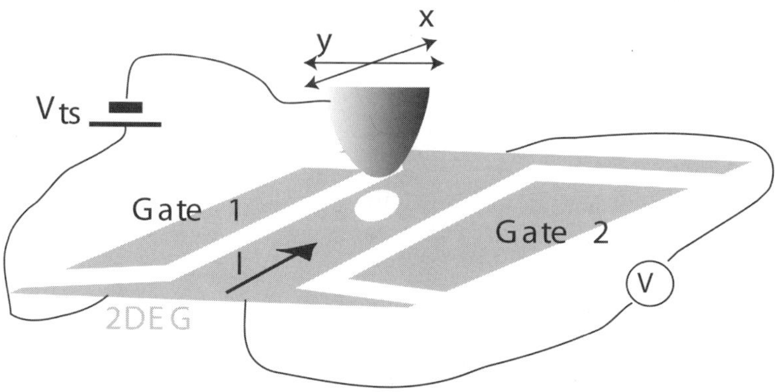

Fig. 17.2. Principle of scanning gate measurements

17.3 Results of low-temperature scanning gate measurements on a quantum wire

The sample was mounted in the sample holder of a low-temperature scanning force microscope (SFM) [407]. This microscope utilizes piezoelectric quartz tuning forks with a metallic tip attached to one prong as the force sensor. Operation characteristics at low temperatures and sensor calibration have been described above and reported in Refs. [410, 414]. The microscope was cooled to $T = 1.7\,\mathrm{K}$ in the He gas flow of a variable temperature insert. The resistance of the wire was measured at an AC current of 20 nA in a four-probe configuration. The measurement frequency was 421 Hz and a lock-in time constant of 10 ms was used in order to obtain a reasonably large output bandwidth. The tungsten tip of the SFM was kept grounded while it was oscillating with an amplitude of about 1 nm normal to the surface at the tuning fork resonance frequency of about 32 kHz. The basic idea of the measurement is shown in Fig. 17.2. The tip induces a local scattering potential in the electron gas. The wire resistance is changed depending on the position of the tip. Resistance images of 256×256 points were taken at scan speeds of 500 nm/s.

Figure 17.3 shows the resistance image of a $5 \times 5\,\mu\mathrm{m}^2$ area obtained with a voltage of +200 mV on one of the in-plane gates. The fact that the wire can be electronically imaged indicates that, when the tip is above the wire, the resistance is enhanced. This resistance contrast is due to the fact that tungsten and the heterostructure have a work function difference of the order of 100 mV that causes the tip to induce a local repulsive electrostatic potential in the 2DEG which is scanned along when the tip moves. The stripe of enhanced resistance is not visible in the topmost quarter of the image. From the simultaneously measured topography we know that the wire ends there.

Cross sections through this image as indicated by three horizontal lines in Fig. 17.3 are shown in Fig. 17.4. As the tip moves across the wire, the resistance

[1] AFM stands for Atomic Force Microscope, synonym for Scanning Force Microscope, SFM

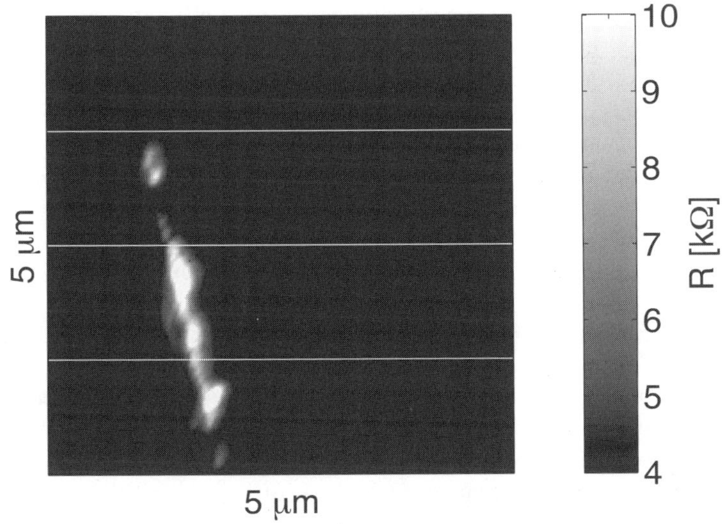

Fig. 17.3. Resistance image of the quantum wire measured at $T = 1.7\,K$

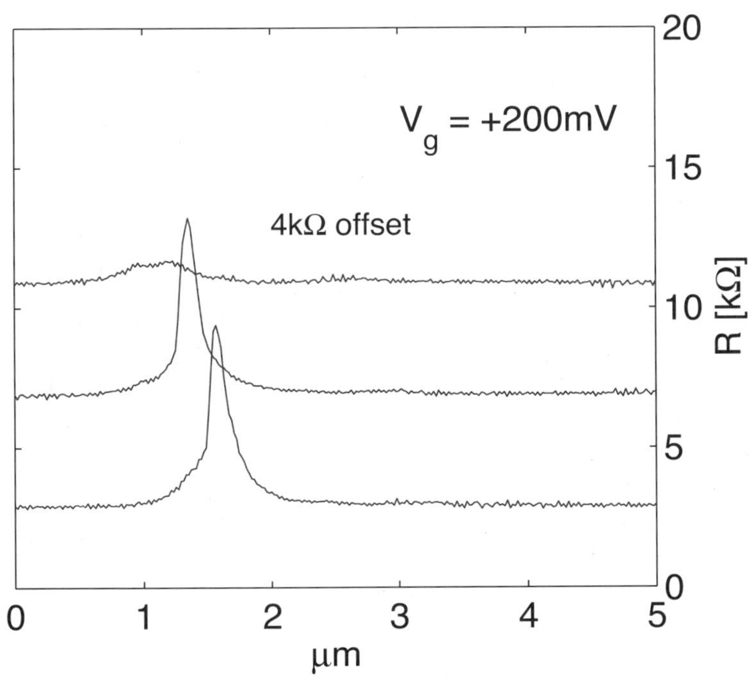

Fig. 17.4. Cross sections through the resistance image shown in Fig. 17.3.

goes from its base value of $3\,\mathrm{k}\Omega$ through a pronounced peak of $9.4\,\mathrm{k}\Omega$. We estimate the number of modes that become locally depleted, ΔN, with a simple model that regards the repulsive tip potential as the cause of an additional resistance ΔR which adds to the background wire resistance R_0. In the simplest case ΔR can be modeled by a short piece of wire with transmission $T = 1$, which is not mode-matched to the rest of the wire. In this case $\Delta R = h/(2e^2)(1/(N_0 - \Delta N) - 1/N_0)$ with $N_0 = 20$ being the number of modes in the wire in the absence of the tip. This gives a substantial depletion of $\Delta N = 18$ modes if the tip is centered above the wire.

From the width of the two lower peaks of about 400 nm, which is close to the geometrical width of the wire, we estimate that the tip-induced potential perturbation cannot be much wider than that.

A very striking feature of the image in Fig. 17.3 is the variation of the resistance along the wire direction. In a perfect wire with no potential fluctuations and perfectly smooth boundaries, no such variation would be expected irrespective of the exact geometrical shape of the tip-induced potential. From a comparison of the resistance variations with the surface topography, we find no correlation between the two that could explain the former as the result of a varying tip-2DEG separation. We therefore conclude that the variations of the resistance along the wire direction reflect the roughness of the potential landscape in the electron gas.

There are two distinct effects in mesoscopic systems that could cause the observed resistance fluctuations as the perturbation is moved along the wire. The first is the ballistic chaotic motion of *classical* electrons in a spatially varying potential landscape. In order to develop a better understanding of how this mechanism would appear in our measurements, we have calculated the classical transmission through wire structures with a given potential landscape as a function of the position of an additional external perturbation. The results which will be discussed in the next section show that the complex electron dynamics does indeed lead to fluctuations in the resistance image.

The second effect that could produce resistance fluctuations along the wire is the *quantum interference* of phase coherent paths, which leads to fluctuations in the transmission as a function of tip position. Theories exist that predict conductance fluctuations in a disordered sample when the position of a single impurity is moved [461–464]. Given our sample parameters and the measurement temperature the phase coherence length can be estimated to be $\ell_\varphi \approx 8\,\mu\mathrm{m}$ but the observability of phase coherence effects is limited by energy averaging through the Fermi-distribution function, i.e.,, the thermal length $\ell_T = \sqrt{\hbar D/(kT)} \approx 2\,\mu\mathrm{m}$. Fluctuations will have the order of magnitude $\Delta R = R^2 \Delta G \approx R^2 e^2/h(\ell_T/\ell_\varphi)(\ell_\varphi/L)^{3/2} \approx 100\,\Omega$. The characteristic length scale for these fluctuations is the Fermi wavelength of the electrons.

Figure 17.5 shows an image of the wire measured in a different measurement configuration, where a DC current was applied and the wire resistance was measured at the resonance frequency of the oscillating tip effectively resulting in a local transconductance measurement. This image taken with increased spatial resolution exhibits amazingly regular stripe-like patterns on a length scale of less than 100 nm. Given the above considerations about phase coherence, the tentative explanation in

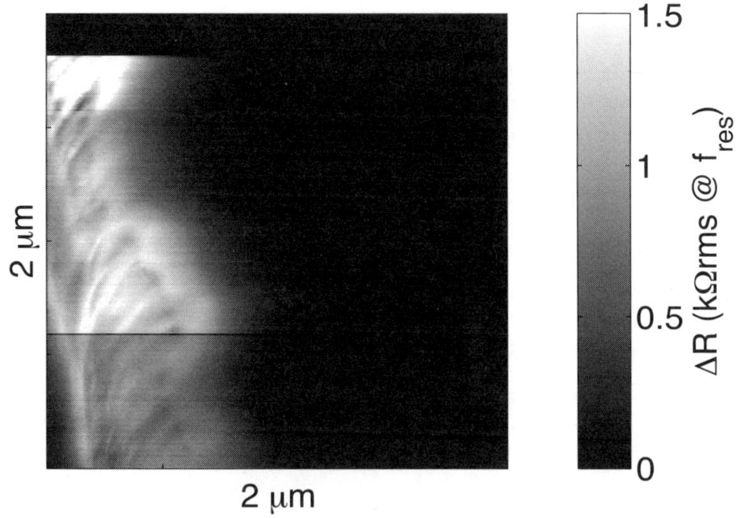

Fig. 17.5. Interference effects in the quantum wire.

terms of interference effects seems reasonable. Such quantum interference effects in scanning gate measurements will be further discussed in the next section.

Summarizing the experiments, we have investigated the mesoscopic transport through a quantum wire as a function of a local SFM-tip-induced perturbation. The resistance image shows fluctuations along the wire which are due to the classical ballistic motion of electrons and quantum interference effects.

17.4 Modeling scanning gate measurements: Classical and quantum effects

Models for conductance fluctuations have been based on different approaches. Among them are the Anderson tight-binding Hamiltonian approach [464, 465], one-dimensional models incorporating randomly spaced delta scatterers [466, 467], and a scattering matrix approach [468].

In order to obtain some insight into the physical processes leading to the observed resistance images, we applied two types of theoretical models. The first model aims at classical effects in the resistance. It is based on a calculation of the classical transmission of a wire within a billiard model following the general approach introduced by Beenakker and van Houten in Ref. [469]. The second model is a fully quantum mechanical model based on the picture of non-interacting electrons in a multimoded wire. It is closely related to a scattering matrix approach used by Cahay and coworkers [468] for calculating the fluctuations in the resistance of a multimoded wire with randomly placed delta scatterers.

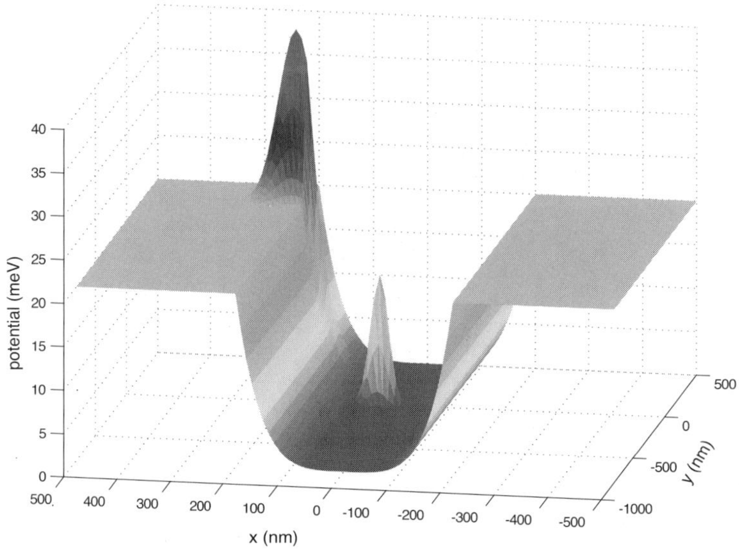

Fig. 17.6. Potential used for the billiard simulation. The fixed Gaussian scatterer is centered at $(x = 30\,\mathrm{nm}, y = 0)$. The tip-induced potential can be seen just at the edge of the wire potential.

17.4.1 Classical billiard model

Within the billiard model we describe the quantum wire with a confinement potential $V(y)$ in y-direction. Electron transport is in x-direction. We consider a single repulsive Gaussian scatterer with potential $V_{\mathrm{s}}(x, y)$ in the wire and model the tip-induced potential with a second Gaussian scatterer [potential $V_{\mathrm{t}}(x, y; x_0, y_0)$] whose position (x_0, y_0) can be varied corresponding to the tip motion. The corresponding classical Hamiltonian function is

$$H(p_x, p_y, x, y) = \frac{p_x^2 + p_y^2}{2m^\star} + V(y) + V_{\mathrm{s}}(x, y) + V_{\mathrm{t}}(x, y; x_0, y_0).$$

The total potential for a certain (x_0, y_0)-position of the tip is depicted in Fig. 17.6.

The strictly two-dimensional classical electron dynamics are described by Newton's equations of motion

$$m^\star \frac{\partial^2 x}{\partial t^2} = -\frac{\partial V_{\mathrm{s}}(x, y)}{\partial x} - \frac{\partial V_{\mathrm{t}}(x, y; x_0, y_0)}{\partial x},$$
$$m^\star \frac{\partial^2 y}{\partial t^2} = -\frac{\partial V(y)}{\partial y} - \frac{\partial V_{\mathrm{s}}(x, y)}{\partial y} - \frac{\partial V_{\mathrm{t}}(x, y; x_0, y_0)}{\partial y}.$$

We solve these equations numerically for given initial conditions at time $t = 0$, i.e., for given starting point $(x_{\mathrm{s}}, y_{\mathrm{s}})$ of the electron and for given injection angle θ. The magnitude of the velocity is the Fermi velocity v_{F}. The trajectory of a single electron is calculated until it leaves the structure, i.e., it crosses the line $x_s = const.$, where it started, or it crosses a line $x_s + L = const.$, where L is chosen large enough to ensure the absence of back-scattering. From the calculation of typically $10^4 - 10^5$ trajectories injected from a line of constant x_s at all angles $-\pi < \theta < \pi$ with a distribution $p(\theta) = \cos(\theta)/2$, we determine the classical transmission and reflection probability T and $1 - T$, respectively. The resistance of the wire, which depends on the position x_0, y_0 of the tip-induced potential, is then given by the formula

$$R(x_0, y_0) = \frac{h}{2e^2} \frac{1}{N} \frac{1 - T(x_0, y_0)}{T(x_0, y_0)},$$

where N is the number of channels in the unperturbed wire. This relation for the resistance is derived from the Landauer–Büttiker conductance formula (2.4)

$$G = \frac{2e^2}{h} NT,$$

by inversion and subtraction of the "contact resistance" $R_c = (h/2e^2)(1/N)$ [31].

Figure 17.7 shows the result of such a simulation. The wire potential has a width of 342 nm at the Fermi energy of 10 meV. The tip-induced potential is 20 meV high and has a half maximum diameter of 75 nm. The fixed scatterer has a potential height of 10 meV and a half maximum diameter of 37 nm. It is centered at $(x = 0, y = 30\,\text{nm})$. It can be seen that, similar to the experiment and to our expectation, the resistance is generally enhanced when the tip-induced potential is within the wire. At the position of the fixed scatterer there is a dip (darker region) in the resistance map, while left and right of the scatterer there are brighter regions of higher resistance. This result is what is intuitively expected: when the tip-induced potential sits on top of the fixed scatterer, its effect is reduced because it just slightly enhances the scattering potential that is already there. However, left (right) of the scatterer, the tip-induced potential blocks the left (right) channel passing the fixed scatterer thus leading to a further reduction of the effective wire width and therefore to a stronger resistance increase than elsewhere. The resistance image even appears to be wider at this x-position although there is no boundary roughness in the system. In the presence of impurity scattering in the wire, the width of the resistance image can therefore not be taken as a direct local measure of the wire width.

Upstream and downstream of the fixed impurity the situation is less intuitive. The picture shows a fluctuating resistance as a function of tip position, which can be called "classical resistance fluctuations." The origin of these fluctuations is the chaotic nature of electron motion in such a two-impurity problem. However, their size is of the order of a few percent of the average resistance, which is way too small to account for the experimentally observed fluctuations. The occurrence of such fluctuations means that even a classical description of electron motion in such a wire system cannot necessarily be interpreted in terms of *local* properties (e.g., local scattering sites) unambiguously. Strong local scattering potentials will nevertheless tend to have strong local signatures in the resistance image.

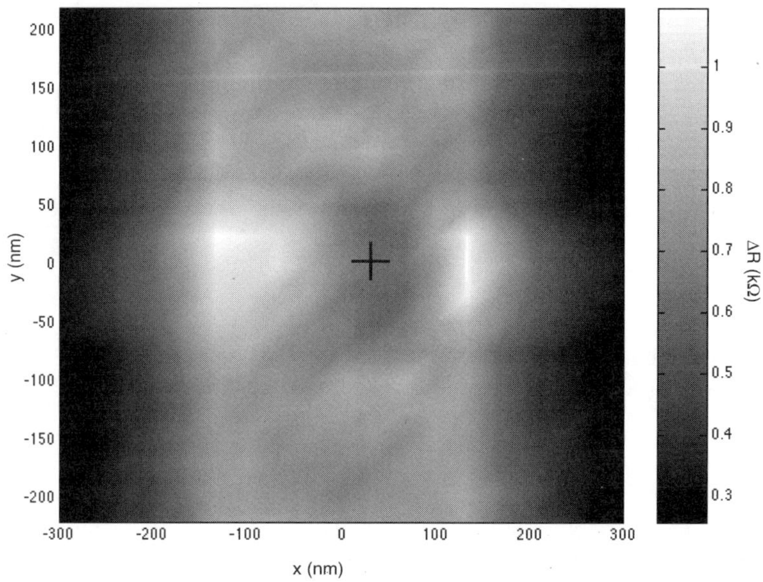

Fig. 17.7. Resistance image of a quantum wire with a width of 342 nm and a fixed local Gaussian impurity potential centered at ($x = 30\,\mathrm{nm}, y = 0$). See text for more detailed parameters of the simulation.

17.4.2 Quantum description of scattering in wires

A key feature missing in the classical description of the scanning gate measurement in a wire is quantum interference. In order to incorporate this effect we describe in the following the scanning gate measurement based on a quantum description.

We consider the Hamiltonian of a two-dimensional quantum wire with confinement potential $V_{\mathrm{c}}(y)$ and some scattering potential $V_{\mathrm{s}}(x, y)$. The Schrödinger equation reads

$$\left[-\frac{\hbar^2}{2m^\star}\left(\frac{\partial^2}{\partial x^2}+\frac{\partial^2}{\partial y^2}\right)+V_{\mathrm{c}}(y)+V_{\mathrm{s}}(x,y)\right]\psi(x,y)=E\psi(x,y).$$

We assume that we have solved the pure wire problem without disorder, i.e.,

$$\left[-\frac{\hbar^2}{2m^\star}\frac{\partial^2}{\partial y^2}+V_{\mathrm{c}}(y)\right]\chi_n(y)=\epsilon_n\chi_n(y).$$

We can then expand the solutions of the perturbed problem:

$$\psi(x,y)=\sum_n C_n(x)\chi_n(y).$$

Inserting this into the equation of motion, multiplying with $\chi_m^\star(y)$, and integrating over y leads to

$$\frac{\partial^2 C_m(x)}{\partial x^2} + k_m^2(E)C_m(x) = \sum_n \Gamma_{mn}(x)C_n(x), \tag{17.1}$$

where

$$\Gamma_{mn}(x) = \frac{2m^\star}{\hbar^2}\int \chi_m^\star(y)V_\mathrm{s}(x,y)\chi_n(y)\mathrm{d}y$$

and

$$k_m^2(E) = \frac{2m^\star(E-\epsilon_m)}{\hbar^2}.$$

The tip-induced potential in the scanning gate experiment has a finite extent in the plane large compared to the Fermi wavelength. As a first step we consider the scattering potential

$$V_\mathrm{s}(x,y) = \delta(x-x_\mathrm{i})\frac{\gamma}{b\sqrt{\pi}}e^{-[(y-y_\mathrm{i})/b]^2}, \tag{17.2}$$

which takes into account the finite extent in y-direction, but neglects the extent in x-direction. This will allow us to illustrate some key features of the quantum system without going through the much more elaborate theory for the potential extended in two directions. In our simplified case the matrix elements take the form

$$\Gamma_{mn}(x) = \delta(x-x_\mathrm{i})\frac{\gamma}{b\sqrt{\pi}}\frac{2m^\star}{\hbar^2}\int \chi_m^\star(y)e^{-[(y-y_\mathrm{i})/b]^2}\chi_n(y)\mathrm{d}y := \delta(x-x_\mathrm{i})\tilde{\Gamma}_{mn}$$

and eq. (17.1) simplifies to

$$\frac{\partial^2 C_m(x)}{\partial x^2} + k_m^2(E)C_m(x) = \delta(x-x_\mathrm{i})\sum_n \tilde{\Gamma}_{mn}(x)C_n(x).$$

The solutions of this system of equations for $x \neq x_\mathrm{i}$ are left- and right-going plane-wave states that have to be matched at the position x_i just as in the case of the familiar one-dimensional delta-scattering problem. It can be shown (see Appendix C) that the transmission and reflection matrixes for the case of a single delta scatterer in the wire, t and r, can be determined from the equations

$$\mathsf{t} = (1-\mathsf{M})^{-1},$$
$$\mathsf{r} = (1-\mathsf{M})^{-1}\mathsf{M},$$

where the matrix M is given by

$$M_{mn} = \frac{\tilde{\Gamma}_{mn}}{2\mathrm{i}k_m}.$$

In the quantum case the resistance of the wire induced by the scatterer can again be calculated from the conductance formula of Landauer and Büttiker [30–32]

$$G = \frac{e^2}{h}\mathrm{Tr}(\mathsf{t}\mathsf{t}^\dagger).$$

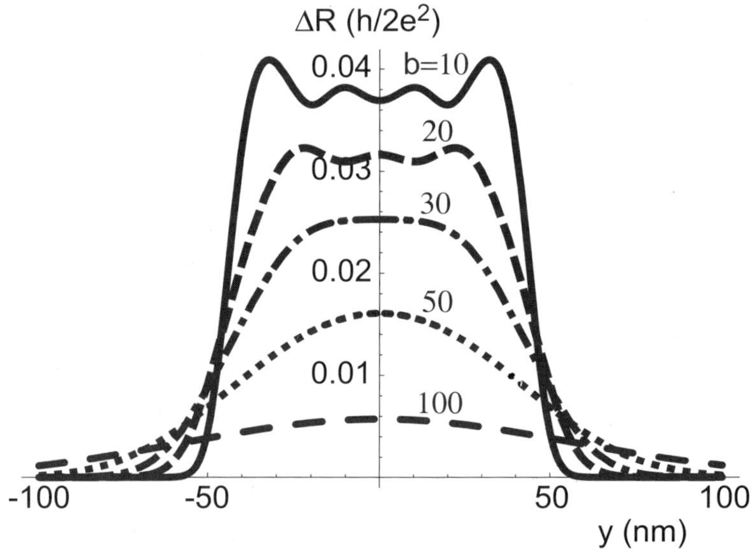

Fig. 17.8. Calculated resistance change in a homogeneous wire due to scatterers with different width $b = 100, 50, 30, 20, 10\,\mathrm{nm}$ in y-direction and constant strength $\gamma = 4000\,\mathrm{meVnm}^2$ as a function of the y-position of the scatterer.

For specific calculations we have used an infinite square well potential modeling the wire potential and a tip-induced potential with a Gaussian shape [eq. (17.2)] in the y-direction. Within this framework we first discuss the influence of the width of the tip-induced potential in the y-direction on cross sections through a resistance image along constant x (cf., Fig. 17.4). The two extreme limits are intuitively clear: when the tip-induced potential is much wider than the wire, the wire will act as the probe for the tip-induced potential and the cross section will have the width and shape of the tip-induced potential. On the other hand, if the wire width is large compared to the tip-induced potential, details of the internal electronic structure of the wire will become visible. Figure 17.8 shows the result of a calculation for a wire with width $W = 100\,\mathrm{nm}$ in which four modes are occupied, corresponding to a Fermi energy $E_\mathrm{F} = 14\,\mathrm{meV}$. For $b = 100\,\mathrm{nm}$ the resistance shows a broad unstructured Gaussian peak when the tip crosses the wire. This peak becomes narrower at $b = 50\,\mathrm{nm}$ and develops a flat top at 30 nm, i.e., when the width of the tip-induced potential becomes smaller than the wire width. Reducing b further results in additional fine structure reflecting the mode structure of the wire. With four occupied modes the highest mode has four maxima of the probability amplitude, which is reflected in the resistance cross section. The condition for the observability of the mode structure in the wire is that the width of the tip-induced potential has to be smaller than the Fermi wavelength λ_F. Comparing these results to the cross sections through the resistance image shown in Fig. 17.4, we conclude that the width of the tip-induced potential in the experiment was of the order of the wire width, but certainly not much larger.

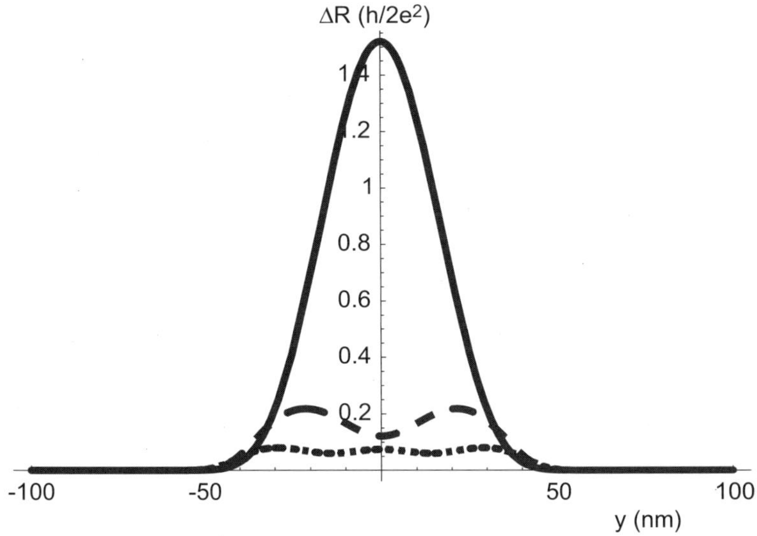

Fig. 17.9. Calculated resistance change in a homogeneous wire with different Fermi energies $E_{\mathrm{F}} = 2.24, 5, 9\,\mathrm{meV}$ and hence different numbers of modes ($N = 1, 2, 3$, respectively). Here the width of the scatterer has been taken to be $b = 10\,\mathrm{nm}$, its strength is $\gamma = 4000\,\mathrm{meVnm}^2$.

We briefly want to use our model calculation to illustrate the possibility of imaging individual modes in a quantum wire or quantum point contact. Corresponding experiments on a quantum point contact have been published recently by Crook and coworkers [371]. However, in their experiments, the two-dimensional electron gas with the quantum point contact are buried 100 nm below the surface and a model for screening by mobile carriers in the remote doping plane had to be invoked in order to understand the experimental resolution of about 40 nm. Figure 17.9 shows calculated resistance changes due to a gaussian tip potential ($b = 10\,\mathrm{nm}$) for a 100 nm wide wire with Fermi energies $E_{\mathrm{F}} = 2.24, 5, 9\,\mathrm{meV}$ corresponding to the occupation of one, two and three modes. It can be seen that the number of local maxima in the resistance cross sections corresponds to the number of maxima in the probability density of the highest occupied mode, confirming the above statement that the mode structure of narrow constrictions can be mapped with scanning gate techniques.

We continue the discussion of quantum effects in scanning gate measurements by considering the case of many scatterers in the quantum wire. For simplicity, we again use δ-shaped potentials in x-direction with Gaussian shape in y. In this case the above formalism can easily be extended to the case of an arbitrary number of impurities. Qualitatively we now expect interference effects between electron waves to arise due to multiple scattering. Figure 17.10 shows a gray scale resistance image calculated for two fixed impurities in addition to the moving tip. In some sense, all three "perturbing potentials" form two neighboring cavities with transmission resonances changing when the sizes or shapes of the cavities are changed by moving the tip. Indeed, interference patterns can be seen along the wire axis. In contrast to the expectation for a very wide two-dimensional system, where interference effects have been predicted to occur on the scale of the Fermi wavelength λ_{F} [461–464], in a wire with a relatively small number of modes more than one frequency contributes to these fluctuations because the individual modes contribute with different Fermi wavelengths. In addition, it can be seen that not all interference stripes are exactly normal to the wire axis. Both findings are also in agreement with the measurement shown in Fig. 17.5 where different frequencies of the fluctuations are observed as well and the stripes' orientation deviates from being normal to the wire axis. All this supports the interpretation of the measurements in terms of interference effects. In contrast to the classical fluctuations found in the billiard model (cf., Fig. 17.7), the quantum interference effects look more regular, i.e., they come in clearly discernible stripes, and they occur at a smaller length scale.

The aspect of a finite phase coherence length was not taken into account in the above model. The obvious question is whether phase coherence is needed over the whole length of the wire to make the observation of interference effects possible. While in a conventional transport experiment — where the conductance is measured as a function of magnetic field —, the answer would certainly be "yes" because otherwise, the connection of many phase coherent sections in series fluctuating independently would average sufficiently. In *local* scanning gate experiments this argument is no longer true. There is always only *one* fluctuating segment, i.e., the one in which the tip is scanning, and no averaging takes place. Phase coherence effects can therefore be observed even in systems with characteristic lengths $L \gg \min(l_\varphi, l_T)$, provided that temperature is low enough to maintain phase coherence over length scales larger than the mean free path. This argument is also of relevance for the ob-

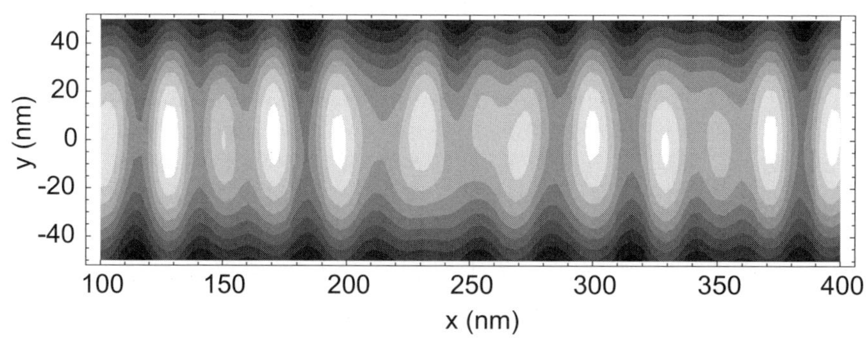

Fig. 17.10. Calculated resistance change ΔR in a homogeneous 100 nm wide wire with four occupied modes ($E_{\mathrm{F}} = 14$ meV). One fixed Gaussian impurity with $b = 50$ nm and strength $\gamma = 500\,\mathrm{meVnm}^2$ was positioned at $(x = 0, y = 40\,\mathrm{nm})$. A second Gaussian impurity had the parameters $b = 30$ nm, $\gamma = 500\,\mathrm{meVnm}^2$ and $(x = 500, y = -20)$. The scanned tip-induced potential was modeled with a second Gaussian potential with $b = 40$ nm and $\gamma = 4000\,\mathrm{meVnm}^2$.

servations in Ref. [373], where scanning gate measurements were performed in the vicinity of a quantum point contact. It was observed that interference fringes do not disappear with increasing separation from the point contact. The explanation given in Ref. [373] is essentially identical to the argument presented above. It implies that the phase coherence length cannot be read from a spatial resistance image obtained from scanning gate measurements. However, interference effects should disappear with increasing temperature. Scanning gate measurements at various temperatures could therefore allow one to determine the phase coherence length and to discriminate between fluctuations of the resistance due to classically chaotic motion and phase coherent effects.

Another aspect neglected in the above model are interaction effects. In principle it is conceivable that due to potential inhomogeneities, electrons localize in the wire and form — in the extreme case — quantum dots exhibiting the Coulomb-blockade effect, similar to the situation in carbon nanotubes in the experiments of Ref. [457]. However, from the over-all conductance of our wires, which is smaller than e^2/h, and the excellent sample quality, we can conclude that interaction effects will play a negligible role in our experiments.

A

Formal solution of the electrostatic problem with Green's functions

A.1 The electrostatic problem

We wish to solve Poisson's equation

$$\nabla\left[\varepsilon(\boldsymbol{r})\varepsilon_0\nabla\phi(\boldsymbol{r})\right]=-\rho(\boldsymbol{r}) \tag{A.1}$$

with constant voltages ϕ_i on the metallic surfaces S_i:

$$\phi(\boldsymbol{r})|_{\boldsymbol{r}\in S_i}=\phi_i.$$

Making use of the superposition principle we split the potential according to

$$\phi(\boldsymbol{r})=\phi_{\text{ion}}(\boldsymbol{r})+\sum_i\phi_i\alpha_i(\boldsymbol{r}), \tag{A.2}$$

where $\phi_{\text{ion}}(\boldsymbol{r})$ obeys the equation

$$\nabla\left[\varepsilon(\boldsymbol{r})\varepsilon_0\nabla\phi_{\text{ion}}(\boldsymbol{r})\right]=-\rho_{\text{ion}}(\boldsymbol{r}) \tag{A.3}$$

with boundary conditions

$$\phi_{\text{ion}}(\boldsymbol{r})|_{\boldsymbol{r}\in S_i}=0$$

and the characteristic functions $\alpha_i(\boldsymbol{r})$ obey Laplace's equation

$$\nabla\left[\varepsilon(\boldsymbol{r})\varepsilon_0\nabla\alpha_i(\boldsymbol{r})\right]=0 \tag{A.4}$$

with boundary conditions

$$\alpha_i(\boldsymbol{r})|_{\boldsymbol{r}\in S_j}=\delta_{ij}=\begin{cases}1 \text{ for } j=i\\ 0 \text{ for } j\neq i.\end{cases}$$

A.2 Formal solution with Green's functions

In order to solve eq. (A.1) formally, we need an extended version of one of Green's identities, which will be derived in the following: let ϕ, ε, and ψ be any scalar functions of the spatial coordinates $\boldsymbol{r}$. Then

$$\nabla\left[\phi\varepsilon\nabla\psi\right] = \phi\nabla\left[\varepsilon\nabla\psi\right] + \nabla\phi\varepsilon\nabla\psi.$$

Application of Gauss' integral identity leads to

$$\oint_S \mathrm{d}s\,\left[\phi\varepsilon\nabla\psi\right]\boldsymbol{n} = \int_V \mathrm{d}V\,\left\{\phi\nabla\left[\varepsilon\nabla\psi\right] + \nabla\phi\varepsilon\nabla\psi\right\}.$$

Exchanging ψ and ϕ gives the identity

$$\oint_S \mathrm{d}s\,\left[\psi\varepsilon\nabla\phi\right]\boldsymbol{n} = \int_V \mathrm{d}V\,\left\{\psi\nabla\left[\varepsilon\nabla\phi\right] + \nabla\psi\varepsilon\nabla\phi\right\}.$$

The difference of the two latter equations is

$$\oint_S \mathrm{d}s\,\left[\psi\varepsilon\nabla\phi - \phi\varepsilon\nabla\psi\right]\boldsymbol{n} = \int_V \mathrm{d}V\,\left\{\psi\nabla\left[\varepsilon\nabla\phi\right] - \phi\nabla\left[\varepsilon\nabla\psi\right]\right\}, \tag{A.5}$$

which is the extended version of Green's integral identities we were looking for.

We now define Green's function $G(\boldsymbol{r},\boldsymbol{r}')$ being a solution of the equation

$$\nabla\left[\varepsilon(\boldsymbol{r})\varepsilon_0\nabla G(\boldsymbol{r},\boldsymbol{r}')\right] = -\delta(\boldsymbol{r}-\boldsymbol{r}')$$

with boundary conditions

$$G(\boldsymbol{r},\boldsymbol{r}')|_{\boldsymbol{r}\in S_i} = 0.$$

Green's function is the potential at $\boldsymbol{r}$ created by a unit point charge in $\boldsymbol{r}'$, if all electrodes are grounded. We replace in eq. (A.5) the quantity ψ with Green's function $G(\boldsymbol{r},\boldsymbol{r}')$, the quantity ε with $\varepsilon(\boldsymbol{r})\varepsilon_0$, and the quantity ϕ with $\phi(\boldsymbol{r})$ from eq. (A.1) and obtain

$$\phi(\boldsymbol{r}') = \int_V \mathrm{d}V\,G(\boldsymbol{r},\boldsymbol{r}')\rho_{\text{ion}}(\boldsymbol{r}) - \sum_i \phi_i \int\!\!\int_{S_i} \mathrm{d}s_i\,\left[\varepsilon(\boldsymbol{r})\varepsilon_0\nabla G(\boldsymbol{r},\boldsymbol{r}')\right]\boldsymbol{n}_i.$$

Comparing to eq. (A.2) we identify

$$\phi_{\text{ion}}(\boldsymbol{r}') = \int_V \mathrm{d}V\,G(\boldsymbol{r},\boldsymbol{r}')\rho_{\text{ion}}(\boldsymbol{r})$$

and

$$\alpha_i(\boldsymbol{r}) = -\int\!\!\int_{S_i} \mathrm{d}s_i\,\left[\varepsilon(\boldsymbol{r})\varepsilon_0\nabla G(\boldsymbol{r},\boldsymbol{r}')\right]\boldsymbol{n}_i.$$

A.3 Induced charges on the electrodes

The surface charge density induced on one of the electrodes is given by

$$\begin{aligned}\sigma_i(\boldsymbol{r}') &= -\varepsilon(\boldsymbol{r}')\varepsilon_0\nabla'\phi(\boldsymbol{r}')\boldsymbol{n}_i' \\ &= -\varepsilon(\boldsymbol{r}')\varepsilon_0\nabla'\left[\int_V \mathrm{d}V\, G(\boldsymbol{r},\boldsymbol{r}')\rho_{\mathrm{ion}}(\boldsymbol{r})\right. \\ &\qquad \left. -\sum_j \phi_j \int\!\!\int_{S_j} \mathrm{d}s_j\, \left[\varepsilon(\boldsymbol{r})\varepsilon_0\nabla G(\boldsymbol{r},\boldsymbol{r}')\right]\boldsymbol{n}_j\right]\boldsymbol{n}_i' \\ &= \sigma_i^{(0)}(\boldsymbol{r}) + \sum_j c_{ij}(\boldsymbol{r})\phi_j.\end{aligned}$$

The total charge on electrode i is therefore given by

$$Q_i = Q_i^{(0)} + \sum_j C_{ij}\phi_j, \tag{A.6}$$

where

$$\begin{aligned}Q_i^{(0)} &= -\int_V \mathrm{d}V\, \rho_{\mathrm{ion}}(\boldsymbol{r}) \int\!\!\int_{S_i} \mathrm{d}s_i'\, \varepsilon(\boldsymbol{r}')\varepsilon_0\nabla' G(\boldsymbol{r},\boldsymbol{r}')\boldsymbol{n}_i' \\ &= \int_V \mathrm{d}V\, \rho_{\mathrm{ion}}(\boldsymbol{r})\alpha_i(\boldsymbol{r})\end{aligned}$$

is the charge induced on electrode i if all other electrodes are grounded and the

$$C_{ij} = \int\!\!\int_{S_i} \mathrm{d}s_i'\, \varepsilon(\boldsymbol{r}')\varepsilon_0 \int\!\!\int_{S_j} \mathrm{d}s_j\, \varepsilon(\boldsymbol{r})\varepsilon_0\nabla'\left\{\left[\nabla G(\boldsymbol{r},\boldsymbol{r}')\right]\boldsymbol{n}_j\right\}\boldsymbol{n}_i'$$

are the elements of the capacitance matrix of the system. Because the charge on electrode i must not change if all potentials ϕ_i are increased by the same constant, the capacitance matrix obeys

$$\sum_j C_{ij} = 0.$$

From the above definition of the capacitance matrix follows the symmetry relation

$$C_{ij} = C_{ji},$$

meaning that the number of independent matrix elements in a problem with N electrodes reduces to $N(N-1)/2$. With these properties of the capacitance matrix we obtain for the total charge on electrode i:

$$Q_i = Q_i^{(0)} + \sum_j C_{ij}\left(\phi_j - \phi_i\right).$$

A.4 Total electrostatic energy of the system

The energy of the charge distribution in the system is given by

$$\begin{aligned} W &= \frac{1}{2}\int_V \mathrm{d}V\, \rho(\boldsymbol{r})\phi(\boldsymbol{r}) \\ &= \frac{1}{2}\int_V \mathrm{d}V\, \rho_{\mathrm{ion}}(\boldsymbol{r})\phi(\boldsymbol{r}) + \frac{1}{2}\sum_i \int\int_{S_i} \mathrm{d}s_i\, \sigma_i(\boldsymbol{r})\phi(\boldsymbol{r}) \\ &:= W_1 + W_2. \end{aligned}$$

For the energy W_1 we obtain

$$\begin{aligned} W_1 &= \frac{1}{2}\int_V \mathrm{d}V\, \rho_{\mathrm{ion}}(\boldsymbol{r})\phi_{\mathrm{ion}}(\boldsymbol{r}) + \frac{1}{2}\sum_i \phi_i \int_V \mathrm{d}V\, \rho_{\mathrm{ion}}(\boldsymbol{r})\alpha_i(\boldsymbol{r}) \\ &= W_{\mathrm{ion}} + \frac{1}{2}\sum_i \phi_i Q_i^{(0)}. \end{aligned}$$

For the energy W_2 we get

$$W_2 = \frac{1}{2}\sum_j \phi_j Q_j^{(0)} + \frac{1}{2}\sum_{ij} \phi_j C_{ji}\phi_i.$$

Therefore the total energy can be written as

$$W = \frac{1}{2}\sum_{ij} \phi_j C_{ji}\phi_i + \sum_j \phi_j Q_j^{(0)} + W_{\mathrm{ion}}.$$

A.5 Force gradient acting on an electrode

The force gradient F' acting on an electrode is the second derivative of the system's energy with respect to the displacement of the electrode:

$$F' = \frac{1}{2}\sum_{ij} \phi_j C''_{ji}\phi_i + \sum_j \phi_j Q_j^{(0)\prime\prime} + W''_{\mathrm{ion}}. \tag{A.7}$$

B

Screened addition energy of an electron to a quantum ring

In Ref. [326] the energy of adding an electron to a system of interacting electrons is calculated in the Hartree–Fock approximation. The only assumption is that the addition of one electron does not change the wave functions of all the other electrons. The result is

$$\begin{aligned}\Delta E = &\int \varphi_i^\star(\boldsymbol{q}_1) H_i \varphi_i(\boldsymbol{q}_1) \mathrm{d}\tau_1 + e^2 \int \sum_{k(\neq i)} \frac{|\varphi_i(\boldsymbol{q}_1)|^2 \, |\varphi_k(\boldsymbol{q}_2)|^2}{|\boldsymbol{r}_1 - \boldsymbol{r}_2|} \mathrm{d}\tau_1 \mathrm{d}\tau_2 \\ &- e^2 \sum_{k(\neq i)} \int \frac{\varphi_i^\star(\boldsymbol{q}_1)\varphi_i(\boldsymbol{q}_2)\varphi_k^\star(\boldsymbol{q}_2)\varphi_k(\boldsymbol{q}_1)}{|\boldsymbol{r}_1 - \boldsymbol{r}_2|} \mathrm{d}\tau_1 \mathrm{d}\tau_2 .\end{aligned}$$

The first term is the single-particle energy of state i where the electron is added. The second term is the Hartree interaction, which gives the main contribution to the charging energy in most quantum dots. The third term is the exchange interaction. Integrals over τ_i include the summation over the spin degree of freedom. The coordinate $\boldsymbol{q}_i$ comprises the space coordinate $\boldsymbol{r}_i$ and the spin s_i.

The Hartree-interaction term has a straightforward interpretation. We identify the charge density as

$$\rho(\mathbf{r}) = \sum_k |\varphi_k(\mathbf{r})|^2$$

and define the Hartree potential

$$V_{\mathrm{H}}(\boldsymbol{r}) := \int \mathrm{d}^3 r' \, \frac{\rho(\boldsymbol{r}')}{|\boldsymbol{r} - \boldsymbol{r}'|} .$$

The Hartree contribution to the addition energy can then be written as

$$E_{\mathrm{H}} = e \int \mathrm{d}^3 r \, |\varphi_i(\boldsymbol{r})|^2 \, V_{\mathrm{H}}(\boldsymbol{r}) .$$

Expansion of the Coulomb interaction in spherical harmonics

The Coulomb interaction potential can be expanded in spherical harmonics. The result is [284]:

$$\frac{1}{|\boldsymbol{r}-\boldsymbol{r}'|} = 4\pi \sum_{\ell=0}^{\infty} \sum_{m=-\ell}^{+\ell} \frac{1}{2\ell+1} \frac{r_<^\ell}{r_>^{\ell+1}} Y_{\ell m}^\star(\vartheta', \varphi') Y_{\ell m}(\vartheta, \varphi).$$

If $r > r'$, then $r_> = r$ and $r_< = r'$, while for $r' > r$ we have $r_> = r'$ and $r_< = r$.

2D expansion of the Coulomb interaction in Bessel functions

In the following all vectors are considered to be two dimensional. We start from the identities

$$J_0(qr) = \frac{1}{2\pi} \int_0^{2\pi} \mathrm{d}\varphi \, e^{\mathrm{i}qr\cos\varphi},$$

$$\frac{1}{r} = \int_0^\infty \mathrm{d}q \, J_0(qr).$$

Inserting the first into the second identity we find

$$\frac{1}{r} = \frac{1}{2\pi} \int_0^\infty \mathrm{d}q \int_0^{2\pi} \mathrm{d}\varphi \, e^{\mathrm{i}qr\cos\varphi},$$

which implies that

$$\frac{1}{|\boldsymbol{r}-\boldsymbol{r}'|} = \frac{1}{2\pi} \int_0^\infty \mathrm{d}q \int_0^{2\pi} \mathrm{d}\varphi \, e^{\mathrm{i}\boldsymbol{q}(\boldsymbol{r}-\boldsymbol{r}')}.$$

We now use the expansion of plane waves into Bessel functions [470, 471]

$$e^{\mathrm{i}\boldsymbol{q}\boldsymbol{r}} = e^{\mathrm{i}qr\cos\theta} = \sum_{m=0}^{\infty} \varepsilon_m \mathrm{i}^m \cos(m\theta) J_m(qr),$$

where $\varepsilon_m = 2$ for $m \neq 0$ and $\varepsilon_0 = 1$ and write

$$\begin{aligned}\frac{1}{|\boldsymbol{r}-\boldsymbol{r}'|} &= \frac{1}{2\pi} \int_0^\infty \mathrm{d}q \int_0^{2\pi} \mathrm{d}\varphi \sum_{m=0}^{\infty} \varepsilon_m \mathrm{i}^m \cos[m(\theta-\varphi)] J_m(qr) \\ &\quad \times \sum_{\ell=0}^{\infty} \varepsilon_\ell (-\mathrm{i})^\ell \cos[\ell(\theta'-\varphi)] J_\ell(qr') \\ &= \frac{1}{2\pi} \sum_{m=0}^{\infty} \sum_{\ell=0}^{\infty} \varepsilon_m \varepsilon_\ell \mathrm{i}^m (-\mathrm{i})^\ell \int_0^\infty \mathrm{d}q \, J_m(qr) J_\ell(qr') \\ &\quad \times \int_0^{2\pi} \mathrm{d}\varphi \, \cos[m(\theta-\varphi)] \cos[\ell(\theta'-\varphi)]\end{aligned}$$

$$
\begin{aligned}
&= \frac{1}{4\pi} \sum_{m=0}^{\infty} \sum_{\ell=0}^{\infty} \varepsilon_m \varepsilon_\ell \mathrm{i}^m (-\mathrm{i})^\ell \int_0^\infty \mathrm{d}q \, J_m(qr) J_\ell(qr') \\
&\quad \times \int_0^{2\pi} \mathrm{d}\varphi \, \{\cos[m\theta + \ell\theta' - (m+\ell)\varphi] + \cos[(\ell - m)\varphi + m\theta - \ell\theta']\} \\
&= \frac{1}{2} \sum_{m=0}^{\infty} \sum_{\ell=0}^{\infty} \varepsilon_m \varepsilon_\ell \mathrm{i}^m (-\mathrm{i})^\ell \int_0^\infty \mathrm{d}q \, J_m(qr) J_\ell(qr') \\
&\quad \times \{\delta_{\ell,-m} + \delta_{\ell,m}\} \cos[m(\theta - \theta')] \\
&= \frac{1}{2} \left\{ \sum_{m=0}^{\infty} \varepsilon_m^2 \int_0^\infty \mathrm{d}q \, J_m(qr) J_m(qr') \cos[m(\theta - \theta')] \right. \\
&\quad \left. + \int_0^\infty \mathrm{d}q \, J_0(qr) J_0(qr') \right\} \\
\frac{1}{|\boldsymbol{r} - \boldsymbol{r}'|} &= \sum_{m=0}^{\infty} \varepsilon_m \int_0^\infty \mathrm{d}q \, J_m(qr) J_m(qr') \cos[m(\theta - \theta')].
\end{aligned}
$$

Ring with finite extent in radial direction

The ring has an inner radius of r_1 and an outer radius of r_2. In this case the charge density is

$$
\rho(\boldsymbol{r}) = \begin{cases} -\frac{eN}{\pi(r_2^2 - r_1^2)} \delta(z) & \text{for } r_1 < r < r_2 \\ 0 & \text{elsewhere.} \end{cases}
$$

We further assume that

$$
|\varphi_i(\boldsymbol{r})|^2 = \begin{cases} \frac{1}{\pi(r_2^2 - r_1^2)} \delta(z) & \text{for } r_1 < r < r_2 \\ 0 & \text{elsewhere.} \end{cases}
$$

The Hartree contribution to the addition energy is then

$$
\begin{aligned}
E_\mathrm{H} &= e \int \mathrm{d}^3 r \int \mathrm{d}^3 r' \, \frac{|\varphi_i(\boldsymbol{r})|^2 \rho(\boldsymbol{r}')}{|\boldsymbol{r} - \boldsymbol{r}'|} \\
&= \frac{e^2 N}{\pi^2 (r_2^2 - r_1^2)^2} \int \mathrm{d}^2 r \int \mathrm{d}^2 r' \, \frac{1}{|\boldsymbol{r} - \boldsymbol{r}'|} \\
&= \frac{e^2 N}{\pi^2 (r_2^2 - r_1^2)^2} \int \mathrm{d}^2 r \int_{r_0}^{r_1} \mathrm{d}r' \, r' \int_0^{2\pi} \mathrm{d}\theta' \\
&\quad \times \sum_{m=0}^{\infty} \varepsilon_m \int_0^\infty \mathrm{d}q \, J_m(qr) J_m(qr') \cos[m(\theta - \theta')] \\
&= \frac{2\pi e^2 N}{\pi^2 (r_2^2 - r_1^2)^2} \int_0^\infty \mathrm{d}q \int_{r_0}^{r_1} \mathrm{d}r \, r J_0(qr) \int_{r_0}^{r_1} \mathrm{d}r' \, r' J_0(qr') \int_0^{2\pi} \mathrm{d}\theta
\end{aligned}
$$

$$
\begin{aligned}
&= \frac{(2\pi)^2 e^2 N}{\pi^2 \left(r_2^2 - r_1^2\right)^2} \int_0^\infty \mathrm{d}q \left[\int_{r_0}^{r_1} \mathrm{d}r\, r J_0(qr) \right]^2 \\
&= \frac{(2\pi)^2 e^2 N}{\pi^2 \left(r_2^2 - r_1^2\right)^2} \int_0^\infty \mathrm{d}q \frac{1}{q^2} \left[r_2 J_1(qr_2) - r_1 J_1(qr_1) \right]^2 \\
&= \frac{(2\pi)^2 e^2 N}{\pi^2 \left(r_2^2 - r_1^2\right)^2} \int_0^\infty \mathrm{d}q \frac{1}{q^2} \left[r_2^2 (J_1(qr_2))^2 + r_1^2 (J_1(qr_1))^2 \right. \\
&\left. -2 r_1 r_2 J_1(qr_2) J_1(qr_1) \right] \\
&= \frac{(2\pi)^2 e^2 N}{\pi^2 \left(r_2^2 - r_1^2\right)^2} \left[\frac{4r_2^3}{3\pi} + \frac{4r_1^3}{3\pi} - r_1^2 r_2\, {}_2F_1\left(-\frac{1}{2}, \frac{1}{2}; 2; \frac{r_1^2}{r_2^2}\right) \right] \\
E_{\mathrm{H}} &= \frac{16 e^2 N}{3\pi r_2} \frac{1}{\left[1 - (r_1/r_2)^2\right]^2} \left[1 + \left(\frac{r_1}{r_2}\right)^3 \right. \\
&\left. - \frac{3\pi}{4} \left(\frac{r_1}{r_2}\right)^2 {}_2F_1\left(-\frac{1}{2}, \frac{1}{2}; 2; \frac{r_1^2}{r_2^2}\right) \right].
\end{aligned}
$$

Screening of the interaction by a metallic top gate

In the presence of a metallic top gate the electron–electron interaction potential is screened by image charges. We obtain for the Hartree contribution to the addition energy

$$
\begin{aligned}
E_{\mathrm{H}} &= e \int \mathrm{d}^3 r \int \mathrm{d}^3 r' \left|\varphi_i(\boldsymbol{r})\right|^2 \rho(\boldsymbol{r}') \left(\frac{1}{|\boldsymbol{r} - \boldsymbol{r}'|} \right. \\
&\left. - \frac{1}{\sqrt{(x - x')^2 + (y - y')^2 + (z + z')^2}} \right) \\
&= \frac{e^2 N}{\pi^2 \left(r_2^2 - r_1^2\right)^2} \int \mathrm{d}^2 r \int \mathrm{d}^2 r' \left(\frac{1}{|\boldsymbol{r} - \boldsymbol{r}'|} - \frac{1}{\sqrt{(x - x')^2 + (y - y')^2 + 4d^2}} \right) \\
\bar{E}_{\mathrm{H}} &:= \frac{e^2 N}{\pi^2 \left(r_2^2 - r_1^2\right)^2} \int \mathrm{d}^2 r \int \mathrm{d}^2 r' \frac{1}{|\boldsymbol{r} - \boldsymbol{r}'|} \\
&= \frac{16 e^2 N}{3\pi r_2} \frac{1}{\left[1 - (r_1/r_2)^2\right]^2} \left[1 + \left(\frac{r_1}{r_2}\right)^3 - \frac{3\pi}{4} \left(\frac{r_1}{r_2}\right)^2 {}_2F_1\left(-\frac{1}{2}, \frac{1}{2}; 2; \frac{r_1^2}{r_2^2}\right) \right]^2 \\
&= \frac{16 e^2 N}{3\pi r_2} \frac{1}{\left[1 - (r_1/r_2)^2\right]^2} E_{\mathrm{u}} \left(\frac{r_1}{r_2}\right) \\
\tilde{E}_{\mathrm{H}} &:= \frac{e^2 N}{\pi^2 \left(r_2^2 - r_1^2\right)^2} \int \mathrm{d}^2 r \int \mathrm{d}^2 r' \frac{1}{\sqrt{(\boldsymbol{r} - \boldsymbol{r}')^2 + 4d^2}} \\
&= \frac{e^2 N}{\pi^2 \left(r_2^2 - r_1^2\right)^2} \int \mathrm{d}r\, r \mathrm{d}\theta \int \mathrm{d}r'\, r' \mathrm{d}\theta' \frac{1}{\sqrt{r^2 + r'^2 - 2rr' \cos(\theta - \theta') + 4d^2}}
\end{aligned}
$$

$$
\begin{aligned}
&= \frac{e^2 N}{\pi^2 \left(r_2^2 - r_1^2\right)^2} \int \mathrm{d}r\, r \mathrm{d}\theta \int \mathrm{d}r'\, r' \mathrm{d}\varphi \, \frac{1}{\sqrt{(r-r')^2 + 4d^2 + 2rr'(1-\cos\varphi)}} \\
&= \frac{2\pi e^2 N}{\pi^2 \left(r_2^2 - r_1^2\right)^2} \int \mathrm{d}r\, r \int \mathrm{d}r'\, r' \frac{1}{\sqrt{(r-r')^2 + 4d^2}} \\
&\quad \times \int_0^{2\pi} \mathrm{d}\varphi \, \frac{1}{\sqrt{1 + \frac{4rr'}{(r-r')^2+4d^2} \sin^2 \varphi/2}} \\
&= \frac{8\pi e^2 N}{\pi^2 \left(r_2^2 - r_1^2\right)^2} \int \mathrm{d}r\, r \int \mathrm{d}r'\, r' \frac{1}{\sqrt{(r-r')^2 + 4d^2}} K\left(-\frac{4rr'}{(r-r')^2 + 4d^2}\right) \\
&= \frac{16 e^2 N}{3\pi r_2 \left[1 - (r_1/r_2)^2\right]^2} \frac{3}{2} \int_{r_1/r_2}^{1} \mathrm{d}x\, x \int_{r_1/r_2}^{1} \mathrm{d}x'\, x' \\
&\quad \times \frac{1}{\sqrt{(x-x')^2 + \eta^2}} K\left(-\frac{4xx'}{(x-x')^2 + \eta^2}\right) \\
&:= \frac{16 e^2 N}{3\pi r_2 \left[1 - (r_1/r_2)^2\right]^2} E_\mathrm{s}\left(\frac{r_1}{r_2}, \eta\right),
\end{aligned}
$$

where $\eta = 2d/r_2$. The double integral for the determination of S has no divergences and can be solved numerically. The total Hartree energy is then

$$
E_\mathrm{H} = \frac{16 e^2 N}{3\pi r_2} \frac{1}{\left[1 - (r_1/r_2)^2\right]^2} \left[E_\mathrm{u}\left(\frac{r_1}{r_2}\right) - E_\mathrm{s}\left(\frac{r_1}{r_2}, \eta\right)\right].
$$

C

Scattering in quantum wires

C.1 Single δ scatterer

We consider the Hamiltonian of a two-dimensional quantum wire with confinement potential $V_{\mathrm{c}}(y)$ and some scattering potential $V_{\mathrm{s}}(x,y)$. The Schrödinger equation reads

$$\left[-\frac{\hbar^2}{2m^\star}\left(\frac{\partial^2}{\partial x^2}+\frac{\partial^2}{\partial y^2}\right)+V_{\mathrm{c}}(y)+V_{\mathrm{s}}(x,y)\right]\psi(x,y)=E\psi(x,y).$$

We assume that we have solved the pure wire problem without disorder, i.e.,

$$\left[-\frac{\hbar^2}{2m^\star}\frac{\partial^2}{\partial y^2}+V_{\mathrm{c}}(y)\right]\chi_n(y)=\epsilon_n\chi_n(y).$$

We can then expand the solutions of the perturbed problem:

$$\psi(x,y)=\sum_n c_n(x)\chi_n(y).$$

Inserting this into the equation of motion, multiplying with $\chi^\star_m(y)$, and integrating over y leads to

$$\frac{\partial^2 C_m(x)}{\partial x^2}+\underbrace{\frac{2m^\star(E-\epsilon_m)}{\hbar^2}}_{k_m^2}C_m(x)=\sum_n\underbrace{\frac{2m^\star U_{mn}(x)}{\hbar^2}}_{\Gamma_{mn}(x)}c_n(x),$$

where

$$\Gamma_{mn}(x)=\frac{2m^\star}{\hbar^2}\int\chi^\star_m(y)V_{\mathrm{s}}(x,y)\chi_n(y)dy.$$

As a specific example we consider a δ-function scatterer

$$V_{\mathrm{s}}(x,y)=\gamma\delta(x-x_{\mathrm{i}})\delta(y-y_{\mathrm{i}}).$$

In this case

$$\Gamma_{mn}(x) = \underbrace{\frac{2m^\star\gamma}{\hbar^2}\chi^\star_m(y_\mathrm{i})\chi_n(y_\mathrm{i})}_{\tilde{\Gamma}_{mn}}\,\delta(x - x_\mathrm{i})$$

and the Schrödinger equation reads

$$\frac{\partial^2 C_m(x)}{\partial x^2} + k_m^2 C_m(x) = \delta(x - x_\mathrm{i}) \sum_n \tilde{\Gamma}_{mn} c_n(x).$$

We know the solution of this set of equations for $x \neq x_\mathrm{i}$. They are plane wave states that can be written as

$$C_m(x) = \begin{cases} \alpha_m \mathrm{e}^{\mathrm{i}k_m(x-x_\mathrm{i})} + \beta_m \mathrm{e}^{-\mathrm{i}k_m(x-x_\mathrm{i})} & \text{for } x > x_\mathrm{i} \\ \gamma_m \mathrm{e}^{\mathrm{i}k_m(x-x_\mathrm{i})} + \delta_m \mathrm{e}^{-\mathrm{i}k_m(x-x_\mathrm{i})} & \text{for } x < x_\mathrm{i}. \end{cases} \tag{C.1}$$

In order to match the two solutions for $x > x_\mathrm{i}$ and $x < x_\mathrm{i}$, we proceed as in the case of a one-dimensional problem. Integration of the Schrödinger equation over x between $x_\mathrm{i} - \varepsilon$ and $x_\mathrm{i} + \varepsilon$ and letting $\varepsilon \to 0$ leads to

$$\frac{\mathrm{d}C_m(x_\mathrm{i}^+)}{\mathrm{d}x} - \frac{\mathrm{d}C_m(x_\mathrm{i}^-)}{\mathrm{d}x} = \sum_n \tilde{\Gamma}_{mn} c_n(x_\mathrm{i}).$$

At the same time, the $C_m(x)$ have to be steady, leading to

$$C_m(x_\mathrm{i}^+) = C_m(x_\mathrm{i}^-).$$

These two equations connect the amplitudes of left- and right-going waves on both sides of the scatterer. Inserting the wave function in eq. (C.1) gives

$$\begin{aligned} \alpha_m - \beta_m &= \gamma_m - \delta_m + \sum_n \frac{\tilde{\Gamma}_{mn}}{\mathrm{i}k_m}(\gamma_m + \delta_m) \\ \alpha_m + \beta_m &= \gamma_m + \delta_m. \end{aligned}$$

In vector notation these equations can be written as

$$\begin{aligned} \boldsymbol{\alpha} - \boldsymbol{\beta} &= \boldsymbol{\gamma} - \boldsymbol{\delta} + 2\mathsf{M}(\boldsymbol{\gamma} + \boldsymbol{\delta}) \\ \boldsymbol{\alpha} + \boldsymbol{\beta} &= \boldsymbol{\gamma} + \boldsymbol{\delta}, \end{aligned}$$

where

$$M_{mn} = \frac{\tilde{\Gamma}_{mn}}{2\mathrm{i}k_m} = -\mathrm{i}\frac{\gamma}{\hbar v_m}\chi_m(y_\mathrm{i})\chi_n(y_\mathrm{i}).$$

Here we have introduced the velocity $v_m = \hbar k_m / m^\star$. We now regard the amplitudes of the wave on the left-hand side, $\boldsymbol{\alpha}$ and $\boldsymbol{\beta}$, as given and solve for the amplitudes on the right hand side. The result is

$$\begin{aligned} \boldsymbol{\alpha} &= (1 + \mathsf{M})\boldsymbol{\gamma} + \mathsf{M}\boldsymbol{\delta} \\ \boldsymbol{\beta} &= -\mathsf{M}\boldsymbol{\gamma} + (1 - \mathsf{M})\boldsymbol{\delta}. \end{aligned}$$

If we define $\boldsymbol{\alpha}, \boldsymbol{\beta}$ and $\boldsymbol{\gamma}, \boldsymbol{\delta}$ to be the components of vectors $\boldsymbol{a}$ and $\boldsymbol{b}$, respectively, we can define the transfer matrix

$$\mathsf{T} = \begin{pmatrix} 1+\mathsf{M} & \mathsf{M} \\ -\mathsf{M} & 1-\mathsf{M} \end{pmatrix} \tag{C.2}$$

such that $\boldsymbol{a} = \mathsf{T}\boldsymbol{b}$.

We now consider a scattering experiment in which the electron beam incident from the left, i.e., the vector $\boldsymbol{\gamma}$, is given and there is no electron beam incident from the right, i.e., $\boldsymbol{\beta} = 0$. In this case, $\boldsymbol{\delta}$ represents the amplitudes of the reflected beam and $\boldsymbol{\alpha}$ those of the transmitted beam. Simple algebra leads to the results

$$\boldsymbol{\alpha} = (1-\mathsf{M})^{-1}\boldsymbol{\gamma}$$

$$\boldsymbol{\delta} = (1-\mathsf{M})^{-1}\mathsf{M}\boldsymbol{\gamma}.$$

We can therefore identify transmission and reflection matrices

$$\mathsf{t} = (1-\mathsf{M})^{-1}$$
$$\mathsf{r} = (1-\mathsf{M})^{-1}\mathsf{M}.$$

C.2 Multiple δ scatterers

The transfer matrix approach is particularly useful in the case of multiple scatterers in series. In this case, each scatterer s is described by a transfer matrix T_s according to eq. (C.2). For propagation between the scatterers we define the propagator P given by

$$\mathsf{P}_s = \begin{pmatrix} \mathsf{R}^{(s)} & 0 \\ 0 & \mathsf{L}^{(s)} \end{pmatrix}$$

with

$$R^{(s)}_{mn} = \delta_{mn} e^{ik_m(x_s - x_{s-1})}$$
$$L^{(s)}_{mn} = \delta_{mn} e^{-ik_m(x_s - x_{s-1})}.$$

The total transfer matrix for N scatterers is then given by

$$\mathsf{T}_N = \mathsf{T}_1 \prod_{s=2}^{N} \mathsf{P}_s \mathsf{T}_s.$$

From the total transfer matrix, the transmission and reflection matrices can then be determined analog to the above case of a single scatterer.

C.3 Delta scatterer with finite extent in y-direction

The tip-induced potential in a scanning gate experiment has a finite extent in the plane large compared to the Fermi wavelength. As a first step we consider the scattering potential

$$V_{\mathrm{s}}(x,y) = \delta(x - x_{\mathrm{i}})\frac{\gamma}{b\sqrt{\pi}}e^{-[(y-y_{\mathrm{i}})/b]^2}.$$

In this case the matrix elements take the form

$$M_{mn} = -\mathrm{i}\frac{\gamma}{\hbar v_m b\sqrt{\pi}}\int \mathrm{d}y\, \chi_m(y) e^{-[(y-y_{\mathrm{i}})/b]^2}\chi_n(y).$$

The remaining analyis is analogous to the discussion of the delta scatterers.

C.4 Scatterer with finite extent in y-direction and rectangular shape in x-direction

Here we extend the above model for a finite rectangular scattering potential in x-direction

$$V_{\mathrm{s}}(x,y) = \begin{cases} \frac{\gamma}{b\sqrt{\pi}}e^{-[(y-y_{\mathrm{i}})/b]^2} & \text{for } x_{\mathrm{i}} - L/2 < x < x_{\mathrm{i}} + L/2 \\ 0 & \text{elsewhere.} \end{cases}$$

This leads to matrix elements

$$\Gamma_{mn}(x) = \begin{cases} \tilde{\Gamma}_{mn} & \text{for } x_{\mathrm{i}} - L/2 < x < x_{\mathrm{i}} + L/2 \\ 0 & \text{elsewhere} \end{cases},$$

where

$$\tilde{\Gamma}_{mn} = \frac{2m^\star}{\hbar^2}\frac{\gamma}{b\sqrt{\pi}}\int \mathrm{d}y\, \chi_m(y) e^{-[(y-y_{\mathrm{i}})/b]^2}\chi_n(y).$$

In this case we solve Schrödinger's equation in the three regions $x < x_{\mathrm{i}} - L/2$ (region I), $x_{\mathrm{i}} - L/2 < x < x_{\mathrm{i}} + L/2$ (region II), and $x > x_{\mathrm{i}} + L/2$ (region III). In regions I and III different modes m do not mix and the solutions are simply linear combinations of left and right going plane waves. In region II Schrödinger's equation reads

$$\frac{\partial^2 C_m(x)}{\partial x^2} + k_m^2 C_m(x) = \sum_n \tilde{\Gamma}_{mn} c_n(x).$$

It can be rewritten as

$$\frac{\partial^2 C_m(x)}{\partial x^2} = \sum_n \left(\tilde{\Gamma}_{mn} - \delta_{mn}k_m^2\right) c_n(x),$$

or, in vector notation

$$\frac{\partial^2 \boldsymbol{C}(x)}{\partial x^2} = \mathsf{M}\boldsymbol{C}(x),$$

with

$$M_{mn} = \tilde{\Gamma}_{mn} - \delta_{mn} k_m^2.$$

Solutions of this set of coupled differential equations can be found with the *Ansatz*

$$\boldsymbol{C}(x) = \boldsymbol{c}e^{\mathrm{i}\kappa(x-x_\mathrm{i})},$$

which leads to the eigenvalue problem

$$(\mathsf{M} - \kappa^2)\boldsymbol{c} = 0.$$

After determination of all the eigenvalues κ_m^2 and eigenvectors $\boldsymbol{c}_m$, the solution of Schrödinger's equation in region II can be expressed as

$$C_m(x) = \sum_n c_{mn} \left(a_n e^{\mathrm{i}\kappa_n (x-x_\mathrm{i})} + b_n e^{-\mathrm{i}\kappa_n (x-x_\mathrm{i})} \right)$$

with $\kappa_n > 0$ and c_{mn} being the mth component of the eigenvector belonging to the eigenvalue κ_n^2, i.e., the matrix c is the eigenvector-matrix of M.

The solution of Schrödinger's equation in all three regions can therefore be written as

$$C_m(x) = \begin{cases} \alpha_m e^{\mathrm{i}k_m(x-x_\mathrm{i}-L/2)} + \beta_m e^{-\mathrm{i}k_m(x-x_\mathrm{i}-L/2)} & \text{in region III} \\ \sum_n c_{mn} \left(a_n e^{\mathrm{i}\kappa_n(x-x_\mathrm{i})} + b_n e^{-\mathrm{i}\kappa_n(x-x_\mathrm{i})} \right) & \text{in region II} \\ \gamma_m e^{\mathrm{i}k_m(x-x_\mathrm{i}+L/2)} + \delta_m e^{-\mathrm{i}k_m(x-x_\mathrm{i}+L/2)} & \text{in region I.} \end{cases}$$

Four of the six constants $\alpha_m, \beta_m, \gamma_m, \delta_m, a_m, b_m$ are determined by matching the wave functions at the interfaces. This leads to the equations

$$\begin{aligned}
\gamma_m + \delta_m &= \sum_n c_{mn} \left(a_n e^{-\mathrm{i}\kappa_n L/2} + b_n e^{\mathrm{i}\kappa_n L/2} \right), \\
\alpha_m + \beta_m &= \sum_n c_{mn} \left(a_n e^{\mathrm{i}\kappa_n L/2} + b_n e^{-\mathrm{i}\kappa_n L/2} \right), \\
\mathrm{i}k_m (\gamma_m - \delta_m) &= \sum_n c_{mn} \mathrm{i}\kappa_n \left(a_n e^{-\mathrm{i}\kappa_n L/2} - b_n e^{\mathrm{i}\kappa_n L/2} \right), \\
\mathrm{i}k_m (\alpha_m - \beta_m) &= \sum_n c_{mn} \mathrm{i}\kappa_n \left(a_n e^{\mathrm{i}\kappa_n L/2} - b_n e^{-\mathrm{i}\kappa_n L/2} \right).
\end{aligned}$$

In matrix notation these equations can be written as

$$\begin{aligned}
\boldsymbol{\gamma} + \boldsymbol{\delta} &= \mathsf{c}\left(\mathsf{p}^\star \boldsymbol{a} + \mathsf{p}\boldsymbol{b}\right), \\
\boldsymbol{\alpha} + \boldsymbol{\beta} &= \mathsf{c}\left(\mathsf{p}\boldsymbol{a} + \mathsf{p}^\star \boldsymbol{b}\right), \\
\mathsf{k}\left(\boldsymbol{\gamma} - \boldsymbol{\delta}\right) &= \mathsf{c}\kappa\left(\mathsf{p}^\star \boldsymbol{a} - \mathsf{p}\boldsymbol{b}\right), \\
\mathsf{k}\left(\boldsymbol{\alpha} - \boldsymbol{\beta}\right) &= \mathsf{c}\kappa\left(\mathsf{p}\boldsymbol{a} - \mathsf{p}^\star \boldsymbol{b}\right),
\end{aligned}$$

where $p_{mn} = \delta_{mn} e^{\mathrm{i}\kappa_m L/2}$, $\kappa_{mn} = \delta_{mn}\mathrm{i}\kappa_m$ and $k_{mn} = \delta_{mn}\mathrm{i}k_m$. In order to obtain a transfer matrix T we eliminate the vectors $\boldsymbol{a}$ and $\boldsymbol{b}$ and obtain

$$\begin{pmatrix} \alpha \\ \beta \end{pmatrix} = \begin{pmatrix} \mathsf{c}^{-1} + \kappa^{-1}\mathsf{c}^{-1}\mathsf{k} & \mathsf{c}^{-1} - \kappa^{-1}\mathsf{c}^{-1}\mathsf{k} \\ \mathsf{p}^2[\mathsf{c}^{-1} - \kappa^{-1}\mathsf{c}^{-1}\mathsf{k}] & \mathsf{p}^2[\mathsf{c}^{-1} + \kappa^{-1}\mathsf{c}^{-1}\mathsf{k}] \end{pmatrix}^{-1}$$
$$\times \begin{pmatrix} \mathsf{p}^2[\mathsf{c}^{-1} + \kappa^{-1}\mathsf{c}^{-1}\mathsf{k}] & \mathsf{p}^2[\mathsf{c}^{-1} - \kappa^{-1}\mathsf{c}^{-1}\mathsf{k}] \\ \mathsf{c}^{-1} - \kappa^{-1}\mathsf{c}^{-1}\mathsf{k} & \mathsf{c}^{-1} + \kappa^{-1}\mathsf{c}^{-1}\mathsf{k} \end{pmatrix} \begin{pmatrix} \gamma \\ \delta \end{pmatrix}.$$

The matrix T is the product of the two matrices on the right-hand side.

D

Response of a harmonic oscillator to a resonance frequency step

Following eq. (14.4) we assume that the oscillator is at all times driven by the force $2\alpha_{\mathrm{p}}/mU_0 \cos(\omega_{\mathrm{e}}t + \phi)$. Tip-sample interaction forces are neglected. For $t < 0$ we excite on resonance, i.e., $\omega_{\mathrm{e}} = \omega_0$ and assume that the oscillator performs its steady-state oscillations. At $t = 0$ we instantaneously change the resonance frequency of the oscillator to the new frequency $\Omega_0 = \omega_0 + \Delta\omega$ and seek solutions of eq. (10.2), neglecting F_{ts}, of the form

$$\xi(t) = Y(t)\cos(\omega_{\mathrm{e}}t + \phi) + X(t)\sin(\omega_{\mathrm{e}}t + \phi).$$

For $t < 0$ the amplitudes are $X(t) = a_<$ and $Y(t) = 0$. For $t > 0$ we find

$$\begin{aligned} Y(t) &= a_> \cos\phi + be^{-\omega_0 t/(2Q)} \cos[\Delta\omega t + \theta - \phi], \\ X(t) &= -a_> \sin\phi - be^{-\omega_0 t/(2Q)} \sin[\Delta\omega t + \theta - \phi]. \end{aligned}$$

The constants b and $\theta - \phi$, found from the boundary conditions at $t = 0$, are given by

$$\begin{aligned} b^2 &= a_>^2 + a_<^2 + 2a_> a_< \sin\phi, \\ \cot(\theta - \phi) &= \frac{a_> \cos\phi}{a_< + a_> \sin\phi}. \end{aligned}$$

So far the results are generally valid and no approximations have been made. Now we consider approximate behavior. For small frequency steps $\Delta f \ll f_{\mathrm{G}} = f_0/(2Q)$, i.e., in the regime where $a_<$ depends linearly on $\Delta\omega$, one finds after Fourier transformation that

$$Y(t) = 2a_< \frac{Q\Delta\omega}{\omega_0}\left[1 - e^{-\omega_0 t/(2Q)} \cos\left(\Delta\omega t + \frac{2Q\Delta\omega}{\omega_0}\right)\right].$$

It is interesting to note that the response has an oscillatory component with frequency $\Delta\omega$. We have experimentally verified the existence of this oscillatory response by applying frequency steps of varying magnitude $\Delta\omega$ and found excellent agreement with the above expression.

Because the long-time behavior of the response for small enough $\Delta\omega$ will be governed by the exponential decay, we may approximate the response by

$$Y(t) = 2a_< \frac{Q\Delta\omega}{\omega_0} \left[1 - e^{-\omega_0 t/(2Q)}\right].$$

This gives immediately the step response of the demodulated current

$$Y_{res}(t) = -X_{res} \frac{2Q}{\omega_0} \Delta\omega \left[1 - e^{-\omega_0 t/(2Q)}\right],$$

or, after Fourier transformation

$$Y_{res}(\omega) = k(\omega)\Delta f(\omega) = -X_{\mathrm{res}} \frac{2Q}{f_0} \Delta f(\omega) \frac{1}{1 + \mathrm{i}\omega/(2\pi f_{\mathrm{G}})},$$

i.e., a first-order low-pass behavior with bandwidth $f_{\mathrm{G}} = f_0/(2Q)$, in agreement with the result of Ref. [384].

E

Acknowledgements

The scientific work presented in this book would not have come about without essential contributions from many collaborators and colleagues.

I would like to thank Klaus Ensslin, who gives me the great opportunity to work within his research team at ETH. None of the presented results would have been possible without his generous and persistent support, his enthusiastic and effective way of promoting our joint projects, and his openness for all kinds of concerns occurring in everyday life of a research lab.

I thank Thomas Heinzel with whom I shared an office for many years for numberless discussions about physics and many other topics concerning us "senior postdocs." In particular, I wish to thank him and his PhD students Ryan Held and Silvia Lüscher for generously supplying the cryo-SFM crew with the samples nanostructured by AFM lithography.

It was a great pleasure to work with a number of very talented and enthusiastic PhD students, who, as a matter of fact, measured most of the presented data. Volkmar Senz contributed almost all the work on the metal–insulator transition in SiGe presented in part II. His very focused style of approaching new tasks and his joyous personality made our joint project successful and enjoyable.

I am still very grateful to have the opportunity to work with Andreas Fuhrer on the quantum ring experiments which already made a substantial contribution to this review in chapter 12 and continue to make us excited about more and more surprises from week to week. I thank him and Alessandro Pioda for incouraging me to join the academic orchestra, which helps to relax blinkered physicist's brains and brings a lot of fun.

I also have to thank Stefan Lindemann, who contributed the data presented in chapter 13. I will never forget his superb seminar talks about cats and quantum dots or his stories from the world of the stock market.

I am indebted to Jörg Rychen, who was a real pioneer in the field of low-temperature scanning probe microscopy in our group. His dedication, creativity, and persistence lead to the most successful scanning probe experiments in our group, namely, the scanning gate measurements on a quantum wire and the local tunneling experiments on edge channels in the quantum Hall regime (see chapters 16 and 17).

Tobias Vancura deserves many thanks for putting the design of the ^{3}He-scanning force microscope described in chapter 14 into practice. With this microscope we now have a very reliable piece of equipment operating under adverse conditions as demonstrated by the Kelvin probe measurements shown in chapter 15 performed by Tobias and Slavo Kicin. I thank Andreas Baumgartner for his contribution to this work (especially for Fig. 14.12) and his persistent enthusiasm for the physics, sometimes hidden behind the undue hardship of extremely challenging experiments.

Several diploma students contributed directly or indirectly to this work. I thank Tommaso Cilento, Martin Sigrist, Alessandro Pioda, Laura Bremme, Peter Vorburger, and many others for their collaboration.

All the ideas would not have become real and all the experimental achievements would have been so much harder to make without the excellent technical support in our group. I have to thank Paul Studerus for his tireless efforts to improve our measurement setups. He taught me "the art of electronics" from his practical point of view and was always open-minded when physcists had some vague and unconventional idea how things could possibly be improved. Many thanks go to Andreas Herrmann and Cecil Barengo, our lab technicians, who manufactured a great part of the mechanical setups in the labs and contributed all their practical knowledge about materials and workmanship. I also thank Jean Pierre Stucki and the workshop of the physics department for their great support.

Experiments based on semiconductor heterostructures rely on high quality samples tailored for the experimental needs. I am very grateful for the opportunity to collaborate with some leading experts on MBE growth. I am indebted to Werner Wegscheider, formerly from the Walter-Schottky-Institute in Garching, Germany, now at the University of Regensburg, for supplying the excellent shallow heterostructures needed for the AFM lithography and also for the scanning probe experiments. I would like to thank Kevin Maranowski and Art Gossard from the University of Santa Barbara, California, for the fantastic parabolic quantum well samples. Many thanks go to Detlev Grützmacher, Ulf Gennser, and Gabriel Dehlinger for the gated state-of-the-art p-SiGe quantum well structures.

I very much appreciate the collaboration with Denis L'Hote and Renaud Leturcq who performed the heating measurements on the p-type SiGe quantum wells presented in part II.

I am deeply grateful to the many theoreticians who dedicated a lot of time to discussing our results and helped to make progress in the interpretation. Among them, I especially thank Gianni Blatter from ETH Zurich, Sergiu Ulloa from the University of Athens in Ohio, and Wilhelm Zwerger from the University of Munich.

F

List of symbols

F.1 Physics constants

Symbol	Meaning	Value
c	speed of light	299792458 m/s
e	elementary charge	$1.602176462 \times 10^{-19}$ As
$\hbar = h/(2\pi)$	Planck's constant	$1.054571596 \times 10^{-34}$ Js
k	Boltzmann constant	$1.3806503 \times 10^{-23}$ J/K
m_0	rest mass of the electron	$9.10938188 \times 10^{-31}$ kg
ε_0	electric constant	$8.854187817 \times 10^{-12}$ As/(Vm)
Φ_0	magnetic flux quantum h/e	$4.135667272 \times 10^{-15}$ Vs

F.2 Variables

Symbol	Meaning	Units
$a_{\mathrm{B}}^{\star}$	effective Bohr radius	[m]
a	slope of TF Y-signal around resonance	
a	quantum well width	[m]
$\boldsymbol{a}$	vector of plane wave amplitudes	
$a_{>,<}$	oscillation amplitudes before and after a step in resonance frequency	
a_1	plane wave amplitude	
a_2	plane wave amplitude	
a	amplitude of CB oscillations in SET	
a_{T}	tip oscillation amplitude	
A	normalization area	
A	enclosed area	
$\boldsymbol{b}$	vector of plane wave amplitudes	
b	width of scattering potential	
b	amplitude of the transient signal after a frequency step	
b_1	plane wave amplitude	
b_2	plane wave amplitude	
B	magnetic flux density (also referred to as 'magnetic field')	[m^2/(Vs)]
B_{φ}	field scale associated with phase coherence	[T]
$B_{\perp}$	field scale	
$B_{\parallel}$	field scale	
$B_{\parallel}$	parallel magnetic field	
B_{total}	total magnetic field (if tilted)	
$\boldsymbol{B}_{\mathrm{eff}}$	effective spin-orbit coupling field	
c_1	plane wave amplitude	
c_2	plane wave amplitude	
$c_{1\ldots4}$	constants	
$c_n(x)$	wave function coefficients (wire simulation)	
$c(p)$	density-dependent parameter (screening)	
$c_{ij}(\boldsymbol{r})$	characteristic capacitance functions	
$\boldsymbol{c}$	vector of amplitudes of plane waves	
c	matrix in wire scattering problem	
C	self-capacitance	[As/V]
C	capacitance of tuning fork	
C_{Σ}	self-capacitance	[As/V]
C_{ij}	capacitance coefficients	[F]
C_{tg}	ring-top gate capacitance	
C_{pg}	plunger gate-dot capacitance	
C_{ring}	self-capacitance of a ring	
C_{plate}	plate capacitance for ring-topgate	
C_{g}	geometric capacitance	

Symbol	Meaning	Units
C_q	quantum capacitance	
C_0	tuning fork capacitance	
C_K	cable capacitance	
D	diffusion constant	[m^2/s]
$\mathcal{D}(x, y; B)$	(local) density of states	
$\mathcal{D}_k$	density of states (k-dependent)	
d_1	plane wave amplitude	
d_2	plane wave amplitude	
d	sample thickness	[m]
d	dimension of a system	
d	diameter of ring structure	[m]
d	separation from gate to 2DEG	
$d_{1,2}$	thickness of layers	
$e(\omega)$	error signal	
$\boldsymbol{\mathcal{E}}$	vector of the electric field	[V/m]
$\mathcal{E}$	electric field	[V/m]
E	energy	[J]
E_n	quantized energy level	[J]
$E_\mathrm{u}(x)$	functional dependence of unscreened Coulomb interaction	
$E_\mathrm{s}(x, \eta)$	functional dependence of the screening contribution to the interaction	
E_α	quantized energy level	[J]
$E_{\alpha\beta}$	energy difference $E_\alpha - E_\beta$	[J]
E_F	Fermi energy	[J]
ΔE_c	charging energy $e^2/(2C)$	[J]
E_c	electrostatic charging energy ($\propto N!$)	[J]
E_C	Coulomb interaction energy	[J]
E_ee	electron electron interaction energy	[J]
E_H	Hartree energy	[J]
E_r	characteristic energy scale for a quantum ring	
E_N	single-particle energy of a quantum dot	[J]
$E(N)$	total energy of a quantum dot	[J]
$E_\mathrm{elstat}(N)$	electrostatic energy	[J]
E_tot	total energy of a quantum dot	
$\Delta E_\mathrm{Z}(B)$	Zeeman shift of individual conductance peak	

Symbol	Meaning	Units
F	force	[N]
F_{ts}	tip-sample interaction force	
F_z	tip-sample interaction force, z-component	
$F(z)$	electrostatic force	
$F_{\mathrm{AC}}^{\mathrm{tr}}$	AC force caused by AC transport voltage	
$F_{\mathrm{AC}}^{\mathrm{eq}}$	AC force caused by AC voltage	
$\tilde{F}_\sigma$	Fermi-liquid interaction parameter	
$F^\star$	Fermi-liquid interaction parameter	
$F_{\mathrm{eq}}(E_p, N)$	conditional probability for N electrons in the dot and energy E_p occupied	
f_0	resonance frequency of TF	
f_0	center frequency of VCO	
f	frequency	
f_{G}	characteristic frequency	
f_{res}	tuning fork resonance frequency	
f_{d}	tuning fork driving frequency	
f_{PI}	characteristic frequency of P-I controller	
f_{PLL}	characteristic frequency of PLL	
Δf	frequency shift	
$\Delta f(\omega)$	modulated shift of TF	
Δf_{set}	setpoint for frequency shift	
Δf_{meas}	measured frequency shift	
Δf_{fit}	fitted frequency shift	
f_z	characteristic frequency of z-feedback	
f_{FB}	characteristic feedback frequency	
G	electric conductance	$[\Omega^{-1}]$
$G_{p,N}$	conductance contribution of level p with N electrons in the dot	
G_0	Drude conductance	$[\Omega^{-1}]$
ΔG	change in conductance	
$G(\boldsymbol{r}, \boldsymbol{r}')$	Green's function	
g	dimensionless conductance $g/(e^2/\hbar)$	
$\langle g \rangle$	average conductance through quantum dot	
g_0	dimensionless conductance constant	
g_{a}	dimensionless conductance constant	
g_{v}	valley degeneracy factor	
g_{L}	Lande factor	
H_N	Hamilton operator for N-electron system	
H_{H}	Hartree hamiltonian	
I	electric current	[A]
I_{SET}	SET current	
I_{p}	piezoelectric current	
I_{c}	capacitive current	

Symbol	Meaning	Units
I_{T}	tip current	
I_{TF}	tuning fork current	
$I_{\perp}$	numerical parameter for WL	
$I_{\parallel}$	numerical parameter for WL	
i	index: label for gates	
$\boldsymbol{j}$	electric current density	[A/m^2]
$\boldsymbol{j}_{\alpha\beta}(\boldsymbol{x})$	matrix element of the current density operator	[A/m^2]
$\boldsymbol{J}$	total angular momentum vector	
J	total angular momentum quantum number	
k	wave vector	[m^{-1}]
k'	wave vector	
k_{F}	Fermi wave vector	[m^{-1}]
k_1	spring constant	
k_2	spring constant	
$k(E)$	wave vector	
k_{c}	spring constant	
$k(\omega)$	response function	
k	matrix in scattering problem	
K	current to voltage conversion ratio of amplifier	
l	mean free path	[m]
l_{D}	Drude mean free path	[m]
l_{e}	elastic mean free path of electrons	[m]
$l_{\mathrm{e}}^{\mathrm{2D}}$	mean free path of electrons	
l_{φ}	phase coherence length	[m]
l_{eff}	effective length scale in Tan–Inkson model	
l_{c}	magnetic length	
l_{T}	thermal length	
l_y	characteristic length scale of harmonic oscillator	
l_z	characteristic length scale of harmonic oscillator	
L	sample length	[m]
L	separation between voltage probes	[m]
L	characteristic sample size	[m]
L	length of quantum wire	
L	separation of delta scatterers	
L	length of wire attached to tuning fork	
L	inductance of tuning fork	
$L(\omega)$	open loop response function of PLL	
$\mathcal{L}$	parameter of the Tan–Inkson model	
$\mathsf{L}^{(s)}$	part of the propatator matrix P_s	
ℓ	angular momentum quantum number	
ℓ	quantum number	
m	mass	[kg]
m_1	mass	

Symbol	Meaning	Units
m_2	mass	
m	number of flux quanta in a ring	
$m^\star$	effective mass	[kg]
m_J	quantum number for the z-component of total angular momentum	
m_i	magnetic moment of state i	
M	total magnetic moment	
$M(\omega)$	open loop response of z-feedback	
M_k	total transmission matrix	
M	matrix containing scattering information in wire	
$\|n\rangle$	quantum state with quantum number n (Ket)	
n	electron number density	[m^{-2}]
n	index: Landau level quantum number	
n_{H}	Hall density	[m^{-2}]
n_{SdH}	Shubnikov–de Haas density	[m^{-2}]
n_{s}	electron sheet density	[m^{-2}]
n_{S}	electron sheet density	
n_{d}	electron density in the quantum dot	
n_{IUC}	I-U converter output noise	
n_{Y}	lockin output noise	
n_{PLL}	PLL output noise	
n_z	z-feedback noise	
n_i	occupation number	
N	number of modes	
N	number of electrons	
N_0	number of modes in the absence of the tip	
ΔN	change in the number of modes	
$\mathcal{N}$	number of energy levels	
N_{i}	impurity density	[m^{-3}]
p	hole density	m^{-2}
p	matrix in wire problem	
p_{c}	critical hole density	[m^{-2}]
$p_{x,y}$	components of the momentum	
$p(\theta)$	angular probability distribution	
P	gain of P-I controller	
P	power dissipated in TF	
P_z	gain of z-controller	
P_{eff}	effective gain	
$P_{\mathrm{eff}}^{(z)}$ P_{c}	critical gain of PI-controller	
$P_{\mathrm{c}}^{(z)}$	critical gain of z-controller	
P_E	dissipated power	[W]
P_0	power offset	[W]
$PLL(\omega)$	PLL response function	

Symbol	Meaning	Units
$PI(\omega)$	response of z-feedback P-I controller	
P_k	propagation matrix	
P_s	propagation matrix between scatterers s and $s-1$	
$\boldsymbol{p}$	momentum operator	
P_{eq}	equilibrium Gibbs distribution function	
$\boldsymbol{q}$	scattering wave vector	[m^{-1}]
$\boldsymbol{q}_{1,2}$	combined space and spin coordinate	
q	modulus of the scattering wave vector $\lvert\boldsymbol{q}\rvert$	
q_{p}	piezoelectric charge	
Q_i	electric charge on electrode i	[As]
Q_{i}	charge on SET island	
Q	quality factor	
Q_{2DEG}	induced charge in the 2DEG	
$Q_{\mathrm{tip}}^{(0)}$	charge induced on the tip at zero bias	
r	radius of a ring	[m]
r_0	radius of a ring	[m]
r_1	inner ring radius	
r_2	outer ring radius	
$r_{>,<}$	radius coordinates	
$\boldsymbol{r}$	vector in space	[m]
r	distance from the origin	[m]
Δr	radial width of ring	[m]
R	electric resistance	[Ω]
R	resistance of tuning fork	
R_{c}	contact resistance	[Ω]
R_{c}	classical cyclotron radius	[m]
R_{H}	Hall coefficient	[Ω/T]
$R(\omega)$	response function of P-I controller	
$\mathsf{R}^{(s)}$	part of the propatator matrix P_s	
ΔR	resistance change	
δR_{H}	correction to the Hall coefficient	
R_s	pure (series) resistance of a QPC without R_{c}	[Ω]
r_{s}	interaction parameter (conventional definition)	
$\bar{r}_{\mathrm{s}}$	interaction parameter as energy ratio in 2D	
$r_{\mathrm{s}}^{\mathrm{eff}}$	effective interaction parameter for electrons in ring	
r	reflection matrix	
S	ground-state spin of quantum dots	
S_{LI}	lockin sensitivity	
S_N	ground-state spin of N electron dot	
S_i	surface of electrode i	
s	z-component of ground state spin	
T	temperature	[K]
T	transmission probability	

Symbol	Meaning	Units
T_{F}	Fermi temperature	[K]
T_{e}	electron temperature	[K]
T_{min}	minimum temperature	[K]
T_{l}	lattice temperature	[K]
T	transfer matrix	
T_s	transfer matrix for scatterer s	
t	transmission matrix	
t	time	
U	applied or measured voltage	[V]
U	longitudinal voltage	[V]
U_{a}	voltage at Kelvin parabola apex	
U_{H}	Hall voltage, transverse voltage	[V]
U_{G}	gate voltage	[V]
U_{SD}	source-drain voltage	[V]
U_{ts}	tip-sample voltage	
U_{set}	setpoint voltage for feedback	
U_{depl}	onset voltage for depletion	
U_0	voltage applied to tuning fork	
U_{PG}	plunger gate voltage	
U_{TG}	top gate voltage	
U_{BG}	back gate voltage	
U_{G}	gate voltage	
U_{tip}	tip voltage	
U_{2DEG}	2DEG voltage	
$U_{\mathrm{AC}}(x,y)$	local voltage between tip and sample	
U_{ij}	gate voltage difference $U_i - U_j$	[V]
$U_{\Delta f}$	VCO input voltage	
U_i	gate voltage	[V]
$\delta U_{\Delta f}$	noise on VCO input voltage	
ΔU_{HV}	voltage on piezo tube	
u	interaction strength (spin chapter)	
$V(q)$	Fourier transform of scattering potential	
V	volume	
$V(y,z)$	confinement potential	
$V(y)$	confinement potential	
$V_{\mathrm{s}}(x,y)$	scattering potential	
$V_{\mathrm{t}}(x,y)$	tip induced potential	
V_{c}	constant interaction energy	[J]
$V_{\mathrm{c}}(y)$	confinement potential	
$V(r)$	confinement potential in radial direction	
$V(r,\varphi)$	confinement potential	
$V_{\mathrm{H}}(\boldsymbol{r})$	Hartree potential	
V_{δ}	strength of delta scattering potential	

Symbol	Meaning	Units
$v_\ell(r)$	Fourier components of $V(r,\varphi)$	
v_F	Fermi velocity	[m/s]
$\boldsymbol{v}$	velocity	[m/s]
$v(q)$	form factor for scattering	
W	width of a sample	[m]
W	width of quantum wire	
W	electrostatic energy of charge distribution	
$W_{1,2}$	electrostatic energy contributions	
W_ion	electrostatic energy contribution due to ions	
W_gates	work done by voltage sources	[J]
$\boldsymbol{x}$	vector in space	[m]
$\boldsymbol{x}'$	vector in space	[m]
x	aluminum content in Ga[Al]As	
X_res	X-output lockin on resonance	
$X(t)$	sin-component of harmonic oscillator oscillation	
Y	elastic modulus	
Y_res	Y-output lockin on resonance	
$Y(\omega)$	Y-output lockin	
$Y(t)$	cos-component of harmonic oscillator oscillation	
δY_res	integrated lockin output noise	
Z	partition function	
$Z(\omega)$	response function of z-feedback	
Δz_eff	effective surface roughness	
$\langle z^2_\mathrm{2DEG}\rangle$	wave function extent in 2DEG	
$\langle z^2_\mathrm{dot}\rangle$	wave function extent in dot	
δz	integrated z-noise	
$\boldsymbol{\alpha}$	vector of plane wave amplitudes	
$\lvert\alpha\rangle$	quantum mechanical basis state, ket-vector	
α,β	index: (continuous) quantum numbers of basis states	
α	Exponent for heating effects	
α_pr	piezoresistive coupling constant	
α	proportionality constant of frequency shift and U_ts	
α_p	piezoelectric coupling constant of tuning fork	
$\alpha_i(\boldsymbol{r})$	characteristic function of gate electrode	
α_i	lever arm of electrode i	
α'	exponent for heating effects	
α	prefactor for WL	
α_G	lever arm of the gate electrode	
α_D	lever arm of the drain electrode	
α_S	lever arm of the source electrode	
$\boldsymbol{\beta}$	vector of plane wave amplitudes	
β	proportionality constant of frequency shift and Δz	
$\beta(g)$	scaling function	

Symbol	Meaning	Units
$\beta(\boldsymbol{r})$	shape function in quantum dot	
$\varepsilon(q,T)$	dielectric function	
$\varepsilon(\boldsymbol{r})$	dielectric function ($\boldsymbol{r}$ dependent)	
$\varepsilon_{1,2}$	dielectric constant	
ϵ_N	single-particle energy level	
ϵ	small asymmetry parameter in Tan–Inkson	
ϵ_n	single-particle energy levels	
$\boldsymbol{\gamma}$	vector of plane wave amplitudes	
γ	friction constant	
γ	temperature exponent	
γ	normalized strength of delta scattering potential	
γ	strength of scattering potential	
γ_1	damping constant	
γ_2	damping constant	
γ_{c}	damping constant	
Γ	tunneling coupling	
Γ	width of conductance peaks	
$\Gamma_{mn}(x)$	matrix element of scattering potential $V_{\mathrm{s}}(x,y)$	
$\tilde{\Gamma}_{mn}$	matrix element	
Γ_p^{l}	tunneling coupling of state p to left lead	
Γ_p^{r}	tunneling coupling of state p to right lead	
$\bar{\Gamma}_p$	$=\Gamma_p^{\mathrm{l}}\Gamma_p^{\mathrm{r}}/(\Gamma_p^{\mathrm{l}}+\Gamma_p^{\mathrm{r}})$: mean tunneling coupling of level p	
Δ	single-particle level spacing	
Δ_N	single-particle level spacing E_N-E_{N-1}	[J]
$\boldsymbol{\delta}$	vector of plane wave amplitudes	
δ	$=e\alpha_{\mathrm{G}}(U_{\mathrm{G}}-U_{\mathrm{G}}^{\mathrm{res}})$	
κ	coupling constant of harmonic oscillators	
κ	wave vector	
κ	matrix in wire scattering problem	
λ_{F}	Fermi wave length	[m]
λ	ratio of interaction strength to single-particle level spacing	
μ	electron mobility	
μ_{H}	Hall mobility	[m^2/(Vs)]
μ_{elch}	electrochemical potential	[J]
μ_{ch}	chemical potential	[J]
μ_N	addition energy of a quantum dot	[J]
μ_{S}	electrochemical potential of source contact	
μ_{D}	electrochemical potential of drain contact	
μ_{B}	Bohr magneton	
μ	calibration of piezo tube in z direction	
ν	filling factor of Landau levels	
ν	exponent for scaling analysis	

Symbol	Meaning	Units
ν	exponent in wave function (effective angular momentum) in Tan–Inkson model	
$\phi(\boldsymbol{r})$	electrostatic potential	
ϕ_i	electrostatic potentials on gate electrodes	[V]
ϕ_{G}	electrostatic potential on the gate	[V]
$\phi_{\mathrm{ion}}(\boldsymbol{r})$	electrostatic potential of ionic charges	
$\phi(N)$	total electrostatic potential of the dot	
Φ	magnetic flux	[Vs]
φ	phase of electron path	
φ	angle variable	
$\varphi_i(\boldsymbol{r})$	wave function	
$\Pi(q, T, \mu_{\mathrm{ch}})$	polarization function	
ϱ	specific resistivity	[Ωm]
ϱ	resistivity tensor in two dimensions	[Ω]
ϱ_{xx}	longitudinal resistivity	[Ω]
ϱ_{xy}	Hall resistivity	[Ω]
ϱ_{c}	critical resistivity in 2D	[Ω]
ρ	mass density	
ρ_{ion}	density of ionic charges	[As/m^3]
ρ_{e}	density of electronic charges	[As/m^3]
$\rho(\boldsymbol{r})$	density of electronic charges	
σ	specific conductivity	[Ω^{-1}/m]
σ_{D}	Drude conductivity in 2D	Ω^{-1}
$\Delta\sigma_{\mathrm{I}}$	interaction correction to the conductivity	[Ω^{-1}]
$\delta\sigma_{\mathrm{I}}$	interaction correction to the conductivity	[Ω^{-1}]
$\delta\sigma_{\mathrm{WL}}$	weak-localization correction to the conductivity	[Ω^{-1}]
σ	conductivity tensor in two dimensions	[Ω^{-1}]
$\sigma(\boldsymbol{x}, \boldsymbol{x}')$	non-local conductivity	[Ω^{-1}/m]
$\sigma_i(\boldsymbol{r})$	surface charge density	
σ_{d}	charge density in doping plane	
σ_{s}	surface charge density	
$\sigma_{\mathrm{2DEG}}(\boldsymbol{r})$	induced density in the 2DEG	
τ	Drude scattering time	[s]
τ	time constant	
τ_{e}	Drude scattering time	[s]
τ_{q}	quantum life time	[s]
τ_{φ}	phase coherence time	[s]
$\tau_{\parallel}$	relaxation time	[s]
$\tau_{\perp}$	relaxation time	[s]
θ	scattering angle	
θ	angle	
ω	angle frequency $2\pi f$	
ω_{c}	cyclotron frequency	[s^{-1}]

Symbol	Meaning	Units
ω_{c}	coupling frequency of tuning fork arms	
ω_z	harmonic oscillator frequency in z-direction	
ω_z^{2DEG}	h.o. frequency for 2DEG	
ω_z^{dot}	h.o. frequency for dot	
ω_y	harmonic oscillator frequency in y-direction	
ω_0	frequency	
ω_0	resonance frequency of TF	
ω_{e}	excitation frequency of TF	
ω_1	diagonalized harmonic oscillator frequency	
ω_2	diagonalized harmonic oscillator frequency	
$\Delta\omega$	step in resonance frequency	
Ω_d	solid angle in d dimensions	
Ω_1	frequency of Eigenmode	
Ω_2	frequency of Eigenmode	
$\psi_k(x)$	wave function	
$\chi(r)$	radial wave function	
ξ	mean exchange energy	
$\xi(t)$	coordinate of harmonic oscillator	
η	parameter for the charging energy model for the ring	
η	calibration constant for frequency shift	
η	response function VCO	

F.3 Special functions

Function	Meaning	Units
$f(E, \mu_{\mathrm{ch}}, T), f(E)$	Fermi–Dirac distribution function	
$\delta(E)$	Dirac delta function	
$\psi(x)$	digamma function	
$\Gamma(x)$	gamma function	
$L_n^{(\nu)}$	generalized Laguerre polynomials	
${}_2F_1(a, b; c; x)$	hypergeometric function	
$Y_{\ell,m}(\vartheta, \varphi)$	spherical harmonic functions	
$J_n(x)$	Bessel functions	
$K(x)$	elliptic integral	

References

[1] G.S. Ohm. Bestimmung des Gesetzes, nach welchem Metalle die Kontakt-Elektrizität leiten, nebst einem Entwurfe zu einer Theorie des Voltaschen Apparates und des Schweiggerschen Multiplikators. *Schweiggers Journal für Chemie und Physik*, 46:137, 1826.

[2] G.S. Ohm. *Die mathematisch bearbeitete Theorie der Kette*. Riemann, Berlin, 1827.

[3] E.H. Hall. On a new action of the magnet on electric currents. *Am. J. Math.*, 2:287, 1879.

[4] E.H. Hall. On a new action of the magnet on electric currents. *The American Journal of Science*, 19(XXV):200, 1880.

[5] P. Drude. Zur Elektronentheorie. I. *Annalen der Physik*, 1:566, 1900.

[6] P. Drude. Zur Elektronentheorie. II. *Annalen der Physik*, 3:369, 1900.

[7] N. W. Ashcroft and N. D. Mermin. *Solid State Physics*. Saunders College Publishing, 1976.

[8] H.A. Lorentz. *The theory of electrons*. Teubner, 1909.

[9] A. Sommerfeld. Zur Elektronentheorie der Metalle. *Die Naturwissenschaften*, 15:825, 1927.

[10] A. Sommerfeld. Zur Elektronentheorie der Metalle. *Die Naturwissenschaften*, 16:374, 1928.

[11] A. Sommerfeld. Zur Elektronentheorie der Metalle aufgrund der Fermischen Statistik. 2. Thermo-elektrische, galvano-magnetische und thermomagnetische Vorgänge. *Z. Phys.*, 47:43, 1928.

[12] A. Sommerfeld and N.H. Frank. The statistical theory of thermoelectric, galvano- and thermomagnetic phenomena in metals. *Rev. Mod. Phys.*, 3:1, 1931.

[13] J.M. Ziman. *Principles of the theory of solids*. Cambridge University Press, 2nd edition, 1972.

[14] K. Seeger. *Semiconductor Physics: An Introduction*, volume 40 of *Springer Series in Solid-State Sciences*. Springer-Verlag, 5th edition, 1991.

[15] F. Bloch. Über die Quantenmechanik der Elektronen in Kristallgittern. *PhD thesis, University of Leipzig*, 1928.

[16] L.D. Landau. Diamagnetismus der Metalle. *Z. Phys.*, 64:629, 1930.

[17] V. Fock. Bemerkung zur Quantelung des harmonischen Oszillators im Magnetfeld. *Z. Physik*, 47:446, 1928.

[18] C.G. Darwin. The diamagnetism of the free electron. *Proc. Cambridge Philos. Soc.*, 27:86, 1930.

[19] E. Teller. Der Diamagnetismus von freien Elektronen. *Zeitschrift f. Physik*, 67:311, 1931.

[20] L. Shubnikov and W.J. de Haas. Magnetische Widerstandsvergrösserung in Einkristallen von Wismut bei tiefen Temperaturen. *Leiden Commun.*, 207a, 207c, 207d, 210a, 1930.

[21] E.N. Adams and T.D. Holstein. Quantum theory of transverse galvanomagnetic phenomena. *J. Phys. Chem. Solids*, 10:254, 1959.

[22] R. Kubo. A general expression for the conductivity tensor. *Can. J. Phys.*, 34:1274, 1956.

[23] R. Kubo. Statistical-Mechanical Theory of Irreversible Processes. I. *J. Phys. Soc. Japan*, 12:570, 1957.

[24] D.A. Greenwood. The Boltzmann Equation in the Theory of Electrical Conduction in Metals. *Proc. Phys. Soc.*, 71:585, 1958.

[25] W. Zwerger. Theory of coherent transport. In *Quantum Transport and Dissipation*. Wiley-VCH, 1998.

[26] W. Kohn and J.M. Luttinger. Quantum Theory of Electrical Transport Phenomena. *Phys. Rev. B*, 108:590, 1957.

[27] J.H. Davies. *The physics of low-dimensional semiconductors: an introduction*. Cambridge University Press, 1998.

[28] R. Landauer. Spatial Variation of Currents and Fields Due to Localized Scatterers in Metallic Conduction. *IBM J. Res. Dev.*, 1:223, 1957.

[29] R. Landauer. Residual resistivity dipoles. *Z. Physik B*, 21:247, 1975.

[30] C.W.J. Beenakker and H. van Houten. Quantum transport in semiconductor nanostructures. In H. Ehrenreich and D. Turnbull, editors, *Semiconductor Heterostructures and Nanostructures*, volume 44 of *Solid State Physics*, pages 1–228. Academic Press Ltd, 1991.

[31] S. Datta. *Electronic Transport in Mesoscopic Systems*, volume 3 of *Cambridge Studies in Semiconductor Physics and Microelectronic Engineering*. Cambridge University Press, 1995.

[32] D.K. Ferry and S.M. Goodnick. *Transport in Nanostructures*. Number 6 in Cambridge Studies in Semiconductor Physics and Microelectronic Engineering. Cambridge University Press, 1997.

[33] M. Büttiker. Four-Terminal Phase-Coherent Conductance. *Phys. Rev. Lett.*, 57:1761, 1986.

[34] H. U. Baranger and A.D. Stone. Electrical linear-response theory in an arbitrary magnetic field: A new Fermi-surface formation. *Phys. Rev. B*, 40:8169, 1989.

[35] R.H. Fowler and L. Nordheim. Electron Emission in Intense Electric Fields. *Proc. Roy. Soc. London, Series A*, 119:173, 1928.

[36] C. Zener. A Theory of the Electrical Breakdown of Solid Dielectrics. *Proc. Roy. Soc. London, Series A*, 145:523, 1943.

[37] L. Esaki. New Phenomenon in Narrow Germanium p-n Junctions. *Phys. Rev.*, 109:603, 1957.

[38] I. Giaever. Energy Gap in Superconductors Measured by Electron Tunneling. *Phys. Rev. Lett.*, 5:147, 1960.

[39] I. Giaever. Electron Tunneling Between Two Superconductors. *Phys. Rev. Lett.*, 5:464, 1960.

[40] J. Bardeen. Tunnelling from a many-particle point of view. *Phys. Rev. Lett.*, 6:57, 1961.

[41] C.B. Duke. *Tunnelling in Solids*. Solid State Physics, Supplement 10. Academic Press, 1969.

[42] T. Ando, A.B. Fowler, and F. Stern. Electronic properties of two-dimensional systems. *Rev. Mod. Phys.*, 54:437, 1982.

[43] C. Weisbuch and B. Vinter. *Quantum Semiconductor Structures: Fundamentals and Applications*. Academic Press, INC., 1991.

[44] P.T. Coleridge, R. Stoner, and R. Fletcher. Low-field transport coefficients in GaAs/$Ga_{1-x}Al_xAs$ heterostructures. *Phys. Rev. B*, 39:1120, 1989.

[45] B. Laikhtman and E.L. Altshuler. Quasiclassical Theory of Shubnikov-de Haas Effect in 2D Electron Gas. *Annals of Physics*, 232:332, 1994.

[46] R. Tsu and L. Esaki. Tunnelling in a finite superlattice. *Appl. Phys. Lett.*, 22:562, 1973.

[47] L.L. Chang, L. Esaki, and R. Tsu. Resonant tunnelling in semiconductor double barriers. *Appl. Phys. Lett.*, 24:593, 1974.

[48] K. von Klitzing, G. Dorda, and M. Pepper. New Method for High-Accuracy Determination of the Fine-Structure Constant Based on Quantized Hall Resistance. *Phys. Rev. Lett.*, 45:494, 1980.

[49] M. Buttiker. Absence of backscattering in the quantum Hall effect in multiprobe conductors. *Phys. Rev. B*, 38:9375, 1988.

[50] D.B. Chklovskii, B.I. Shklovskii, and L.I. Glazman. Electrostatics of edge channels. *Phys. Rev. B*, 46:4026, 1992.

[51] D.C. Tsui, H.L. Störmer, and A.C. Gossard. Two-Dimensional Magnetotransport in the Extreme Quantum Limit. *Phys. Rev. Lett.*, 48:1559, 1982.

[52] R. B. Laughlin. Anomalous quantum Hall effect: an incompressible quantum fluid with fractionally charged excitations. *Phys. Rev. Lett.*, 50:1395, 1983.

[53] R.E. Prange and S.M. Girvin, editors. *The quantum Hall effect*. Springer, New York, 1988.

[54] T. Chakraborty and P. Pietilainen. *The fractional quantum Hall effect*. Springer, Berlin, 1988.

[55] J. P. Eisenstein and H.L. Störmer. The fractional quantum Hall effect. *Science*, 248:1510, 1990.

[56] O. Heinonen, editor. *Composite Fermions; a unified view of the quantum Hall regime*. World Scientific, Singapore, 1998.

[57] E. Abrahams, P.W. Anderson, D.C. Licciardello, and T.V. Ramakrishnan. Scaling Theory of Localization: Absence of Quantum Diffusion in Two Dimensions. *Phys. Rev. Lett.*, 42:673, 1979.

[58] W. Franz. Elektroneninterferenzen im Magnetfeld. *Verhandlungen der deutschen physikalischen Gesellschaft [Phys. Ber.* **21**, *686 (1940)]*, 2:65, 1940.

[59] Y. Aharonov and D. Bohm. Significance of electromagnetic potentials in the quantum theory. *Phys. Rev.*, 115:485, 1959.

[60] S. Datta, M.R. Melloch, S. Bandyopadhyay, R. Noren, M. Vaziri, M. Miller, and R. Reifenberger. Novel interference effects between parallel quantum wells. *Phys. Rev. Lett.*, 55:2344, 1985.

[61] G. Timp, A.M. Chang, J.E. Cunningham, T.Y. Chang, P. Mankiewich, R. Behringer, and R.E. Howard. Observation of the Aharonov-Bohm effect for $\omega_c\tau > 1$. *Phys. Rev. Lett.*, 58:2814, 1987.

[62] K. Ishibashi, Y. Takagaki, K. Gamo, S. Namba, S. Ishida, K. Murase, Y. Aoyagi, and M. Kawabe. Observation of Aharonov-Bohm magnetoresistance oscillations in selectively doped GaAs-AlGaAs submicron structures. *Solid State Communications*, 64:573, 1987.

[63] C.J.B. Ford and H. Ahmed. Fabrication of GaAs heterojunction ring structures. *Microelectronic Engineering*, 6:169, 1987.

[64] G. Timp, A.M. Chang, P. DeVegvar, R.E. Howard, R. Behringer, J.E. Cunningham, and P. Mankiewich. Quantum transport in one-dimensional GaAs/AlGaAs microstructures. *Suf. Sci.*, 196:68, 1988.

[65] S. Washburn and R.A. Webb. Aharonov-Bohm effect in normal metal. Quantum coherence and transport. *Adv. Phys.*, 35:375, 1986.

[66] A.G. Aronov and Y.V. Sharvin. Magnetic flux effects in disordered conductors. *Rev. Mod. Phys.*, 59:755, 1987.

[67] Y. Imry. *Introduction to Mesoscopic Physics*, volume 1 of *Mesoscopic Physics and Nanotechnology*. Oxford University Press, 2nd edition, 2002.

[68] A. Pioda, V. Senz, T. Ihn, T. Heinzel, K. Ensslin, W. Wegscheider, and M. Bichler. Electron transport through an array of coherently coupled cavities. In *Proceedings of the 26th International Conference on the Physics of Semiconductors, Edinborough 2002*. World Scientific, 2002.

[69] B.J. van Wees, H. van Houten, C.W.J. Beenakker, J.G. Williamson, L.P. Kouwenhoven, D. van der Marel, and C.T. Foxon. Quantized conductance of point contacts in a two-dimensional electron gas. *Phys. Rev. Lett.*, 60:848, 1988.

[70] D.A. Wharam, T.J. Thornton, R. Newbury, M. Pepper, H. Ahmed, J. Frost, D.G. Hasko, D.C. Peacock, D.A. Ritchie, and G.A.C. Jones. One dimensional transport and the quantization of the ballistic resistance. *J. Phys. C*, 21:L209, 1988.

[71] A. Yacoby and Y. Imry. Quantization of the conductance of ballistic point contacts beyond the adiabatic approximation. *Phys. Rev. B*, 41:5341, 1990.

[72] S.M. Cronenwett, H.J. Lynch, D. Goldhaber-Gordon, L.P. Kouwenhoven, C.M. Marcus, K. Hirose, N.S. Wingreen, and V. Umansky. Low-temperature

fate of the 0.7 structure in a point contact: a Kondo-like correlated state in an open system. *Phys. Rev. Lett.*, 88:226805–1, 2002.

[73] R. de Picciotto, H.L. Stormer, L.N. Pfeiffer, K.W. Baldwin, and K.W. West. Four terminal resistance of a ballistic quantum wire. *Nature*, 411:6833, 2001.

[74] C.J. Gorter. A possible explanation of the increase of the electrical resistance of thin metal films at low temperatures and small field strengths. *Physica*, 17:777, 1951.

[75] I. Giaever and H.R. Zeller. Superconductivity of small tin particles measured by tunneling. *Phys. Rev. Lett.*, 20:1504, 1968.

[76] J. Lambe and R.C. Jaklevic. Charge-quantization studies using a tunnel capacitor. *Phys. Rev. Lett.*, 22:1371, 1969.

[77] I.O. Kulik and R.I. Shekhter. Kinetic phenomena and charge discreteness effects in granulated media. *Sov. Phys. JETP*, 41:308, 1975.

[78] T.A. Fulton and G.J. Dolan. Observation of single-electron charging effects in small tunnel junctions. *Phys. Rev. Lett.*, 59:109, 1987.

[79] J.H.F. Scott-Thomas, S.B. Field, M.A. Kastner, H.I. Smith, and D.A. Antoniadis. Conductance oscillations periodic in the density of a one-dimensional electron gas. *Phys. Rev. Lett.*, 62:583, 1989.

[80] S. Tarucha, D.G. Austing, T. Honda, R.J. van der Haage, and L.P. Kouwenhoven. Shell filling and spin effects in a few-electron quantum dot. *Phys. Rev. Lett.*, 77:3613, 1996.

[81] L.P. Kouwenhoven, D.G. Austing, and S. Tarucha. Few electron quantum dots. *Rep. Prog. Phys.*, 64:701, 2001.

[82] D. Goldhaber-Gordon, H. Shtrikman, D. Mahalu, D. Abusch-Magder, U. Meirav, and M.A. Kastner. Kondo effect in a single-electron transistor. *Nature*, 391:156, 1998.

[83] J. Kondo. *Theory of Dilute Magnetic Alloys*, volume 23 of *Solid State Physics*, page 184. Academic Press, 1969.

[84] B.A. Jones, C.M. Varma, and J.W. Wilkins. Low-temperature properties of the two-impurity Kondo Hamiltonian. *Phys. Rev. Lett.*, 61:125, 1988.

[85] L.I. Glazman and M.E. Raikh. Resonant Kondo transparency of a barrier with quasilocal impurity states. *JETP Lett.*, 47:452, 1988.

[86] T.K. Ng and P.A. Lee. On-site Coulomb repulsion and resonant tunneling. *Phys. Rev. Lett.*, 61:1768, 1988.

[87] Grabert and Devoret, editors. *Single charge tunnelling*. Plenum Press, New York, 1991.

[88] T. Ando, Y. Arakawa, K. Furuya, S. Komiyama, and H. Nakashima, editors. *Mesoscopic Physics and Electronics*. Springer, 1998.

[89] P. Hawrylak, L. Jacak, and A. Wojs. *Quantum dots*. Springer Verlag, Berlin, 1998.

[90] T. Chakraborty. *Quantum dots. A survey of the properties of artificial atoms*. Elsevier, 1999.

[91] M.A. Kastner. The single-electron transistor. *Rev. Mod. Phys.*, 64:849, 1992.

[92] L.P. Kouwenhoven, C.M. Marcus, P.L. McEuen, S. Tarucha, R.M. Westervelt, and N.S. Wingreen. Electron transport in quantum dots. In L. P. Kouwen-

hoven, G. Schön, and L.L. Sohn, editors, *Nato ASI conference proceedings*, pages 105–214. Kluwer, Dordrecht, 1997.

[93] C.W.J. Beenakker. Random-matrix theory of quantum transport. *Rev. Mod. Phys.*, 69:731, 1997.

[94] Y. Alhassid. The statistical theory of quantum dots. *Rev. Mod. Phys.*, 72:895, 2000.

[95] I.L. Aleiner, P.W. Brouwer, and L.I. Glazman. Quantum effects in Coulomb blockade. *Physics Reports*, 358:309, 2002.

[96] A.H. Wilson. The theory of electronic semi-conductors. *Proc. R. Soc. London, Series A*, 133:458, 1931.

[97] A.H. Wilson. The theory of electronic semi-conductors. II. *Proc. R. Soc. London, Series A*, 134:277, 1931.

[98] N.F. Mott. *Metal-Insulator Transitions*. Taylor & Francis, London, 2nd edition, 1990.

[99] P.W. Anderson. Absence of diffusion in certain random lattices. *Phys. Rev.*, 109:1492, 1958.

[100] B.I. Shklovskii and A.L. Efros. *Electronic Properties of Doped Semiconductors*. Springer Series in Solid-State Sciences 45. Springer-Verlag, 1984.

[101] P.A. Lee and T.V. Ramakrishnan. Disordered elctronic systems. *Reviews of Modern Physics*, 57:287, 1985.

[102] H. Fukuyama. Theory of weakly localized regime of the Anderson localization in 2-dimensions. *Surface Science*, 113:489, 1982.

[103] L.D. Landau. The theory of a Fermi liquid. *Sov. Phys. JETP*, 3:920, 1957.

[104] D. Pines and P. Nozieres. *The theory of quantum liquids*, volume 1: Normal Fermi Liquids. W.A. Benjamin, Inc, 1966.

[105] B.L. Altshuler and A.G. Aronov. Zero bias anomaly in tunnel resistance and electron-electron interaction. *Solid State Commun.*, 39:115, 1979.

[106] B.L. Altshuler and A.G. Aronov. Theory of disordered metals in strongly doped semiconductors. *Sov. Phys. JETP*, 50:968, 1979.

[107] B.L. Altshuler and A.G. Aronov. *Electron-Electron Interactions in Disordered Systems*. eds. A.L. Efros, M. Pollak (North-Holland, Amsterdam), 1985.

[108] S. Chakravarty and A. Schmid. Weak localization: the quasiclassical theory of electrons in a random potential. *Phys. Reports 140*, 4:194, 1986.

[109] E.D. Siggia and P.C. Kwok. Properties of electrons in semiconductor inversion layers with many occupied electric subbands. I. screening and impurity scattering. *Phys. Rev. B*, 2:1024, 1970.

[110] F. Stern. Calculated temperature dependence of mobility in silicon inversion layers. *Phys. Rev. Lett.*, 44:1469, 1980.

[111] P.W. Anderson, E. Abrahams, and T.V. Ramakrishnan. Possible explanation of nonlinear conductivity in thin-film metal wires. *Phys. Rev. Lett.*, 43:718, 1979.

[112] L.P. Gorkov, A.I. Larkin, and D.E. Khmelnitskii. Particle conductivity in a two-dimensional random potential. *JETP Lett.*, 30:228, 1979.

[113] G. Bergmann. Physical interpretation of weak localisation: A time-of-flight experiment with conduction electrons. *Phys. Rev. B*, 28:2914, 1983.

[114] G. Bergmann. Weak localization in thin-films — a time-of-flight experiment with conduction electrons. *Phys. Rep.*, 107:1, 1984.

[115] A.I. Larkin and D.E. Khmelnitskii. Anderson localization and anomalous magnetoresistance. *Sov. Phys. Usp.*, 25:185, 1982.

[116] D.E. Khmelnitskii. Localization and coherent scattering of electrons. *Physica*, 126B:235, 1984.

[117] G. Zala, B.N. Narozhny, and I.L. Aleiner. Interaction corrections at intermediate temperatures: Longitudinal conductivity and kinetic equation. *Phys. Rev. B*, 64:214204–1, 2001.

[118] F. Stern. Polarizability of a two-dimensional electron gas. *Phys. Rev. Lett.*, 18:546, 1967.

[119] T. Ando. Self-Consistent Results for a GaAs/Al_xGa_{1-x}As Heterojunction. II. Low Temperature Mobility. *J. Phys. Soc. Japan*, 51:3900, 1982.

[120] S. Das Sarma and F. Stern. Single-particle relaxation time versus scattering time in an impure electron gas. *Phys. Rev. B*, 32:8442, 1985.

[121] D. Das Sarma. Theory of finite-temperature screening in a disordered two-dimensional electron gas. *Phys. Rev. B*, 33:5401, 1986.

[122] A.Gold and V.T. Dolgopolov. Temperature dependence of the conductivity for the two-dimensional electron gas: Analytical results for low temperatures. *Phys. Rev. B*, 33:1076, 1986.

[123] S. Das Sarma and E.H. Hwang. Charged impurity-scattering-limited low-temperature resistivity of low-density silicon inversion layers. *Phys. Rev. Lett.*, 83:164, 1999.

[124] S. Das Sarma and E.H. Hwang. Calculated temperature dependent resistance in low-density two-dimensional hole gases in GaAs heterostructures. *Phys. Rev. B*, 61:R7838, 2000.

[125] B. Laikhtman and R.A. Kiehl. Theoretical hole mobility in a narrow Si/SiGe quantum well. *Phys. Rev. B*, 47:10515, 1993.

[126] G.D. Mahan. *Many particle physics*. Plenum Press, New York, 1990.

[127] A. Gold. Linear temperature dependence of mobility in quantum wells and the effects of exchange and correlation. *Journal of Physics: Condensed Matter*, 13:11641, 2001.

[128] E. Wigner. On the interaction of electrons in metals. *Phys. Rev.*, 46:1002, 1934.

[129] E. Wigner. Effects of the electron interaction on the energy levels of electrons in metals. *Tran. Faraday Soc.*, 34:678, 1938.

[130] L. Bonsall and A.A. Maradudin. Some dynamical properties of a two-dimensional wigner crystal. *Surf. Sci.*, 58:312, 1976.

[131] L. Bonsall and A.A. Maradudin. Some static and dynamical properties of a two-dimensional Wigner crystal. *Phys. Rev. B*, 15:1959, 1977.

[132] B. Tanatar and D.M. Ceperley. Ground-state of the two-dimensional electron-gas. *Phys. Rev. B*, 39:5005, 1989.

[133] S.T.Chui and B. Tanatar. Impurity effect on the 2-dimensional electron fluid-solid transition in zero-field. *Phys. Rev. Lett.*, 74:458, 1995.

[134] S. Kawaji. Weak localization and negative magnetoresistance in semiconductor two-dimensional systems. *Surf. Sci.*, 170:682, 1986.

[135] D. Vollhardt. *Localization Effects in Disordered Systems*, volume 27 of *Festkörperprobleme*. Vieweg, 1987.

[136] G.J. Dolan and D.D. Osheroff. Nonmetallic conduction in thin metal films at low temperatures. *Phys. Rev. Lett.*, 43:721, 1979.

[137] N. Giordano. Experimental study of localization in thin wires. *Phys. Rev. B*, 22:5635, 1980.

[138] Y. Kawaguchi and S. Kawaji. Negative magnetoresistance in silicon (100) MOS inversion layers. *J. Phys. Soc. Jpn.*, 48:699, 1980.

[139] D.J. Bishop, D.C. Tsui, and R.C. Dynes. Nonmetallic conduction in electron inversion layers at low temperatures. *Phys. Rev. Lett.*, 44:1153, 1980.

[140] R.G. Wheeler. Magnetoconductance and weak localization in silicon inversion layers. *Phys. Rev. B*, 24:4645, 1981.

[141] M.J. Uren, R.A. Davis, and M. Pepper. The observation of interaction and localisation effects in a two-dimensional electron gas at low temperatures. *J. Phys. C*, 13:L986, 1980.

[142] M.J. Uren, R.A. Davis, M. Kaveh, and M. Pepper. Magnetic delocalisation of a two-dimensional electron gas and the quantum law of electron-electron scattering. *J. Phys. C*, 14:L395, 1981.

[143] R.W. Glew D.A. Poole, M. Pepper. The observation of localisation and interaction effects in the two-dimensional electron gas of a GaAs-GaAlAs heterojunction at low temperatures. *J. Phys. C*, 14:L995, 1981.

[144] B.J.F. Lin, M.A. Paalanen, A.C. Gossard, and D.C. Tsui. Weak localisation of two-dimensional electrons in GaAs-$Al_xGa_{1-x}As$ heterostructures. *Phys. Rev. B*, 29:927, 1984.

[145] D.J. Bishop, R.C. Dynes, and D.C. Tsui. Magnetoresistance in Si metal-oxide-semiconductor field-effect transistors: Evidence of weak localization and correlation. *Phys. Rev. B*, 24:773, 1982.

[146] T. Nambu, S. Kawaji, K. Kuboki, Y. Kawaguchi, J. Yoshino, and H. Sakaki. Negative magnetoresistance and inelastic scattering time in two-dimensional electron systems in GaAs/$Al_xGa_{1-x}As$ heterojunction interfaces. *J. Phys. Soc. Jap.*, 53:682, 1984.

[147] R.A. Davies, M.J. Uren, and M. Pepper. Magnetic separation of localisation and interaction effects in a two-dimensional electron gas at low temperatures. *J. Phys. C: Solid State Phys.*, 14:L531, 1981.

[148] S.V. Kravchenko, G.V. Kravchenko, J.E. Furneaux, V.M. Pudalov, and M D'Iorio. Possible metal-insulator transition at B=0 in two dimensions. *Phys. Rev. B*, 50:8039, 1994.

[149] S.V. Kravchenko, W. E. Mason, G.E. Bowker, J.E. Furneaux, and V.M. Pudalovand M. D'Iorio. Scaling of an anomalous metal-insulator transition in a two-dimensional system in silicon at B=0. *Phys. Rev. B*, 51:7038, 1995.

[150] X. Waintal abd G. Benenti and J.-L. Pichard. Delocalized Coulomb phase in two dimensions. *Europhys. Lett.*, 49:466, 2000.

[151] E. Abrahams and G. Kotliar. The metal-insulator transition in correlated disordered systems. *Science*, 274:1853, 1996.

[152] M. Vojta. Quantenphasenübergänge – 'Schmelzen' am Temperaturnullpunkt. *Physik Journal*, 1(3):55, 2002.

[153] S.L. Sondhi, S.M. Girvin, J.P. Carini, and D. Shahar. Continuous quantum phase transitions. *Rev. Mod. Phys.*, 69:315, 1997.

[154] D. Popovic, A.B. Fowler, and S. Washburn. Metal-insulator transition in two dimensions: effects of disorder and magnetic field. *Phys. Rev. Lett.*, 79:1543, 1997.

[155] R. Heemskerk and T.M. Klapwijk. Nonlinear resistivity at the metal-insulator transition in a two-dimensional electron gas. *Phys. Rev. B*, 58:R1754, 1998.

[156] S.S. Safonov, S.H. Roshko, A.K. Savchenko, A.G. Pogosov, and Z.D. Kvon. "Metallic" and "insulating" behaviour of the two-dimensional electron gas on a vicinal surface of Si MOSFETs. *Phys. Rev. Lett.*, 86:272, 2001.

[157] Y. Hanein, U. Meirav, D. Shahar, C.C. Li, D.C. Tsui, and H. Shtrikman. The Metalliclike Conductivity of a Two-Dimensional Hole System. *Phys. Rev. Lett.*, 80:1288, 1998.

[158] J. Yoon, C.C. Li, D. Shanar, D.C.Tsui, and M. Shayegan. Wigner crystallization and metal–insulator transition of two-dimensional holes in GaAs at B=0. *Phys. Rev. Lett.*, 82:1744, 1999.

[159] M.Y. Simmons, A.R. Hamilton, M. Pepper, E.H. Linfield, P.D. Rose, and D. A. Ritchie. Weak localization, hole–hole interactions, and the "metal"–insulator transition in two dimensions. *Phys. Rev. Lett.*, 84:2489, 2000.

[160] Y. Yaish, O. Prus, E. Buchstab, S. Shapira, G. Ben Yoseph, U. Sivan, and A. Stern. Interband scattering and the Metallic Phase" of two-dimensional holes in GaAs/AlGaAs. *Phys. Rev. Lett.*, 84:4954, 2000.

[161] A.R. Hamilton, M.Y. Simmons, M. Pepper, and D.A. Ritchie. Metallic behavior in dilute two-dimensional hole systems. *Phys. Rev. Lett.*, 87:126802–1, 2001.

[162] Y.Y. Proskuryakov, A.K. Savchenko, Safonov S.S, M. Pepper, M.Y. Simmons, and D.A. Ritchie. Fermi-liquid behavior of the low-density 2D hole gas in a GaAs/AlGaAs heterostructure at large values of r_s. *Phys. Rev. Lett.*, 86:4895, 2001.

[163] E. Ribeiro, R.D. Jäggi, T. Heinzel, and K. Ensslin. Metal–insulator transition in a disordered two-dimensional electron gas in GaAs-AlGaAs at zero magnetic field. *Phys. Rev. Lett.*, 82:996, 1999.

[164] S.J. Papadakis and M. Shayegan. Apparent metallic behavior at B=0 of a two-dimensional electron system in AlAs. *Phys. Rev. B*, 57:R15068, 1998.

[165] J. Lam, M. D'Iorio, D. Brown, and H. Lafontaine. Scaling and the metal-insulator transition in Si/SiGe quantum wells. *Phys. Rev. B*, 56:R12741, 1997.

[166] P.T. Coleridge, R. L. Williams, Y. Feng, and P. Zawadzki. Metal-insulator transition at B=0 in p-type SiGe. *Phys. Rev. B*, 56:R12764, 1997.

[167] V. Senz, U. Dötsch, U. Gennser, T. Ihn, T. Heinzel, K. Ensslin, R. Hartmann, and D. Grützmacher. Metal-insulator transition in a 2-dimensional system with an easy spin axis. *Annalen der Physik*, 8:237, 1999.

[168] V. Senz, T. Heinzel, T. Ihn, K. Ensslin, G. Dehlinger, D. Grützmacher, and U. Gennser. Coexistence of weak localization and a metallic phase in Si/SiGe quantum wells. *Phys. Rev. B*, 61:R5082, 2000.

[169] E. Abrahams, S.V. Kravchenko, and M.P. Sarachik. Metallic behavior and related phenomena in two dimensions. *Rev. Mod. Phys.*, 73:251, 2001.

[170] S.V. Kravchenko. Metal-insulator transition in two dimensions. *Braz. J. Phys.*, 29:623, 1999.

[171] D. Simonian, S.V. Kravchenko, and M.P. Sarachik. Reflection symmetry at a $B = 0$ metal-insulator transition in two dimensions. *Phys. Rev. B*, 55:R13421, 1997.

[172] V. Senz, U. Dötsch, U. Gennser, T. Ihn, T. Heinzel, K. Ensslin, R. Hartmann, and D. Grützmacher. Temperature and electric field scaling in a two-dimensional hole gas. *Proceedings of the XXXIV Rencontres de Moriand, France*, 1999.

[173] M.Y. Simmons, A. R. Hamilton, M. Pepper, E. H. Linfield, P. D. Rose, D. A. Ritchie, A. K. Savchenko, and T. G. Griffiths. Metal–insulator transition at B=0 in a dilute two-dimensional GaAs-AlGaAs hole gas. *Phys. Rev. Lett.*, 80:1292, 1998.

[174] Y. Hanein, D. Sahar, J. Yoon, C.C. Li, D.C. Tsui, and H. Shtrikman. Properties of the apparent metal-insulator transition in two-dimensional systems. *Phys. Rev. B*, 58:R7520, 1998.

[175] D. Popovic, A.B. Fowler, and S. Washburn. Metal-insulator transition in a low-mobility two-dimensional electron system. *Physica B*, 249-251:701, 1998.

[176] V. M. Pudalov, G. Brunthaler, A. Prinz, and G. Bauer. Maximum matallic conductivity in Si-MOS structures. *Phys. Rev. B*, 60:R2154, 1999.

[177] S. Washburn, N.J. Kim, X.G. Feng, and D. Popovic. Scaling laws, phase diagram, localized magnetic moments and Kondo effect in two-dimensional metals. *Ann. Phys. (Leipzig)*, 7-9:569, 1999.

[178] S. J. Papadakis, E.P. De Poortere, H.C. Manoharan, M. Shayegan, and R. Winkler. The effect of spin-splitting on the metallic behavior of a two-dimensional system. *Science*, 283:2056, 1999.

[179] A.P. Mills, A.P. Ramirez, L.N. Pfeiffer, and K.W. West. Nonmonotonic temperature-dependent resistance in low density 2D hole gases. *Phys. Rev. Lett.*, 83:2805, 1999.

[180] X.G. Feng, D. Popovic, S. Washburn, and V. Dobrosavljevic. Novel metallic behavior in two dimensions. *Phys. Rev. Lett.*, 86:2625, 2001.

[181] S.V. Kravchenko and T.M. Klapwijk. Metallic low-temperature resistivity in a 2D electron system over an extended temperature range. *Phys. Rev. Lett.*, 84:2909, 2000.

[182] A.R. Hamilton, M.Y. Simmons, M. Pepper, E.H. Linfield, P.D. Rose, and D.A. Ritchie. Reentrant insulator-metal trasition at B=0 in a two-dimensional hole gas. *Phys. Rev. Lett.*, 82:1542, 1999.

[183] S.V. Kravchenko, D. Simonian, and M.P. Sarchik. Electric field scaling at B=0 metal–insulator transition in two dimensions. *Phys. Rev. Lett.*, 77:4938, 1996.

[184] A.A. Shashkin, S.V. Kravchenko, and T.M. Klapwijk. Metal-insulator transition in a 2D electron gas: equaivalence of two approaches for determining the critical point. *Phys. Rev. Lett.*, 87:266402–1, 2001.

[185] X. G. Feng, D. Popovic, S. Washburm, and V. Dobrosavljevic. Effect of local magnetic moments on the metallic behavior in two dimensions. *Phys. Rev. Lett*, 83:368, 1999.

[186] D. Simonian, S. V. Kravchenko, M.P. Sarachik, and V.M. Pudalov. Magnetic field suppression of the conducting phase in two dimensions. *Phys. Rev. Lett.*, 79:2304, 1997.

[187] V.M. Pudalov, G. Brunthaler, A. Prinz, and G. Bauer. Breakdown of the anomalous two-dimensional metallic phase in a parallel magnetic field. *Physica B*, 249-251:697, 1998.

[188] K.M. Mertes, D. Simonian, M.P. Sarachik, S.V. Kravchenko, and T.M. Klapwijk. Response to parallel magnetic field of a dilute two-dimensional electron system across the metal-insulator transition. *Phys. Rev. B*, 60:R5093, 1999.

[189] X.P.A. Gao, A.P. Mills, A.P. Ramirez, L.N. Pfeiffer, and K.W. West. Two-dimensional metal in a parallel magnetic field. *Phys. Rev. Lett.*, 88:166803–1, 2002.

[190] K. Eng, X.G. Feng, D. Popovic, and S. Washburn. Effects of a parallel magnetic field on the metal-insulator transition in a dilute two-dimensional electron system. *Phys. Rev. Lett.*, 88:136402–1, 2002.

[191] A.A. Shashkin, S.V. Kravchenko, V.T. Dolgopolov, and T.M. Klapwijk. Indication of the ferromagnetic instability in a dilute two-dimensional electron system. *Phys. Rev. Lett.*, 87:086801–1, 2001.

[192] S.A. Vitkalov, H. Zheng, K.M. Mertes, M.P. Sarachik, and T.M. Klapwijk. Hall coefficient of a dilute two-dimensional electron system in a parallel magnetic field. *Phys. Rev. B*, 63:193304–1, 2001.

[193] S.A. Vitkalov, H. Zheng, K.M. Mertes, M.P. Sarachik, and T.M. Klapwijk. Scaling of the Magnetoconductivity of Silicon MOSFETs: evidence for a quantum phase transition in two dimensions. *Phys. Rev. Lett.*, 87:086401–1, 2001.

[194] S.V. Kravchenko, D. Simonian, K. Mertes, M.P. Sarachik, and T.M.Klapwijk. Classical versus quantum effects in the B=0 conducting phase in two dimensions. *Phys. Rev. B*, 59:R12740, 1999.

[195] G. Brunthaler, A. Prinz, G. Bauer, V. M. Pudalov, E. M. Dizhur, J. Jaroszynski, P. Glod, and T. Dietl. Weak localization in the 2D metallic regime of Si-MOS. *Ann. Phys. (Leipzig)*, 8:579, 1999.

[196] V. Senz, T. Ihn, T. Heinzel, K. Ensslin, G. Dehlinger, D. Grützmacher, and U. Gennser. Analysis of the Metallic Phase of Two-Dimensional Holes in SiGe in Terms of Temperature Dependent Screening. *Phys. Rev. Lett.*, 85:4357, 2000.

[197] P.T. Coleridge, J. Lapointe, R.L. Williams, and P. Zawadzki. Coulomb interactions and the metal-insulator transition in p-SiGe. *Physica E*, 11:167, 2001.

[198] M.Y. Simmons, A.R. Hamilton, T.G. Griffiths, A.K. Savchenko, M. Pepper, and D.A. Ritchie. Metal-insulator transition at $B = 0$ in an ultra-low density two-dimensional hole gas. *Physica B*, 249-251:705, 1998.

[199] Y. Hanein, N. Nenadovic, D. Shahar, H. Shtrikman, J. Yoon, C.C. Li, and D.C. Tsui. Linking insulator-to-metal transitions at zero and finite magnetic fields. *Nature*, 400:735, 1999.

[200] V.M. Pudalov, G. Brunthaler, and A. Prinz and. Weak-field Hall resistance and effective carrier density measurements across the metal-insulator transition in Si-MOS structures. *JETP Lett.*, 70:48, 1999.

[201] G. Brunthaler, A. Prinz, G. Bauer, and V.M. Pudalov. Exclusion of Quantum Coherence as the Origin of the 2D Metallic State in High-Mobility Silicon Inversion Layers. *Phys. Rev. Lett.*, 87:6802, 2001.

[202] S.V. Kravchenko, A.A. Shashkin, D.A. Bloore, and T.M. Klapwijk. Shubnikov-de Haas oscillations near the metal-insulator transition in a two-dimensional electron system in Silicon. *Solid State Communications*, 116:495, 2000.

[203] P.T. Coleridge, A.S. Sachrajda, and P. Zawadzki. Weak localisation, interaction effects and the metallic phase in p-SiGe. *Phys. Rev. B*, 65:125328–1, 2002.

[204] D. Popovic and S. Washburn. Conductance fluctuations near the two-dimensional metal-insulator transition. *Physica B*, 249-251:504, 1998.

[205] S.V. Kravchenko, D. Simonian, M.P. Sarachik, A.D. Kent, and V.M. Pudalov. Effect of a tilted magnetic field on the anomalous $H = 0$ conducting phase in high-mobility Si MOSFET's. *Phys. Rev. B*, 58:3553, 1998.

[206] T. Okamoto, K. Hosoya, S. Kawaji, and A. Yagi. Spin Degree of Freedom in a Two-Dimensional Electron Liquid. *Phys. Rev. Lett*, 82:3875, 1999.

[207] S. Ilani, A. Yacoby, D. Mahalu, and H. Shtrikman. Unexpected behavior of the local compressibility near the $B = 0$ metal-insulator transition. *Phys. Rev. Lett.*, 84:3133, 2000.

[208] S. C. Dultz and H. W. Jiang. Thermodynamic Signature of a Two-dimensional Metal-Insulator Transition. *Phys. Rev. Lett.*, 84:4689, 2000.

[209] S. Ilani, A. Yacoby, D. Mahalu, and H. Shtrikman. Microscopic Strucutre of the Metal-Insulator Transition in Two Diemsnsions. *Science*, 292:1354, 2001.

[210] Y. Hanein, D. Shahar, J. Yoon, C.C. Li, C. Tsui, and H. Shtrikman. Observation of the metal-insulator transition in two-dimensional n-type GaAs. *Phys. Rev. B*, 58:R13338, 1998.

[211] E. Abrahams. Theoretical approaches to the metal-insulator transition in 2D. *Ann. Phys. (Leipzig)*, 8:539, 1999.

[212] B.L. Altshuler and D.L. Maslov. Theory of metal-insulator transitions in gated semiconductors. *Phys. Rev. Lett.*, 82:145, 1999.

[213] B.L. Altshuler, D.L. Maslov, and V.M. Pudalov. Metal-insulator transition in 2d: Anderson localisation in temperature-dependent disorder? *Physica status solidi B – Basic research*, 218:193, 2000.

[214] B.L. Altshuler, D.L. Maslov, and V.M. Pudalov. Metal-insulator transition in 2D: resistance in the critical region. *Physica E*, 9:209, 2001.

[215] Y. Meir. Percolation-type description of the metal–insulator transition in two dimensions. *Phys. Rev. Lett.*, 83:3506, 1999.

[216] Y. Meir. Two-species percolation and scaling theory of the metal-insultor tranition in two dimensions. *Phys. Rev. B*, 61:16470, 2000.

[217] Y. Meir. Percolation type description of the metal-insulator transition in two dimensions. *Physica A*, 302:391, 2001.

[218] S. Das Sarma and E.H. Hwang. Parallel magnetic field induced giant magnetoresistance in low density quasi-two-dimensional layers. *Phys. Rev. Lett.*, 84:5596, 2000.

[219] T.M. Klapwijk and S. Das Sarma. A few electrons per ion scenario for the $B = 0$ metal-insulator transition in two dimensions. *Solid state communications*, 110:581, 1999.

[220] V. T. Dolgopolov and A. Gold. Magnetoresistance of a two-dimensional electron gas in a parallel magnetic field. *JETP Lett.*, 71:27, 2000.

[221] G. Kastrinakis. Metal-insulator transition in 2D: due to interactions and disorder? *Physica B*, 296:36, 2001.

[222] V. Dobrosavljevic, E. Abrahams, E. Miranda, and S. Chakravarty. Scaling theory of two-dimensional metal-insulator transitions. *Phys. Rev. Lett.*, 79:455, 1997.

[223] S. Chakravarty, L. Yin, and E. Abrahams. Interactions and scaling in a disordered two-dimensional metal. *Phys. Rev. B*, 58:559, 1998.

[224] C. Castellani, C. Di Castro, and P.A. Lee. Metallic phase and metal-insulator transition in two-dimensional electronic systems. *Phys. Rev. B*, 57:R9381, 1998.

[225] Q. Si and C.M. Varma. Metal-insulator transition of disordered interacting electrons. *Phys. Rev. Lett.*, 81:4951, 1998.

[226] D.N. Sheng and Z.Y. Weng. Universal scaling, beta function and metal-insulator transition. *Phys. Rev. Lett.*, 83:144, 1999.

[227] S. Chakravarty, S. Kivelson, C. Nayak, and K. Voelker. Wigner glass, spin liquids and the metal-insulator transition. *Phil. Mag. B*, 79:859, 1999.

[228] B. Spivak. Properties of the strongly correlated two-dimensional electron gas in Si MOSFETs. *Phys. Rev. B*, 64:085317–1, 2001.

[229] D. Belitz and T.R. Kirkpatrick. Possible triplet superconductivity in MOSFETs. *Phys. Rev. B*, 58:8214, 1998.

[230] P. Phillips, Y. Wan, I. Martin, S. Knysh, and D. Dalidovich. Superconductivity in a two-dimensional electron gas. *Nature*, 395:253, 1998.

[231] J.S. Thakur and D. Neilson. Phase diagram of the metal-insulator transition in two-dimensional electronic systems. *Phys. Rev. B*, 59:R5280, 1999.

[232] S. He and X.C. Xie. New Liquid Phase and Metal-Insulator Transition in Si MOSFETs. *Phys. Rev. Lett.*, 80:3324, 1998.

[233] T. Vojta, F. Epperlein, and M. Schreiber. Do interactions increase or reduce the conductance of disordered electrons? It depends! *Phys. Rev. Lett.*, 81:4212, 1998.

[234] G. Benenti, X. Waintal, and J.L. Pichard. New quantum phase between the Fermi glass and the Wigner crystal in two dimensions. *Phys. Rev. Lett.*, 83:1826, 1999.

[235] P.J.H. Denteneer, R.T. Scalettar, and N. Trivedi. Conducting phase in the two-dimensional disordered Hubbard model. *Phys. Rev. Lett.*, 83:4610, 1999.

[236] D.L. Shepelyansky and P.H. Song. Quantum ergodicity for electrons in two dimensions. *Ann. Phys. (Leipzig)*, 8:665, 1999.

[237] C.J. Emeleus, T.E. Whall, D.W. Smith, R.A. Kubiak, E.H.C. Parker, and M.J. Kearney. Scattering mechanisms affecting hole transport in remote-doped Si/SiGe heterostructures. *J. Appl. Phys.*, 73:3852, 1993.

[238] T.E. Whall. p-channel SiGe heterostructures for field effect applications. *Appl. Surf. Sci.*, 102:221, 1996.

[239] P.J. Price. Hot-electrons in a GaAs heterolayer at low-temperature. *J. Appl. Phys.*, 53:6863, 1982.

[240] Y.H. Xie, R. People, J. C. Bean, and K. W. Wecht. Power loss by two-dimensional holes in coherently strained $Ge_{0.2}Si_{0.8}$/Si heterostructures: Evidence for weak screening. *Appl. Phys. Lett.*, 49:283, 1986.

[241] E. Chow, H. P. Wei, S. M. Girvin, and M. Shayegan. Phonon emission from a 2D electron gas: evidence of transition to the hydrodynamic regime. *Phys. Rev. Lett.*, 77:1143, 1996.

[242] E. Chow, H. P. Wei, S. M. Girvin, W. Jan, and J. E. Cunningham. Effect of disorder on phonon emissions from a two-dimensional electron gas in GaAs/Al_xGa_{1-x}As heterostructures. *Phys. Rev. B*, 56:R1676, 1997.

[243] R. Leturcq, D. L'Hote, R. Tourbot, V.Senz, U. Gennser, T. Ihn, K. Ensslin, G. Dehlinger, and D. Grützmacher. Heating of two-dimensional holes in SiGe and the B = 0 metal–insulator transition. *cond-mat/0107457*, 2001.

[244] T. Leturcq, D. L'Hote, R. Tourbot, V. Senz, U. Gennser, T. Ihn, K. Ensslin, G. Dehlinger, and D. Grützmacher. Hot-hole effects in a dilute two-dimensional gas in SiGe. *Europhys. Lett.*, 61:499, 2003.

[245] V. I. Khizhny, O. A. Mironov, E. H. C. Parker, P. J. Phillips, T. E. Whall, and M. J. Kearney. Direct evidence for a piezoelectriclike effect in coherently strained SiGe/Si heterostructures. *Appl. Phys. Lett.*, 69:960, 1996.

[246] S. Hikami, A. I. Larkin, and Y. Nagaoka. Spin-orbit interaction and magnetoresistance in the 2 dimensional random system. *Prog. Theor. Phys.*, 63:707, 1980.

[247] B.L. Altshuler and. On the anomalous magnetoresistance in semiconductors. *Sov. Phys. JETP*, 54:411, 1981.

[248] S.V. Iordanskii, Y.B. Lyanda-Geller, and G.E. Pikus. Weak-localization in quantum-wells with spin-orbit interaction. *JETP Lett.*, 60:206, 1994.

[249] F.G. Pikus and G.E. Pikus. Conduction-band splitting and negative magnetoresistance in A_3B_5 heterostructures. *Phys. Rev. B*, 51:16928, 1995.

[250] W. Knap, C. Skierbiszewski, A. Zduniak, E. Litwin-Staszewska, D. Bertho, F. Kobbi, and J. L. Robert. Weak antilocalization and spin precession in quantum wells. *Phys. Rev. B*, 53:3912, 1996.

[251] T. Hassenkam, S. Pedersen, K. Baklanov, A. Kristensen, C. B. Sorensen, , and P. E. Lindelof. Spin splitting and weak localization in (110) GaAs/Al$_x$Ga$_{1-x}$As quantum wells . *Phys. Rev. B*, 55:9298, 1997.

[252] S. Pedersen, C.B. Sorensen, A. Kristensen, P.E. Lindelof, L.E. Golub, and N.S. Averkiev. Weak localization in Al$_{0.5}$Ga$_{0.5}$As/GaAs p-type quantum wells. *Phys. Rev. B*, 60:4880, 1999.

[253] N.S. Averkiev, L.E. Golub, and G.E. Pikus. Weak localization in semiconductor structures with strong spin-orbit coupling. *JETP*, 86:780, 1998.

[254] G.M. Minkov, A.V. Germanenko, V.A. Larionova, and S.A. Negashev. Analysis of negative magnetoresistance: Statistics of closed paths. I. Theory. *Phys. Rev. B*, 61:13164, 2000.

[255] C.J.C.J. Emeleus, T.E. Whall, D.W. Smith, N.L. Mattey, R.A. Kubiak, E.H.C. Parker, and M.J. Kearney. Observation of novel transport phenomena in a Si$_{0.8}$Ge$_{0.2}$ two-dimensional hole gas. *Phys. Rev. B*, 47:10016, 1993.

[256] G. Zala, B.N. Narozhny, and I.L. Aleiner. Interaction corrections to the Hall coefficient at intermediate temperatures. *Phys. Rev. B*, 64:201201–1, 2001.

[257] T. Kawamura and S. Das Sarma. Temperature dependence of the low-temperature mobility in ultrapure Al$_x$Ga$_{1-x}$As/GaAs heterojunctions: Acoustic phonon scattering. *Phys. Rev. B*, 42:3725, 1990.

[258] T. Kawamura and S. Das Sarma. Phonon-scattering-limited electron mobilities in Al$_x$Ga$_{1-x}$As/GaAs heterojunctions. *Phys. Rev. B*, 45:3612, 1992.

[259] P.F. Maldague. Many-body corrections to polarizability of 2-dimensional electron-gas. *Surf. Sci.*, 73:296, 1978.

[260] A. Gold. Peak Mobility of Silicon Metal-Oxide-Semiconductor Systems. *Phys. Rev. Lett.*, 54:1079, 1985.

[261] A.Gold. Electronic transport properties of a two-dimensional electron gas in a silicon quantum-well structure at low temperature. *Phys. Rev. B*, 35:723, 1987.

[262] A. Gold. Scattering time and single-particle relaxation time in a disordered two-dimensional electron gas. *Phys. Rev. B*, 38:10798, 1988.

[263] L.D. Hallam, J. Weis, and P.A. Maksym. Screening of the electron-electon interaction by gate electrodes in semiconductor quantum dots. *Phys. Rev. B*, 53:1452, 1996.

[264] G.W. Bryant. Electronic structure of ultrasmall quantum-well boxes. *Phys. Rev. Lett.*, 59:1140, 1987.

[265] P.A. Maksym and T. Chakraborty. Quantum dots in a magnetic field: Role of electron-electron interactions. *Phys. Rev. Lett.*, 65:108, 1990.

[266] U. Merkt, J. Huser, and M. Wagner. Energy spectra of two electrons in a harmonic quantum dot. *Phys. Rev. B*, 43:7320, 1991.

[267] D. Pfannkuche, V. Gudmundsson, and P.A. Maksym. Comparison of a Hartree, a Hartree-Fock, and an exact treatment of quantum-dot helium. *Phys. Rev. B*, 47:2244, 1993.

[268] U. Sivan, R. Berkovits, Y. Aloni, O. Prus, A. Auerbach, and G. Ben-Yoseph. Mesoscopic fluctuations in the ground state energy of disordered quantum dots. *Phys. Rev. Lett.*, 77:1123, 1996.

[269] R. Berkovits. Absence of Bimodal Peak Spacing Distribution in the Coulomb Blockade Regime. *Phys. Rev. Lett.*, 81:2128, 1998.

[270] A. Kumar, S.E. Laux, and F. Stern. Electron states in a GaAs quantum dot in a magnetic field. *Phys. Rev. B*, 42:5166, 1990.

[271] K.-H. Ahn, K. Richter, and I.-H. Lee. Addition spectra of chaotic quantum dots: interplay between interactions and geometry. *Phys. Rev. Lett.*, 83:4144, 1999.

[272] L. Bonci and R. Berkovits. The addition spectrum of interacting electrons: Parametric dependence. *Europhys. Lett.*, 47:708, 1999.

[273] A. Cohen, K. Richter, and R. Berkovits. Statistics of Coulomb-blockade peak spacings within the Hartree-Fock approximation. *Phys. Rev. B*, 60:2536, 1999.

[274] S. Levit and D. Orgad. Statistics of Hartree-Fock levels in small disordered systems. *Phys. Rev. B*, 60:5549, 1999.

[275] P.N. Walker, G. Montambaux, and Y. Gefen. Addition spectrum and Koopmans' theorem for disordered quantum dots. *Phys. Rev. B*, 60:2541, 1999.

[276] M. Stopa. Quantum dot self-consistent electronic structure and the Coulomb blockade. *Phys. Rev. B*, 54:13767, 1996.

[277] K. Hirose and N.S. Wingreen. Ground-state energy and spin in disordered quantum dots. *Phys. Rev. B*, 65:193305, 2002.

[278] H. Jiang, H.U. Baranger, and W. Yang. Spin and Conductance-Peak-Spacing Distributions in Large Quantum Dots: A Density-Functional Theory Study. *Phys. Rev. Lett.*, 90:026806, 2003.

[279] F. Pederiva, C.J. Umrigar, and E. Lipparini. Diffusion Monte Carlo Study of Circular Quantum Dots. *Phys. Rev. B*, 62:8120, 2000.

[280] R. Egger, W. Häusler, C.H. Mak, and H. Grabert. Crossover from Fermi Liquid to Wigner Molecule Behavior in Quantum Dots. *Phys. Rev. Lett.*, 82:3320, 1999.

[281] F. Schwabl. *Quantenmechanik*. Springer Verlag, 4th edition, 1993.

[282] C.W.J. Beenakker. Theory of Coulomb-blockade oscillations in the conductance of a quantum dot. *Phys. Rev. B*, 44:1646, 1991.

[283] J.C. Maxwell. *A Treatise on Electricity and Magnetism*, volume I. Clarendon Press, 1891.

[284] J.D. Jackson. *Klassische Elektrodynamik*. de Gruyter, 2nd edition, 1983.

[285] M. Field, C.G. Smith, M. Pepper, D.A. Ritchie, J.E.F. Frost, G.A.C. Jones, and D.G. Hasko. Measurements of Coulomb blockade with a noninvasive voltage probe. *Phys. Rev. Lett.*, 70:1311, 1993.

[286] R.G. Parr and W. Yang. *Density funcitonal theory of atoms and molecules*. Oxford University Press, 1989.

[287] D.V. Averin and Y.V. Nazarov. Virtual electron diffusion during quantum tunnelling of the electric charge. *Phys. Rev. Lett.*, 65:2446, 1990.

[288] D.V. Averin and Y.V. Nazarov. *Single charge tunnelling*, chapter 6: Macroscopic Quantum Tunnelling of Charge and Co-Tunnelling. Plenum Press, 1991.

[289] S.M. Cronenwett, T.H. Oosterkamp, and L.P. Kouwenhoven. A tunable Kondo effect in quantum dots. *Science*, 281:540, 1998.

[290] D. Goldhaber-Gordon, J. Göres, M.A. Kastner, H. Shtrikman, D. Mahalu, and U. Meirav. From the Kondo regime to the mixed-valence regime in a single-electron transistor. *Phys. Rev. Lett.*, 81:5225, 1998.

[291] J. Schmid, J. Weis, K. Eberl, and K.v. Klitzing. A quantum dot in the limit of strong coupling to reservoirs. *Physica B*, 256-258:182, 1998.

[292] J. Hamilton. *Aharonov-Bohm and other cyclic phenomena.* Number 139 in Springer Tracts in Modern Physics. Springer, 1997.

[293] N. Byers and C.N. Yang. Theoretical considerations concerning quantized magnetic flux in superconducting cylinders. *Phys. Rev. Lett.*, 7:46, 1961.

[294] F. Bloch. Josephson effect in a superconducting ring. *Phys. Rev. B*, 2:109, 1970.

[295] M. Büttiker, Y. Imry, and R. Landauer. Josephson behavior in small normal one-dimensional rings. *Phys. Lett.*, 96A:365, 1983.

[296] M.V. Berry. Quantal phase factors accompanying adiabatic changes. *Proc. Roy. Soc. London, Ser. A*, 392:45, 1984.

[297] J.J. Sakurai. *Modern Quantum Mechanics.* Addison-Wesley, 1995.

[298] A. Stern. Berry's phase, motive forces, and mesoscopic conductivity. *Phys. Rev. Lett.*, 68:1022, 1992.

[299] A.G. Aronov and Y.B. Lyanda-Geller. Spin-orbit Berry phase in conducting rings. *Phys. Rev. Lett.*, 70:343, 1993.

[300] D. Loss, H. Schoeller, and P.M. Goldbart. Weak-localization effects and conductance fluctuations: Implications of inhomogeneous magnetic fields. *Phys. Rev. B*, 48:15218, 1993.

[301] Y.B. Lyanda-Geller. Berry phase, conductance oscillations and phase breaking in mesoscopic transport. *Surf. Sci.*, 361/362:692, 1996.

[302] R.A. Webb, S. Washburn, C.P. Umbach, and R.B. Laibowitz. Observation of h/e Aharonov-Bohm oscillations in normal-metal rings. *Phys. Rev. Lett.*, 54:2696, 1985.

[303] A. E. Hansen, A. Kristensen, S. Pedersen, C. B. Sørensen, and P. E. Lindelof. Mesoscopic decoherence in Aharonov-Bohm rings. *Phys. Rev. B*, 64:45327, 2001.

[304] T.M. Jacobs and N. Giordano. Small Au rings with a magnet in the center. *Superlattices and Microstructures*, 23:635, 1998.

[305] P.D. Ye, S. Tarucha, and D. Weiss. Magnetoresistance oscillations in a ballistic Aharonov–Bohm ring with a micromagnet embedded at the center. In *Proceedings of the 24th International Conference on the Physics of Semiconductors.* World Scientific, Singapore, 1999. CD-ROM,ISBN 981-02-3613-1.

[306] A.F. Morpurgo, J.P. Heida, T.M. Klapwijk, B.J. van Wees, and G. Borghs. Ensemble-average spectrum of Aharonov–Bohm conductance oscillations: evidence for spin-orbit-induced Berry's phase. *Phys. Rev. Lett.*, 80:1050, 1998.

[307] J.-B. Yau, E.P. De Poortere, and M. Shayegan. Aharonov–Bohm oscillations with spin: evidence for Berry's phase. *Phys. Rev. Lett.*, 88:146801–1, 2002.

[308] A. Yacoby, M. Heiblum, D. Mahalu, and H. Shtrikman. Coherence and phase-sensitive measurements in a quantum dot. *Phys. Rev. Lett.*, 74:4047, 1995.

[309] R. Schuster, E. Buks, M. Heiblum, D. Mahalu, V. Umansky, and H. Shtrikman. Phase measurement in a quantum dot via a double-slit interference experiment. *Nature*, 385:417, 1997.

[310] L. P. Lévy, G. Dolan, J. Dunsmuir, and H. Bouchiat. Magnetization of mesoscopic copper rings: Evidence for persistent currents. *Phys. Rev. Lett.*, 64:2074, 1990.

[311] V. Chandrasekhar, R.A. Webb, M.J. Brady, M.B. Ketchen, W.J. Gallagher, and A. Kleinsasser. Magnetic response of a single, isolated gold loop. *Phys. Rev. Lett.*, 67:3578, 1991.

[312] D. Mailly, C. Chapelier, and A. Benoit. Experimental observation of persistent currents in GaAs-AlGaAs single loop. *Phys. Rev. Lett.*, 70:2020, 1993.

[313] A. Lorke, R.J. Luyken, A.O. Govorov, J.P. Kotthaus, J.M. Garcia, and P.M. Petroff. Spectroscopy of nanoscopic semiconductor rings. *Phys. Rev. Lett.*, 84:2223, 2000.

[314] R.J. Warburton, C. Schäflein, D. Haft, F. Bickel, A. Lorke, K. Karrai, J.M. Garcia, W. Schoenfeld, and P.M. Petroff. Optical emission from a charge-tunable quantum ring. *Nature*, 405:926, 2000.

[315] A. Fuhrer, S. Lüscher, T. Ihn, T. Heinzel, K. Ensslin, W. Wegscheider, and M. Bichler. Energy spectra of quantum rings,. *Nature*, 413:822, 2001.

[316] R. Held, T. Vancura, T. Heinzel, K. Ensslin, M. Holland, and W. Wegscheider. In-plane gates and nanostructures fabricated by direct oxidation of semiconductor heterostructures with an atomic force microscope. *Appl. Phys. Lett.*, 73:262, 1998.

[317] S. Lüscher, T. Heinzel, K. Ensslin, W. Wegscheider, and M. Bichler. Signatures of spin pairing in a quantum dot in the coulomb blockade regime. *Phys. Rev. Lett.*, 86:2118, 2001.

[318] T. Heinzel, R. Held, S. Lüscher, K. Ensslin, W. Wegscheider, and M. Bichler. Electronic properties of semiconductor nanostructures patterned by AFM lithography. *Physica E*, 9:84, 2001.

[319] S. Pedersen, A.E. Hansen, A. Kristensen, S.B. Sørensen, and P.E. Lindelof. Observation of quantum asymmetry in an Aharonov-Bohm ring. *Phys. Rev. B*, 61:5457, 2000.

[320] M. Cassé, Z.D. Kvon, G.M. Gusev, E.B. Olshanetskii, L.V. Litvin, A.V. Plotnikov, D.K. Maude, and J.C. Portal. Temperature dependence of the Aharonov-Bohm oscillations and the energy spectrum in a single-mode ballistic ring. *Phys. Rev. B*, 62:2624, 2000.

[321] B.L. Altshuler, A.G. Aronov, and B.Z. Spivak. The Aharonov-Bohm effect in disordered conductors. *JETP Lett.*, 33:94, 1981.

[322] W.-C. Tan and J.C. Inkson. Electron states in a two-dimensional ring. An exactly soluble model. *Semicond. Sci. Technol.*, 11:1635, 1996.

[323] E.B. Foxman, P.L. McEuen, U. Meirav, N.S. Wingreen, Y. Meir, P.A. Belk, N.R. Belk, M.A. Kastner, and S.J. Wind. Effects of quantum levels on transport through a Coulomb island. *Phys. Rev. B*, 47:10020, 1993.

[324] T. Chakraborty and P. Pietiläinen. Persistent currents in a quantum ring: Effects of impurities and interactions. *Phys. Rev. B*, 52:1932, 1995.

[325] D. Berman, O. Entin-Wohlman, and M.Y. Azbel. Diamagnetic spectrum and oscillations in an elliptic shell. *Phys. Rev. B*, 42:9299, 1990.

[326] O. Madelung. *Festkörpertheorie I: Elementare Anregungen*, volume 104 of *Heidelberger Taschenbücher*. Springer Verlag, revised reprint edition, 1988.

[327] Ya. M. Blanter, A.D. Mirlin, and B.A. Muzykantskii. Fluctuations of conductance peak spacings in the Coulomb-blockade regime: role of electron–electron interaction. *Phys. Rev. Lett.*, 78:2449, 1997.

[328] P.W. Brouwer, Y. Oreg, and B.I. Halperin. Mesoscopic fluctuations of the ground-state spin of a small metal particle. *Phys. Rev. B*, 60:R13977, 1999.

[329] H.U. Baranger, D. Ullmo, and L.I. Glazman. Interactions and interference in quantum dots:Kinks in Coulomb-blockade peak positions. *Phys. Rev. B*, 61:R2425, 2000.

[330] K. Ensslin R. Schuster. *Antidot superlattices: classical chaos and quantum transport*, page 195. Advances in Solid State Physics (Festkörperprobleme). Vieweg Braunschweig, 1994.

[331] D. Loss and P. Goldbart. Period and amplitude halving in mesoscopic rings with spin. *Phys. Rev. B*, 43:13762, 1991.

[332] F. Simmel, T. Heinzel, and D.A. Wharam. Statistics of conductance oscillations of a quantum dot in the Coulomb-blockade regime. *Europhys. Lett.*, 38:123, 1997.

[333] S. R. Patel, S. M. Cronenwett, D. R. Stewart, A. G. Huibers, C. M. Marcus, C. I. Duruz, Jr. J. S. Harris, K. Campman, and A. C. Gossard. Statistics of Coulomb Blockade Peak Spacings. *Phys. Rev. Lett.*, 80:4522, 1998.

[334] A. Kamenev A.V. Andreev. Itinerant ferromagnetism in disordered metals: a mean-field theory. *Phys. Rev. Lett.*, 81:3199, 1998.

[335] P. Jacquod and A.D. Stone. Suppression of ground-state magnetization in finite-size systems due to off-diagonal interaction fluctuations. *Phys. Rev. Lett*, 84:3938, 2000.

[336] K.A. Matveev, L.I. Glazman, and A.I. Larkin. g-factors of discrete levels in nanoparticles. *Phys. Rev. Lett.*, 85:2789, 2000.

[337] P.W. Brouwer, X. Waintal, and B.I. Halperin. Fluctuation spin g-tensor in small metal grains. *Phys. Rev. Lett.*, 85:369, 2000.

[338] S. Tarucha, D. G. Austing, Y. Tokura, W. G. van der Wiel, and L. P. Kouwenhoven. Direct Coulomb and exchange interaction in artificial atoms. *Phys. Rev. Lett.*, 84:2485, 2000.

[339] P. Hawrylak, C. Gould, A. Sachrajda, Y. Feng, and Z. Wasilewski. Collapse of the Zeeman gap in quantum dots due to electronic correlations. *Phys. Rev. B*, 59:2801, 1999.

[340] R. C. Ashoori, H. L. Stormer, J. S. Weiner, L. N. Pfeiffer, K. W. Baldwin, and K. W. West. N-electron ground state energies of a quantum dot in magnetic field. *Phys. Rev. Lett.*, 71:613, 1993.

[341] A.S.G. Thornton, T. Ihn, P.C. Main, L. Eaves, and M. Henini. Observation of spin splitting in single InAs self-assembled quantum dots in AlAs. *Appl. Phys. Lett.*, 73:354, 1998.

[342] A.C. Gossard. Growth of microstructures by molecular-beam epitaxy. *IEEE J. Quant. Electron.*, 22:1649, 1986.

[343] K.D. Maranowski, J.P. Ibbetson, K.L. Campman, and A.C. Gossard. Interface between low-temperature grown GaAs and undoped GaAs as a conduction barrier for back gates. *Appl. Phys. Lett.*, 66:3459, 1995.

[344] G. Salis, B. Graf, K. Ensslin, K. Campman, K. Maranowski, and A. C. Gossard. Wave-function spectroscopy in quantum wells with tunable electron density. *Phys. Rev. Lett.*, 79:5106, 1997.

[345] S. Lindemann, T. Ihn, T. Heinzel, K. Ensslin, K. Maranowski, and A. C. Gossard. From two-dimensional to three-dimensional quantum dots. *Physica E*, in press, 2002.

[346] P. L. McEuen, E. B. Foxman, Jari Kinaret, U. Meirav, M. A. Kastner, N.S. Wingreen, and S.J. Wind. Self-consistent addition spectrum of a Coulomb island in the quantum Hall regime. *Phys. Rev. B*, 45:11419, 1992.

[347] F. Simmel, D. Abdusch-Magder, D.A. Wharam, M.A. Kastner, and J.P. Kotthaus. Statistics of the Coulomb-blockade peak spacings of a silicon quantum dot. *Phys. Rev. B*, 59:10441, 1999.

[348] D. S. Duncan, D. Goldhaber-Gordon, R. M. Westervelt, K. D. Maranowski, and A. C. Gossard. Coulomb-blockade spectroscopy on a small quantum dot in a parallel magnetic field. *Appl. Phys. Lett.*, 77:2183, 2000.

[349] J.A. Folk, C.M. Marcus, R. Berkovits, I.L. Kurland, I.L. Aleiner, and B.L. Altshuler. Ground state spin and Coulomb blockade peak motion in chaotic quantum dots. *Physica Scripta T*, 90:26, 2001.

[350] A. A. M. Staring, B. W. Alphenaar, H. van Houten, L. W. Molenkamp, O. J. A. Buyk, M. A. A. Mabesoone, and C.T. Foxon. Periodic envelope of Coulomb-blockade oscillations in the quantum Hall regime. *Phys. Rev. B*, 46:12869, 1992.

[351] Landoldt-Börnstein. *Semiconductors*, volume 17a of *Numerical Data and Functionals Relationships in Science and Technology*. Springer Verlag, 1982.

[352] G. Salis, Y. Kato, K. Ensslin, D.C. Driscoll, A.C. Gossard, and D.D. Awshalom. Electrical control of spin coherence in semiconductor nanostructures. *Nature*, 414:619, 2001.

[353] L.P. Rokhinson, L.J. Guo, S.Y. Chou, and D.C. Tsui. Spin transitions in a small Si quantum dot. *Phys. Rev. B*, 63:035321–1, 2001.

[354] M. Ciorga, A. S. Sachrajda, P. Hawrylak, C. Gould, P. Zawadzki, S. Jullian, Y. Feng, and Z. Wasilewski. Addition spectrum of a lateral dot from Coulomb and spin-blockade spectroscopy. *Phys. Rev. B*, 61:R16315, 2000.

[355] W. Beinvogl, A. Kamgar, and J.F. Koch. Influence of a magnetic field on electron subbands in a surface space-charge layer. *Phys. Rev. B*, 14:4274, 1976.

[356] K. Ensslin, A. Wixforth, M. Sundaram, P. F. Hopkins, J. H. English, and A. C. Gossard. Single-particle subband spectroscopy in a parabolic quantum well via transport experiments. *Phys. Rev. B*, 47:1366, 1993.

[357] P. W. Brouwer and I.L. Aleiner. Effects of electron–electron interaction on the conductance of open quantum dots. *Phys. Rev. Lett.*, 82:390, 1999.

[358] D. Weinmann, W. Häusler, and B. Kramer. Spin blockades in linear and non-linear transport through quantum dots. *Phys. Rev. Lett.*, 74:984, 1995.

[359] L.I. Glazman, F.W.J. Hekking, and A.I. Larkin. Spin-charge separation and the Kondo effect in an open quantum dot. *Phys. Rev. Lett.*, 83:1830, 1999.

[360] G. Binnig, C.F. Quate, and Ch. Gerber. Atomic force microscope. *Phys. Rev. Lett.*, 56:930, 1986.

[361] H.J. Hug, B. Stiefel, P.J.A. van Schendel, A. Moser, S. Martin, and H.-J. Güntherodt. A low temperature ultrahigh vacuum scanning force microscope. *Rev. Sci. Instrum.*, 70:3625, 1999.

[362] M.A. Eriksson, R.G. Beck, M. Topinka, J.A. Katine, R.M. Westervelt, K.L. Campman, and A.C. Gossard. Cryogenic scanning probe characterization of semiconductor nanostructures. *Appl. Phys. Lett.*, 69:671, 1996.

[363] K.L. McCormick, M.T. Woodside, M. Huang, P.L. McEuen, C.I. Duruoz, and Jr. J.S. Harris. Scanned potential microscopy of a two-dimensional electron gas. *Physica B*, 249-251:79, 1998.

[364] D.V. Pelekhov, J.B. Becker, and Jr. G. Nunes. Ultralow-temperature atomic force microscopy for the investigation of mesoscopic systems. *Appl. Phys. Lett.*, 72:993, 1998.

[365] R. Crook, C.G. Smith, C.H.W. Barnes, M.Y. Simmons, and D.A. Ritchie. Imaging electron and conduction-band-hole trajectories through one and two series constrictions. *Physica E*, 6:234, 2000.

[366] P. Weitz, E. Ahlswede, J. Weis, K.v. Klitzing, and K. Eberl. A low-temperature scanning force microscope for investigating buried two-dimensional electron systems under quantum Hall conditions. *Appl. Surf. Sci.*, 157:349, 2000.

[367] M. Tortonese, R.C. Barrett, and C.F. Quate. Atomic resolution with an atomic force microscope using piezoresistive detection. *Appl. Phys. Lett.*, 62:834, 1993.

[368] P. Weitz, E. Ahlswede, J. Weis, K.v. Klitzing, and K. Eberl. Hall-potential investigations under quantum Hall conditions using scanning force microscopy. *Physica E*, 6:247, 2000.

[369] R. Crook, C.G. Smith, C.H.W. Barnes, M.Y. Simmons, and D.A. Ritchie. Imaging diffraction-limited electronic collimation from a non-equilibrium one-dimensional ballistic constriction. *J. Phys.: Cond. Matter*, 12:L167, 2000.

[370] R. Crook, C.G. Smith, M.Y. Simmons, and D.A. Ritchie. Imaging cyclotron orbits and scattering sites in a high-mobility two-dimensional electron gas. *Phys. Rev. B*, 62:5174, 2000.

[371] R. Crook R, C.G. Smith, M.Y. Simmons, and D.A. Ritchie. One-dimensional probability density observed using scanned gate microscopy. *J. Phys.: Cond. Matter*, 12:L735, 2000.

[372] R. Crook, C.G. Smith, M.Y. Simmons, and D.A. Ritchie. Imaging random telegraph signal sites near a quasi 1D electron system. *J. Phys.: Cond. Matter*, 13:L249, 2001.

[373] M.A. Topinka, B.J. LeRoy, R.M. Westervelt, S.E.J. Shaw, R. Fleischmann, E.J. Heller, K.D. Maranowski, and A.C. Gossard. Coherent branched flow in a two-dimensional electron gas. *Nature*, 410:183, 2001.

[374] E. Ahlswede, P. Weitz, J. Weis, K.v. Klitzing, and K. Eberl. Hall potential profiles in the quantum Hall regime measured by a scanning force microscope. *Physica B*, 298:562, 2001.

[375] K. Karrai and R.D. Grober. Piezoelectric tip-sample distance control for near field optical microscopes. *Appl. Phys. Lett.*, 66:1842, 1995.

[376] H. Edwards, L. Taylor, W. Duncan, and A.J. Melmed. Fast, high-resolution atomic force microscopy using a quartz tuning fork as actuator and sensor. *J. Appl. Phys.*, 82:980, 1997.

[377] A.G.T. Ruiter, K.O. van der Werf, J.A. Veerman, M.F. Garcia-Parajo, W.H.J. Rensen, and N.F. van Hulst. Tuning fork shear-force feedback. *Ultramicroscopy*, 71:149, 1998.

[378] W.A. Atia and Ch.C. Davis. Phase-locked shear-force microscope for distance regulation in near-field optical microscopy. *Appl. Phys. Lett.*, 70:405, 1997.

[379] J. Salvi, P. Chevassus, A. Mouflard, S. Davy, M. Spajer, and D. Courjon. Piezoelectric shear force detection: A geometry avoiding critical tip/tuning fork gluing. *Rev. Sci. Instrum.*, 69:1744, 1998.

[380] D.P. Tsai and Y.Y. Lu. Tapping-mode tuning fork force sensing for near-field scanning optical microscopy. *Appl. Phys. Lett.*, 73:2724, 1998.

[381] M. Todorovic and S. Schultz. Magnetic Force Microscopy using nonoptical piezoelectric quartz tuning fork detection design with applications to magnetic recording studies. *J. Appl. Phys.*, 83:6229, 1998.

[382] R. Steinke, M. Hoffmann, M. Böhmisch, J. Eisenmenger, K. Dransfeld, and P. Leiderer. Potentiometry with the acoustic near field microscope: A new method for microscopy of surface potentials. *Appl. Phys. A: Mater. Sci. Process.*, 64:19, 1997.

[383] W. Allers, A. Schwarz, U.D. Schwarz, and R. Wiesendanger. A scanning force microscope with atomic resolution in ultrahigh vacuum and at low temperatures. *Rev. Sci. Instrum.*, 69:221, 1998.

[384] T.R. Albrecht, P. Grütter, D. Horne, and D. Rugar. Frequency modulation detection using high-Q cantilevers for enhanced force microscope sensitivity. *J. Appl. Phys.*, 69:668, 1991.

[385] F.J. Giessibl. Atomic-resolution of the silicon (111)-(7X7) surface by atomic-force microscopy. *Science*, 267:68, 1995.

[386] S. Kitamura and M. Iwatsuki. Observation of 7 x 7 reconstructed structure on the silicon (111) surface using ultrahigh vacuum noncontact atomic force microscopy . *Jpn. J. Appl. Phys.*, 34:L145, 1995.

[387] Y. Sugawara, M. Ohta, and H. Ueyama. Defect Motion on an InP(110) Surface Observed with Noncontact Atomic Force Microscopy. *Science*, 270:1648, 1995.

[388] S.H. Tessmer, P.I. Glicofridis, R.C. Ashoori, L.S. Levitov, and M.R. Melloch. Subsurface charge accumulation imaging of a quantum Hall liquid. *Nature*, 392:51, 1998.

[389] M.J. Yoo, T.A. Fulton, H.F. Hess, R.L. Willett, L.N. Dunkleberger, R.J. Chichester, L.N. Pfeiffer, and K.W. West. Scanning Single-Electron Transistor Microscopy: Imaging Individual Charges. *Science*, 276:579, 1997.

[390] R.C. Ashoori, H.L. Stormer, J.S. Weiner, L.N. Pfeiffer, S.J. Pearton, K.W. Baldwin, and K.W. West. Single-electron capacitance spectroscopy of discrete quantum levels. *Phys. Rev. Lett.*, 68:3088, 1992.

[391] H. Birk, K. Oostveen, and C. Schönenberger. Preamplifier for electric-current noise measurements at low temperatures. *Rev. Sci. Instrum.*, 67:2977, 1996.

[392] J.A. Harley and T.W. Kenny. High-sensitivity piezoresistive cantilevers under 1000 Å thick. *Appl. Phys. Lett.*, 75:289, 1999.

[393] C.W. Yuan, E. Batalla, M. Zacher, A.L. de Lozanne, M.D. Kirk, and M. Tortonese. Low temperature magnetic force microscope utilizing a piezoresistive cantilever. *Appl. Phys. Lett.*, 65:1308, 1994.

[394] P.I. Oden, P.G. Datskos, T. Thundat, and R.J. Warmack. Uncooled thermal imaging using a piezoresistive microcantilever. *Appl. Phys. Lett.*, 69:3277, 1996.

[395] D.R. Baselt, G.U. Lee, and R.J. Colton. Biosensor based on force microscope technology. *J. Vac. Sci. Technol. B*, 14:789, 1996.

[396] S.C. Minne, Ph. Flueckiger, H.T. Soh, and C.F. Quate. Atomic-force microscope lithography using amorphous-silicon as a resist and advances in parallel operation. *J. Vac. Sci. Technol. B*, 13:1380, 1995.

[397] S.C. Minne, G. Yaralioglu, S.R. Manalis, J.D. Adams, J. Zesch, A. Atalar, and C.F. Quate. Automated parallel high-speed atomic force microscopy. *Appl. Phys. Lett.*, 72:2340, 1998.

[398] R.G. Beck, M.A. Eriksson, M.A. Topinka, R.M. Westervelt, K.D. Maranowski, and A.C. Gossard. GaAs/AlGaAs Self-Sensing Cantilever for Low Temperature Scanning Probe Microscopy. *Appl. Phys. Lett.*, 73:1149, 1998.

[399] Y.Y. Wei, J. Weis, K.v. Klitzing, and K. Eberl. Single-electron transistor as an electrometer measuring chemical potential variations. *Appl. Phys. Lett.*, 71:2514, 1997.

[400] H.F. Hess, T.A. Fulton, M.J. Yoo, and A. Yacoby. Microscopy with a single electron transistor probe. *Solid State Commun.*, 107:657, 1998.

[401] A. Yacoby, H.F. Hess, T.A. Fulton, L.N. Pfeiffer, and K.W. West. Electrical imaging of the quantum Hall state. *Solid State Commun.*, 111:1, 1999.

[402] N.B. Zhitenev, T.A. Fulton, A. Yacoby, H.F. Hess, L.N. Pfeiffer, and K.W. West. Imaging of localized electronic states in the quantum Hall regime. *Nature*, 404:473, 2000.

[403] N.B. Zhitenev, T.A. Fulton, A. Yacoby, H.F. Hess, L.N. Pfeiffer, and K.W. West. Single-electron states in the quantum Hall effect. *Physica E*, 6:242, 2000.

[404] E. Betzig, J.K. Trautmann, T.D. Harris, J.S. Weiner, and R.L. Kostelak. Breaking the diffraction barrier — optical microscopy on the nanometer scale. *Science*, 251:1468, 1991.

[405] P. Günther, U. Ch. Fischer, and K. Dransfeld. Scanning near-field acoustic microscopy. *Appl. Phys. B*, 48:89, 1989.

[406] F.J. Giessibl. High-speed force sensor for force microscopy and profilometry utilizing a quartz tuning fork. *Appl. Phys. Lett.*, 73:3956, 1998.

[407] J. Rychen, T. Ihn, P. Studerus, A. Herrmann, and K. Ensslin. A low-temperature dynamic mode scanning force microscope operating in high magnetic fields. *Rev. Sci. Instrum.*, 70:2765, 1999.

[408] F.J. Giessibl. Atomic resolution on Si(111)-(7 x 7) by noncontact atomic force microscopy with a force sensor based on a quartz tuning fork. *Appl. Phys. Lett.*, 76:1470, 2000.

[409] R.D. Grober, J. Acimovic, J. Schuck, D. Hessmann, P.J. Kindlemann, J. Hespanha, A.S. Morse, K. Karrai, I. Tiemann, and S. Manus. Fundamental limits to force detection using quartz tuning forks. *Rev. Sci. Instrum.*, 71:2776, 2000.

[410] J. Rychen, T. Ihn, P. Studerus, A. Herrmann, and K. Ensslin. Operation characteristics of piezoelectric quartz tuning forks in high magnetic fields at liquid Helium temperatures. *Rev. Sci. Instrum.*, 71:1695, 2000.

[411] K.B. Shelimov, D.N. Davydov, and M. Moskovits. Dynamics of a piezoelectric tuning fork/optical fiber assembly in a near-field scanning optical microscope. *Rev. Sci. Instrum.*, 71:437, 2000.

[412] D. Sarid. *Scanning force microscopy.* Oxford University Press, 1994.

[413] K. Karrai and R.D. Grober. Tip-sample distance control for near-field scanning optical microscopes. In M.A. Paesler and P.J. Moyer, editors, *Near Field Optics*, volume 2535 of *SPIE Proceedings Series*, page 69, 1995.

[414] J. Rychen, T. Ihn, P. Studerus, A. Herrmann, K. Ensslin, H.J. Hug, P.J.A. van Schendel, and H.J. Güntherodt. Force-distance studies with piezoelectric tuning forks below 4.2 K. *Appl. Surf. Sci.*, 157:290, 2000.

[415] W. Allers, A. Schwarz, U.D. Schwarz, and R. Wiesendanger. Dynamic scanning force microscopy at low temperatures on a van der Waals surface: graphite (0001). *Appl. Surf. Sci.*, 140:247, 1999.

[416] U. Dürig, O. Züger, and A. Stalder. Interaction force detection in scanning probe microscopy: Methods and applications. *J. Appl. Phys.*, 72:1778, 1992.

[417] U. Tietze and Ch. Schenk. *Halbleiter Schaltungstechnik.* Springer Verlag, 1978.

[418] R. Best. *Phase-locked loops: Theory, Design and Applications.* McGraw-Hill, 1984.

[419] D.H. Wolaver. *Phase-locked loop circuit design.* Prentice Hall, 1991.

[420] F.M. Gardner. *Phaselock techniques.* Wiley, 1979.

[421] A. Blanchard. *Phase-locked loops: Application to Coherent Receiver Design.* Wiley, 1976.

[422] Ch. Loppacher, M. Bammerlin, F. Battiston, M. Guggisberg, D. Müller, H.R. Hidber, R. Lüthi, E. Meyer, and H.J. Güntherodt. Fast digital electronics for application in dynamic force microscopy using high-Q cantilevers. *Appl. Phys. A*, 66:S215, 1998.

[423] Ch. Loppacher, M. Bammerlin, M. Guggisberg, F. Battiston, R. Bennewitz, S. Rast, A. Baratoff, E. Meyer, and H.-J. Güntherodt. Phase variation experiments in non-contact dynamic force microscopy using phase locked loop techniques. *Appl. Surf. Sci.*, 140:287, 1999.

[424] U. Dürig, H.R. Steinauer, and N. Blanc. Dynamic force microscopy by means of the phase-controlled oscillator method. *J. Appl. Phys.*, 82:3641, 1997.

[425] F. J. Giessibl. Forces and frequency shifts in atomic-resolution dynamic-force microscopy. *Phys. Rev. B*, 56:16010, 1997.

[426] P. Güthner. Simultaneous imaging of Si(111) 7x7 with atomic resolution in scanning tunneling microscopy, atomic force microscopy, and atomic force microscopy noncontact mode. *J. Vac. Sci. Technol. B*, 14:2428, 1996.

[427] R. Lüthi, E. Meyer, M. Bammerlin, A. Baratoff, T. Lehmann, L. Howald, Ch. Gerber, and H.-J. Güntherodt. Atomic resolution in dynamic force microscopy across steps on Si(111)7 x 7. *Z. Phys. B*, 100:165, 1996.

[428] R. Erlandson, L. Olsson, and P. Martensson. Inequivalent atoms and imaging mechanisms in ac-mode atomic-force microscopy of Si(111)7 x 7. *Phys. Rev. B*, 54:R8309, 1996.

[429] S. Kitamura and M. Iwatsuki. Observation of silicon surfaces using ultrahigh-vacuum noncontact, atomic force microscopy. *Jpn. J. Appl. Phys.*, 35:L668, 1996.

[430] H. Hölscher, U.D. Schwarz, and R. Wiesendanger. Calculation of the frequency shift in dynamic force microscopy. *Appl. Surf. Sci.*, 140:344, 1999.

[431] S. Belaidi, P. Girard, and G. Leveque. Electrostatic forces acting on the tip in atomic force microscopy: Modelization and comparison with analytic expressions. *J. Appl. Phys.*, 81:1023, 1997.

[432] B.I. Halperin. Quantized Hall conductance, current-carrying edge states, and the existence of extended states in a two-dimensional disordered potential. *Phys. Rev. B*, 25:2185, 1982.

[433] D.J. Thouless. Edge voltages and distributed currents in the quantum Hall effect. *Phys. Rev. Lett.*, 71:1879, 1993.

[434] M.R. Geller and G. Vignale. Equilibrium current and orbital magnetization in the quantum Hall fluid. *Physica B*, 212:283, 1995.

[435] L. Bremme, T. Ihn, and K. Ensslin. Magnetization of a two-dimensional electron gas and the role of one-dimensional edge currents. *Phys. Rev. B*, 59:7305, 1999.

[436] A.J. Kent, D.J. McKitterick, P. Hawker, and M. Henini. Imaging the current distribution in the quantum Hall-effect. *Helv. Physica Acta*, 65:331, 1992.

[437] R. Knott, W. Dietsche, K.v. Klitzing, K. Eberl, and K. Ploog. Electrooptic imaging of potential distributions in the quantum Hall regime. *Semicond. Sci. Technol.*, 10:117, 1995.

[438] A.A. Shashkin, A.J. Kent, P.A. Harrison, L. Eaves, and M. Henini. Edge channels and the quantum-Hall-effect breakdown. *Phys. Rev. B*, 49:5379, 1994.

[439] R.J.F. van Haren, F.A.P. Blom, and J.H. Wolter. Direct observation of edge channels in the integer quantum Hall regime. *Phys. Rev. Lett.*, 74:1198, 1995.

[440] E. Yahel, D. Orgad, A. Palevski, and H. Shtrikman. Inductive probing of the integer quantum Hall effect. *Phys. Rev. Lett.*, 76:2149, 1996.

[441] E. Yahel, A. Tsukernik, A. Palevski, and H. Shtrikman. Evidence for bulk current in Hall bar samples and potential screening in the integer quantum Hall effect. *Phys. Rev. Lett.*, 81:5201, 1998.

[442] Y.Y. Wei, J. Weis, K.v. Klitzing, and K. Eberl. Edge strips in the quantum Hall regime imaged by a single-electron transistor. *Phys. Rev. Lett.*, 81:1674, 1998.

[443] K.L. McCormick, M.T. Woodside, M. Huang, M. Wu, P.L. McEuen, C.I. Duruoz, and Jr. J.S. Harris. Scanned potential microscopy of edge and bulk currents in the quantum Hall regime. *Phys. Rev. B*, 59:4654, 1999.

[444] G. Finkelstein, P.I. Glicofridis, S.H. Tessmer, R.C. Ashoori, and M.R. Melloch. Imaging of low-compressibility strips in the quantum Hall liquid. *Phys. Rev. B*, 61:R16323, 2000.

[445] G. Finkelstein, P.I. Glicofridis, R.C. Ashoori, and M. Shayegan. Topographic mapping of the quantum hall liquid using a few-electron bubble. *Science*, 289:90, 2000.

[446] G. Finkelstein, P.I. Glicofridis, S.H. Tessmer, R.C. Ashoori, and M.R. Melloch. Imaging the low compressibility strips formed by the quantum hall liquid in a smooth potential gradient. *Physica E*, 6:251, 2000.

[447] E. Ahlswede, J. Weis, K. v. Klitzing, and K. Eberl. Hall potential distribution in the quantum Hall regime in the vicinity of a potential probe contact. *Physica E*, 12:165, 2002.

[448] K. Lier and R.R. Gerhardts. Self-consistent calculations of edge channels in laterally confined 2-dimensional electron-systems. *Phys. Rev. B*, 50:7757, 1994.

[449] P.I. Glicofridis, G. Finkelstein, R.C. Ashoori, and M. Shayegan. Determination of the resistance across incompressible strips through imaging of charge motion. *Phys. Rev. B*, 65:121312, 2002.

[450] S.H. Tessmer, G. Finkelstein, P.I. Glicofridis, and R.C. Ashoori. Modeling subsurface charge accumulation images of a quantum Hall liquid. *Phys. Rev. B*, 66:125308, 2002.

[451] T. Ihn, J. Rychen, K. Ensslin, W. Wegscheider, and M. Bichler. Local spectroscopy of edge channels in the quantum Hall regime with local probe techniques. *Physica E*, 13:671, 2002.

[452] T. Ihn, J. Rychen, K. Ensslin, W. Wegscheider, and M. Bichler. Tunnelling between edge channels in the quantum Hall regime manipulated with a scanning force microscope. *Microelectron. Eng.*, 63:81, 2002.

[453] I.J. Maasilta and V.J. Goldman. Energetics of quantum antidot states in the quantum Hall regime. *Phys. Rev. B*, 57:R4273, 1998.

[454] P.L. McEuen, A. Szafer, C.A. Richter, B.W. Alphenaar, J.K. Jain, A.D. Stone, R.G. Wheeler, and R.N. Sacks. New resistivity for high-mobility quantum Hall conductors. *Phys. Rev. Lett.*, 64:2062, 1990.

[455] R.J. Haug. Edge-state transport and its experimental consequences in high magnetic-fields. *Semicond. Sci. Technol.*, 8:131, 1993.

[456] B.W. Alphenaar, P.L. McEuen, R.G. Wheeler, and R.N. Sacks. Selective equilibration among the current-carrying states in the quantum Hall regime. *Phys. Rev. Lett.*, 64:677, 1990.

[457] M.T. Woodside and P.L. McEuen. Scanned probe imaging of single-electron charge states in nanotube quantum dots. *Science*, 296:1098, 2002.

[458] S.J. Tans and C. Dekker. Potential modulations along carbon nanotubes. *Nature*, 404:834, 2000.

[459] A. Bachtold, M.S. Fuhrer, S. Plyasunov, M. Forero, E.H. Anderson, A. Zettl, and P.L. McEuen. Scanned probe microscopy of electronic transport in carbon nanotubes. *Phys. Rev. Lett.*, 84:6082, 2000.

[460] M. Bockrath, W. Liang, D. Bozovic, J.H. Hafner, C.M. Lieber, M. Tinkham, and H. Park. Resonant electron scattering by defects in single-walled carbon nanotubes. *Science*, 291:283, 2001.

[461] B.L. Altshuler and B.Z. Spivak. Variation of the random potential and the conductivity of samples of small dimensions. *JETP Lett.*, 42:447, 1985.

[462] S. Feng, P.A. Lee, and A.D. Stone. Sensitivity of the conductance of a disordered metal to the motion of a single atom: implications for 1/f noise. *Phys. Rev. Lett.*, 56:1960, 1986.

[463] S. Feng, P.A. Lee, and A.D. Stone. Sensitivity of the conductance of a disordered metal to the motion of a single atom: implications for 1/f noise. *Phys. Rev. Lett.*, 56:2772, 1986. (Erratum).

[464] N. Giordano. Numerical study of conductance fluctuations in disordered metals. *Phys. Rev. B*, 36:4190, 1987.

[465] A.D. Stone. Magnetoresistance fluctuations in mesoscopic wires and rings. *Phys. Rev. Lett.*, 54:2692, 1985.

[466] J. Sak and B. Kramer. Transmission of particles through a random one-dimensional potential. *Phys. Rev. B*, 24:1761, 1981.

[467] C.J. Lambert and M.F. Thorpe. Random T-matrix approach to one-dimensional localization. *Phys. Rev. B*, 27:715, 1983.

[468] M. Cahay, M. McLennan, and S. Datta. Conductance of an array of elastic scatterers: A scattering matrix approach. *Phys. Rev. B*, 37:10125, 1988.

[469] C.W.J. Beenakker and H. van Houten. Billiard model of a ballistic multiprobe conductor. *Phys. Rev. Lett.*, 63:1857, 1989.

[470] S.K. Adhikari. Quantum scattering in two dimensions. *Am. J. Phys.*, 54:362, 1986.

[471] I.R. Lapidus. Quantum-mechanical scattering in two dimensions. *Am. J. Phys.*, 50:45, 1982.

Index

Springer Tracts in Modern Physics

158 **Nonlinear Optics of Random Media**
Fractal Composites and Metal-Dielectric Films
By V.M. Shalaev 2000. 51 figs. XII, 158 pages

159 **Magnetic Dichroism in Core-Level Photoemission**
By K. Starke 2000. 64 figs. X, 136 pages

160 **Physics with Tau Leptons**
By A. Stahl 2000. 236 figs. VIII, 315 pages

161 **Semiclassical Theory of Mesoscopic Quantum Systems**
By K. Richter 2000. 50 figs. IX, 221 pages

162 **Electroweak Precision Tests at LEP**
By W. Hollik and G. Duckeck 2000. 60 figs. VIII, 161 pages

163 **Symmetries in Intermediate and High Energy Physics**
Ed. by A. Faessler, T.S. Kosmas, and G.K. Leontaris 2000. 96 figs. XVI, 316 pages

164 **Pattern Formation in Granular Materials**
By G.H. Ristow 2000. 83 figs. XIII, 161 pages

165 **Path Integral Quantization and Stochastic Quantization**
By M. Masujima 2000. 0 figs. XII, 282 pages

166 **Probing the Quantum Vacuum**
Pertubative Effective Action Approach in Quantum Electrodynamics and its Application
By W. Dittrich and H. Gies 2000. 16 figs. XI, 241 pages

167 **Photoelectric Properties and Applications of Low-Mobility Semiconductors**
By R. Könenkamp 2000. 57 figs. VIII, 100 pages

168 **Deep Inelastic Positron-Proton Scattering in the High-Momentum-Transfer Regime of HERA**
By U.F. Katz 2000. 96 figs. VIII, 237 pages

169 **Semiconductor Cavity Quantum Electrodynamics**
By Y. Yamamoto, T. Tassone, H. Cao 2000. 67 figs. VIII, 154 pages

170 **d–d Excitations in Transition-Metal Oxides**
A Spin-Polarized Electron Energy-Loss Spectroscopy (SPEELS) Study
By B. Fromme 2001. 53 figs. XII, 143 pages

171 **High-T_c Superconductors for Magnet and Energy Technology**
By B. R. Lehndorff 2001. 139 figs. XII, 209 pages

172 **Dissipative Quantum Chaos and Decoherence**
By D. Braun 2001. 22 figs. XI, 132 pages

173 **Quantum Information**
An Introduction to Basic Theoretical Concepts and Experiments
By G. Alber, T. Beth, M. Horodecki, P. Horodecki, R. Horodecki, M. Rötteler, H. Weinfurter, R. Werner, and A. Zeilinger 2001. 60 figs. XI, 216 pages

174 **Superconductor/Semiconductor Junctions**
By Thomas Schäpers 2001. 91 figs. IX, 145 pages

175 **Ion-Induced Electron Emission from Crystalline Solids**
By Hiroshi Kudo 2002. 85 figs. IX, 161 pages

176 **Infrared Spectroscopy of Molecular Clusters**
An Introduction to Intermolecular Forces
By Martina Havenith 2002. 33 figs. VIII, 120 pages

177 **Applied Asymptotic Expansions in Momenta and Masses**
By Vladimir A. Smirnov 2002. 52 figs. IX, 263 pages

178 **Capillary Surfaces**
Shape – Stability – Dynamics, in Particular Under Weightlessnes
By Dieter Langbein 2002. 182 figs. XVIII, 364 pages

Springer Tracts in Modern Physics

179 **Anomalous X-ray Scattering for Materials Characterization**
Atomic-Scale Structure Determination
By Yoshio Waseda 2002. 132 figs. XIV, 214 pages

180 **Coverings of Discrete Quasiperiodic Sets**
Theory and Applications to Quasicrystals
Edited by P. Kramer and Z. Papadopolos 2002. 128 figs., XIV, 274 pages

181 **Emulsion Science**
Basic Principles. An Overview
By J. Bibette, F. Leal-Calderon, V. Schmitt, and P. Poulin 2002. 50 figs., IX, 140 pages

182 **Transmission Electron Microscopy of Semiconductor Nanostructures**
An Analysis of Composition and Strain State
By A. Rosenauer 2003. 136 figs., XII, 238 pages

183 **Transverse Patterns in Nonlinear Optical Resonators**
By K. Staliūnas, V. J. Sánchez-Morcillo 2003. 132 figs., XII, 226 pages

184 **Statistical Physics and Economics**
Concepts, Tools and Applications
By M. Schulz 2003. 54 figs., XII, 244 pages

185 **Electronic Defect States in Alkali Halides**
Effects of Interaction with Molecular Ions
By V. Dierolf 2003. 80 figs., XII, 196 pages

186 **Electron-Beam Interactions with Solids**
Application of the Monte Carlo Method to Electron Scattering Problems
By M. Dapor 2003. 27 figs., X, 110 pages

187 **High-Field Transport in Semiconductor Superlattices**
By K. Leo 2003. 164 figs.,XIV, 240 pages

188 **Transverse-Pattern Formation in Photorefrative Optical Media**
By C. Denz, M. Schwab, and C. Weilnau 2003. 143 figs., XVIII, 333 pages

189 **Spatio-Temporal Dynamics and Quantum Fluctuations in Semiconductor Lasers**
By O. Hess, E. Gehrig 2003. 91 figs., XIV, 232 pages

190 **Neutrino Mass**
Edited by G. Altarelli, K. Winter 2003. 118 figs., X, 244 pages

191 **Spin-orbit Coupling Effects in Two-dimensional Electron and Hole Systems**
By R. Winkler 2003. 66 figs., XII, 224 pages

192 **Electronic Quantum Transport in Mesoscopic Semiconductor Structures**
By T. Ihn 2003. 90 figs., XII, 280 pages

193 **Spinning Particles - Semiclassics and Spectral Statistics**
By S. Keppeler 2003. 15 figs., X, 190 pages

194 **Light Emitting Silicon for Microphotonics**
By S. Ossicini, L. Pavesi, and F. Priolo 2003. 216 figs., XII, 284 pages

195 **Uncovering *CP* Violation**
Experimental Clarification in the Neutral K Meson and B Meson Systems
By K. Kleinknecht 2003. 67 figs., XII, 144 pages

196 **Ising-type Antiferromagnets**
Model Systems in Statistical Physics and in the Magnetism of Exchange Bias
By C. Binek 2003. 52 figs., X, 120 pages

197 **Electroweak Processes in External Electromagnetic Fields**
By A. Kuznetsov and N. Mikheev 2003. 24 figs., XII, 136 pages

198 **Electroweak Symmetry Breaking**
The Bottom-Up Approach
By W. Kilian 2003. 25 figs., X, 128 pages